Elliptic Boundary Value Problems and Construction of Lᵖ-Strong Feller Processes with Singular Drift and Reflection

T0238489

Benedict Baur

Elliptic Boundary Value Problems and Construction of L^p-Strong Feller Processes with Singular Drift and Reflection

Mit einem Geleitwort von
Professor Dr. Martin Grothaus

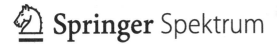 Springer Spektrum

Benedict Baur
Kaiserslautern, Germany

Zugl.: Dissertation, Technische Universität Kaiserslautern, 2013

ISBN 978-3-658-05828-9   ISBN 978-3-658-05829-6 (eBook)
DOI 10.1007/978-3-658-05829-6

The Deutsche Nationalbibliothek lists this publication in the Deutsche Nationalbibliografie;
detailed bibliographic data are available in the Internet at http://dnb.d-nb.de.

Library of Congress Control Number: 2014937835

Springer Spektrum

Printed on acid-free paper

Springer Spektrum is a brand of Springer DE.
Springer DE is part of Springer Science+Business Media.
www.springer-spektrum.de

Preface

The present dissertation of Benedict Baur characterizes a milestone in the the theory of Dirichlet forms. In the last decades the theory of Dirichlet forms emerged to be a very useful concept for the construction and analysis of solutions to stochastic differential equations (SDEs). In particular, this theory was and is of great success in the construction of solutions to equations with singular coefficients, such as they are showing up in many applications from Physics. Moreover, the concepts of Dirichlet forms allow to treat equations in bounded domains with various boundary conditions. Classical existence results from the theory of SDEs in the presence of boundary conditions often are rather limited or, for certain boundary conditions, even not available up to now.

But the general theory of Dirichlet forms has a disadvantage. One can treat many equations, but in general it is not clear for which initial conditions. More precisely, one has a solution for only almost all starting points. Well, the notion "almost all" can even be refined, but in worst case one can not specify even a single starting point for which a solution exists. This disadvantage can be overcome by a combination of Dirichlet from techniques with strong Feller properties. This idea, for example, has been worked out by Masatoshi Fukushima, one of the giants and founders of the theory of Dirichlet forms. Then, approximately 10 years ago, these ideas were refined by Michael Röckner, a further giant of the theory of Dirichlet forms, to be applicable to much more general classes of equations. At that time those concepts were applied to an interesting system of SDEs from Statistical Physics. Later on these concepts were generalized to treat more and more examples.

The impressive contribution of Benedict Baur is the development of a general concept out of the above ideas. In his dissertation he invented a collection of functional analytic conditions. These imply the existence of a solution to a given SDE. The construction of the solution is via Dirichlet form techniques and, nevertheless, the solution process can be started in an explicitly known set of initial points. That these analytic conditions are of practical use, he illustrated by providing several challenging and interesting examples.

In the cases with reflecting boundary conditions, even the corresponding Skorokhod decomposition is provided. Furthermore, as a byproduct, elliptic regularity results up to the boundary were derived.

It is desirable that the present dissertation will serve as a standard reference for constructing solutions to SDEs via Dirichlet forms for an explicitly known set of initial conditions.

Kaiserslautern   Dr. Martin Grothaus

Full Professor of Mathematics
Head of Functional Analysis
and Stochastic Analysis Group
University of Kaiserslautern

Acknowledgement

First of all I would like to thank my supervisor Professor Dr. Martin Grothaus for providing me the possibility to write my doctoral thesis at the University of Kaiserslautern. Especially I am grateful for the fruitful and interesting research topic as well as the granted freedoms. Thank also for the possibility to join interesting conferences in Kiev (MSTA2-Kiev 2010), Cologne (Annual DMV Meeting 2011) and research stays in Moscow (2013).[1]

I would like to express special thanks to Professor Dr. Michael Röckner for helpful discussions on the aforementioned DMV Meeting, for the interest in my research topic and the willingness to examine this thesis as a referee.

Thank is also awarded to all my colleagues Dr. Wolfgang Bock, Patrick Capraro, Dr. Torben Fattler, Florian Jahnert, Patrik Stilgenbauer, Felix Riemann, Herry P. Suryawan, Robert Voßhall and the former colleagues Dr. Florian Conrad and Dr. Tan Thanh Mai. Not to forget our secretaries Claudia Korb and Cornelia Türk.

Furthermore, I would like to thank the (former) members of the Stochastic and real analysis group, namely my secondary supervisor Professor Dr. Heinrich von Weizsäcker, the secretary Beate Siegler, Dr. Benedikt Heinrich, Martin Anders and Dr. Yang Zou. Thank is awarded to all (former) members of the Finanical Mathematics group, especially Dr. Henning Marxen, Dr. Martin Smaga, Junior-Professor Dr. Frank Seifried and Elisabeth Leoff as well as my friends Michael Adam and Daniel Zoufine Bare Contreras.

Warm thanks go to our Russian friends, Professor Dr. Oleg G. Smolyanov, Professor Dr. Yana A. Butko and Dr. Diana Tolstyga for their hospitality in Moscow. Moreover, I would like to thank the ESG Kaiserslautern, my family and further friends who are simply to many to be named explicitly here.

Finally, I would like to thank some of my teachers. They surely had an influence on my wish to study and do research in mathematics: Dr. Dieter Reuße, Herbert Fritsch, Dr. Roland Hoffmann, Peter Schmitz and Peter Staudt.

Kaiserslautern   Benedict Baur

---
[1]Supported by DFG-Projects GR-1809/9-1, GR-1809/8-1 and GR-1809/10-1

Contents

1 Introduction

1.1 Introduction

This thesis consists of three main parts: First, the construction of $\mathcal{L}^p$-strong Feller processes from sub-Markovian strongly continuous contraction semigroups on L^p-spaces that are associated with symmetric regular Dirichlet forms, see Chapter 2.

Second, providing a regularity result for weak solutions to elliptic equations under local assumptions on the coefficients, see Chapter 3 and Section 4.2.

Third, construction of elliptic diffusions with singular drift and reflecting boundary behavior and providing a Skorokhod representation (or semimartingale decomposition). This representation holds for every starting point that is not in the singularity set of the drift term and is either in the interior of the domain or at a C^2-smooth boundary part. See Chapter 4 and Chapter 6 for details.

All results are applied to construct stochastic dynamics for finite particle systems with singular interaction in continuum and for Ginzburg-Landau interface models, see Chapter 5, Section 6.4 and Section 6.5.

Let us now describe the results in more detail.

Construction of $\mathcal{L}^p$-strong Feller processes

We start with the first part, i.e., Chapter 2. In this chapter we provide a general construction scheme for $\mathcal{L}^p$-strong Feller processes that give solutions to a martingale problem for starting points from a known set. With $\mathcal{L}^p$-strong Feller we mean that for some $1 \leq p < \infty$ the semigroup of the process $(P_t)_{t \geq 0}$ maps $\mathcal{L}^p$ (w.r.t. to a specified measure) into $C^0(E_1)$, the space of continuous functions on a given set E_1.

The motivation is the following: Dirichlet form methods allow to construct stochastic processes in a very general setting, see [FOT11] and [MR92]. In particular, the construction of diffusions with very singular drift and general boundary behavior are possible.

However, these methods yield processes that solve the associated martingale problem for the corresponding L^2-generator for starting points outside an exceptional set only. In general, this set cannot be explicitly specified and in particular need not to be empty.

In recent years it turned out that additional techniques allow to refine these results to get a process that can be started from every point in a specified set of admissible starting points. Now the process yields solutions to the martingale problem for starting points from this set and has continuous paths in this set. In applications this specified set is naturally related to coefficients in equations describing the process, like the formulation of the martingale problem.

Albeverio, Kondratiev and Röckner ([AKR03]) construct distorted Brownian motion on $\mathbb{R}^d$, $d \in \mathbb{N}$, with strongly singular drift (see also Theorem 2.4.2). The process can be started from those points where the drift is not singular. Fattler and Grothaus ([FG07] and [Fat08]) generalize these methods to construct Brownian motion with singular drift in the interior and reflecting boundary behavior on domains with certain smoothness assumptions. There one has to exclude all points with singular drift and all non-smooth boundary points. In both cases drifts with very strong (repulsive) singularities are allowed, in particular potentials of Lennard-Jones type can be treated.

Both works make use of an elliptic regularity of Bogachev, Krylov and Röckner, see [BKR97] and [BKR01], and path-regularity techniques of Dohmann, see [Doh05].

The construction method in [FG07] is quite similar to the one of [AKR03]. We generalize this method to an abstract setting in the following way: We start with a regular symmetric strongly local Dirichlet form $(\mathcal{E}, D(\mathcal{E}))$ on $L^2(E, \mu)$ with E being a locally compact separable metric space. Well-known theorems (see Theorem 7.2.3) yield that there exists an associated sub-Markovian strongly continuous contraction semigroup $(T_t^p)_{t \geq 0}$ and resolvent $(G_\lambda^p)_{\lambda > 0}$ on $L^p(E, \mu)$ for $1 \leq p < \infty$ with associated generator $(L_p, D(L_p))$. We assume that there exists a Borel set $E_1 \subset E$ complemented by an exceptional set such that for some $1 < p < \infty$ we have

- $D(L_p) \hookrightarrow C^0(E_1)$ and the embedding is locally continuous.

- $D(L_p)$ is point separating on E_1 in the sense of Condition 2.1.2(ii).

From this we construct a semigroup of $\mathcal{L}^p$-strong Feller transition kernels $(P_t)_{t \geq 0}$ and resolvents of $\mathcal{L}^p$-strong Feller kernels $(R_\lambda)_{\lambda > 0}$. Both give regularized version of the corresponding L^p-semigroup $(T_t^p)_{t \geq 0}$ and L^p-resolvent $(G_\lambda^p)_{\lambda > 0}$, i.e., for $u \in L^p(E, \mu)$

$$P_t u(x) = \widetilde{T_t^p u}(x) \quad \text{and} \quad R_\lambda u(x) = \widetilde{G_\lambda^p u}(x) \quad \text{for } t > 0, \lambda > 0 \text{ and } x \in E_1.$$

Here $\widetilde{T_t^p u}$ $(\widetilde{G_\lambda^p u})$ denotes the continuous version of $T_t^p u$ $(G_\lambda^p u)$ which exists due to the regularity assumption on $D(L_p)$ and the mapping properties of the semigroup and resolvent.

These kernels give rise to an associated process, solving the martingale problem (for functions in a certain space) for starting points in E_1. With techniques of [Doh05] and [AKR03] we get continuity of the paths in $[0, \infty)$. For the right-continuity at $t = 0$ it is crucial to have point separating functions in $D(L_p)$.

So altogether, we obtain a general construction result for processes from symmetric regular Dirichlet forms that can be started from every point in a known set. The generality of the construction scheme is comparable to the construction of classical Feller processes but works under local assumptions. In Section 2.4 we provide concrete examples for the application of the construction scheme.

Elliptic regularity up to the boundary

We aim to apply this general scheme for construction of reflected elliptic diffusions on sets $\overline{\Omega}$ with open interior $\Omega \subset \mathbb{R}^d$, $d \in \mathbb{N}$. Therefore, we have to provide an elliptic regularity result which gives regularity of weak solutions both in the interior Ω and at boundary parts. This is the main part of Chapter 3.

We provide in Chapter 3 an Sobolev space regularity result for weak solutions of elliptic equations. This result is a (partial) generalization of a result of Morrey, see [Mor66, Theo. 5.5.4']. However, therein only a short sketch of the proof is given. Our proof is based on techniques of Shaposhnikov, see [Sha06]. There a detailed proof of an a-priori estimate of Morrey, [Mor66, Theo. 5.5.5'], is given. See also Chapter 3 for a further discussion.

We can prove local regularity at all points where the coefficient matrix is continuous and strictly elliptic, and that are either interior points or located at a C^1-smooth boundary part. Since we do not assume any global assumptions on the matrix, we can handle gradient Dirichlet forms with density, having a non-trivial zero set.

Let us now describe the proof of the regularity result: For interior points we represent a weak solution in terms of potentials containing Green's function. With this representation we can conclude iteratively higher regularity, starting from local $H^{1,2}$-regularity. For boundary points we use a reflection method to reduce this case to the interior point case.

Construction of elliptic diffusions with reflection at the boundary

Combining the results of Chapter 2 and Chapter 3 we construct elliptic diffusions in Chapter 4. They are constructed as $\mathcal{L}^p$-strong Feller processes associated with gradient Dirichlet forms. So let A be a matrix-valued mapping of symmetric and strictly elliptic matrices and ϱ a density on a set $\overline{\Omega} \subset \mathbb{R}^d$, $d \in \mathbb{N}$, with open interior Ω. Our Dirichlet form is constructed as the closure of the pre-Dirichlet form

$$\mathcal{E}(u, v) = \int_{\Omega} (A\nabla u, \nabla v) \, d\mu,$$

$$u, v \in \mathcal{D} := \left\{ u \in C_c(\overline{\Omega}) \,\middle|\, u \in H^{1,1}_{\mathrm{loc}}(\Omega), \mathcal{E}(u, u) < \infty \right\},$$

see (4.1). For the construction of the process we assume that the matrix coefficient A is C^1-smooth, the boundary of Ω is C^2-smooth boundary (except for a set of capacity zero) and certain weak differentiability conditions on the density, see Condition 4.1.1 and Condition 4.1.6. These stronger assumptions on the boundary and matrix are imposed to construct point separating functions in the domain of the L^p-generator. Nevertheless, our assumptions on the density are so weak that very singular drift terms can be handled. In particular, interacting particle systems with Lennard-Jones type potentials can be treated. For the construction we fix a boundary part $\Gamma_2 \subset \partial\Omega$, open in $\partial\Omega$, and complemented in $\partial\Omega$ by a set of capacity zero.

The set of all admissible starting points E_1 consists of all points where the density is non-zero and that are either in the interior or at the smooth boundary part Γ_2, i.e., $E_1 = (\Omega \cup \Gamma_2) \cap \{\varrho > 0\}$.

We can show that the domain of the L^p-generator contains a subspace $\mathcal{D}_{\mathrm{Neu}}$ of C^2-functions with compact support in E_1 and Neumann-type boundary on Γ_2, see (4.4). On this set the generator has the form of an elliptic differential operator of second order with singular drift, denoted by $\hat{L}$. So for $u \in \mathcal{D}_{\mathrm{Neu}}$ it holds

$$L_p u = \hat{L} u := \sum_{i,j=1}^{d} a_{ij} \partial_i \partial_j u + \sum_{j=1}^{d} \left(\sum_{i=1}^{d} \partial_i a_{ij} + \sum_{i=1}^{d} \frac{1}{\varrho} a_{ji} \partial_i \varrho \right) \partial_j u,$$

see (4.5). Using the general construction scheme from the first part together with the elliptic regularity result from the second part we obtain an $\mathcal{L}^p$-strong Feller diffusion

$$\mathbf{M} = (\mathbf{\Omega}, \mathcal{F}, (\mathcal{F}_t)_{t \geq 0}, (\mathbf{X}_t)_{t \geq 0}, (\mathbb{P}_x)_{x \in E \cup \{\Delta\}}),$$

see Chapter 2 or Section 7.3 for the notion. Then the process solves the martingale problem for $u \in D(L_p)$, in particular for $u \in \mathcal{D}_{\text{Neu}}$. So we have that

$$M_t^{[u]} := \widetilde{u}(\mathbf{X}_t) - \widetilde{u}(\mathbf{X}_0) - \int_0^t L_p u(\mathbf{X}_s) \, ds, \ t \geq 0,$$

is an $(\mathcal{F}_t)$-martingale under $\mathbb{P}_x$ for all $x \in E_1$ and $u \in D(L_p)$. Here $\widetilde{u}$ denotes the continuous version of u on E_1 provided by the elliptic regularity result.

Next we aim to investigate the boundary behavior of the constructed diffusion process. We construct the local time of the process on the boundary part $\Gamma_2 \cap \{\varrho > 0\}$. We show that the process solves a martingale problem even for C^2-functions with compact support in E_1 that do not have the Neumann boundary condition. More precisely,

$$M_t^{[u]} := u(\mathbf{X}_t) - u(\mathbf{X}_0) - \int_0^t \hat{L}u(\mathbf{X}_s) \, ds + \int_0^t (A\nabla u, \eta) \, \varrho\left(\mathbf{X}_s\right) d\ell_s, \ t \geq 0,$$

is an $(\mathcal{F}_t)$-martingale under $\mathbb{P}_x$ for all $x \in E_1$ and $u \in C_c^2(E_1)$. Here η denotes the outward unit normal at Γ_2.

We can characterize the quadratic variation process of the martingale $(M_t^{[u]})_{t \geq 0}$ in terms of the matrix coefficient. Altogether, we get a semimartingale decomposition for $(u(\mathbf{X}_t) - u(\mathbf{X}_0))_{t \geq 0}$. Using a localization technique we get such a decomposition (or Skorokhod representation) for the process itself. Denote by $(b_i)_{1 \leq i \leq d}$ the first-order coefficients of $\hat{L}$. Then we have

$$\mathbf{X}_{t \wedge \mathcal{X}}^{(i)} - \mathbf{X}_0^{(i)} = \int_0^{t \wedge \mathcal{X}} b_i(\mathbf{X}_s) \, ds - \int_0^{t \wedge \mathcal{X}} (e_i, A\eta) \varrho\left(\mathbf{X}_s\right) d\ell_s + M_{t \wedge \mathcal{X}}^{(i)}, \ t \geq 0,$$

$\mathbb{P}_x$-a.s. for $x \in E_1$ and $1 \leq i \leq d$. The $(M_t^{(i)})_{t \geq 0}$, $1 \leq i \leq d$, are continuous local martingales (up to the lifetime $\mathcal{X}$) with quadratic variation process (up to $\mathcal{X}$)

$$\langle M^{(i)}, M^{(j)} \rangle_{\cdot \wedge \mathcal{X}} = 2 \left(a_{ij} \cdot t\right)_{\cdot \wedge \mathcal{X}} \quad \text{for } 1 \leq i, j \leq d.$$

Let us emphasize that these decompositions hold under the path measures $\mathbb{P}_x$ for every $x \in E_1$, i.e., we have again a pointwise statement. For conservative processes we can further conclude existence of weak solutions.

The construction of the boundary local time and the semimartingale decomposition for $u \in C_c^2(E_1)$ is based on [FOT11, Ch. 5]. To apply these results to our setting we need the absolute continuity of the semigroup

$(P_t)_{t\geq 0}$ and certain regularity of potentials of the surface measure at compact boundary parts. We first have to refine the construction theorem in [FOT11, Theo. 5.1.6] to our setting since the semigroup $(P_t)_{t\geq 0}$ is in our case absolutely continuous on E_1 only. Then we apply our regularity result from Chapter 3 to potentials of the surface measure at compact boundary parts to conclude the regularity needed to apply the construction of additive functionals. Our regularity result implies even continuity properties of the potentials, but for the construction boundedness properties are already sufficient. Additional care has to be taken in our setting due to the singularity of the drifts.

We apply our results to concrete models in Mathematical Physics. We construct stochastic dynamics for finite particle systems with singular interaction in continuum and reflection at the boundary of the state space. Our approach allows very singular interaction potentials of Lennard-Jones type.

Furthermore, we construct stochastic dynamics for Ginzburg-Landau interface models with reflection (also called: entropic repulsion) at a hard wall. There we can also handle general potentials. These dynamics describe the random evolution of an interface, e.g., the surface of a liquid that is conditioned to stay above a hard wall.

Let us mention other works concerning the construction of reflected diffusions, see also the beginning of Chapter 6 for a more detailed comparison with other works. Strong solutions are constructed by Lions and Sznitman, see [LS84]. Strook and Varadhan construct reflected diffusions as solutions to the sub-martingale problem, see [SV71]. Moreover, there are several works on reflected diffusions and Dirichlet forms: Let us mention [FT95] and [FT96] where classical Feller processes associated with gradient Dirichlet forms with uniformly elliptic coefficient matrix, but without density are constructed. Using the results of [FOT94] a semimartingale decomposition is given. Note, however, that their setting does not cover the case of diffusions with singular drift term. Trutnau (see [Tru03]) constructs diffusions with reflection and singular drift using generalized non-symmetric Dirichlet forms, admitting a more general class of drift terms. The derived semimartingale decomposition, however, holds for quasi-every starting point only.

Furthermore, one can use operator semigroups with Feller(-type) regularizing properties also in the infinite dimensional setting for constructing martingale (or even weak) solutions to stochastic partial differential equations, see e.g. [PR02] and [RS06].

Finally, let us summarize the core results and progress achieved by this work:
- We obtain a general construction result for $\mathcal{L}^p$-strong Feller processes from analytic assumptions, see Theorem 2.1.3.

- We prove a local Sobolev space regularity result up to the boundary for elliptic equations under local assumptions on the coefficients and boundary, see Theorem 3.1.1.
- We construct $\mathcal{L}^p$-strong Feller elliptic diffusions with singular drift and reflection at the boundary, see Theorem 4.1.14.
- We provide a pointwise Skorokhod decomposition for the constructed diffusions, see Theorem 6.2.9 and Theorem 6.3.2, and obtain weak solutions, see Theorem 6.3.5.

In the Appendix we provide several auxiliary results. Most of them are well-known but we needed from time to time modified versions which apply for our specific settings.

1.2 Notation

The notation we use in this work is quite standard. By $\mathbb{N}, \mathbb{Q}, \mathbb{R}, \mathbb{C}$ we denote the set of all natural, rational, real and complex numbers, respectively. Notions like *positive* or *increasing* are meant in the non-strict sense. By $|\cdot|$ we denote the euclidean norm on $\mathbb{R}^d$, $d \in \mathbb{N}$, which yields in one dimension just the modulus. By $(\cdot, \cdot)$ we always denote the Euclidean scalar product on $\mathbb{R}^d$, $d \in \mathbb{N}$. All other scalar products will be explicitly distinguished. For real numbers a and b we denote by $a \wedge b$ and $a \vee b$ the minimum and maximum, respectively.

For a topological space (E, τ) we denote by $\mathcal{B}(E)$ the Borel σ-algebra generated by the open sets. By $\mathcal{B}_b(E)$, respectively $\mathcal{B}^+(E)$, we denote the set of Borel-measurable real-valued bounded, respectively positive, functions. For a subset $A \subset E$ we denote by $\overset{\circ}{A}$, $\overline{A}$ and ∂A the interior, closure and boundary of A, respectively. For $A \subset B$ we say A is open (closed) in B if A is open (closed) w.r.t. the trace topology of E on B. For a matrix A we denote by $A^\top$ the transpose of A.

By $C^0(E)$, $E \subset \mathbb{R}^d$, we denote the space of all continuous functions on E, by $C^{0,\alpha}(E)$, $0 < \alpha < 1$, the space of all Hölder continuous functions of order α. By $C^{m,\alpha}(E)$, $E \subset \mathbb{R}^d$, $d, m \in \mathbb{N}$, $0 < \alpha < 1$, we denote the set of all m-times continuously differentiable functions in $\overset{\circ}{E}$ such that the derivatives up to order $m - 1$ admit a continuous continuation to $\partial E \cap E$ (possibly empty, e.g. if E is open) and the m-th derivatives admit a Hölder continuous extension of order α. The subindex c marks that the functions are supposed to have compact support in E. The subindex b marks that the function and its derivatives up to order m are bounded on E. By *support of a function* u we denote the closure of all points where u is not zero, denoted by $\mathrm{supp}[u]$. Denote by $C^\infty(E)$ the intersection of all $C^m(E)$, $m \in \mathbb{N}$. With *cutoff* for A in B we mean a $C_c^\infty(E)$-function that is constantly equal 1 on A and has compact support in B. We call an open subset of $\mathbb{R}^d$ sometimes also a *domain*. For a differentiable function u on an open set $\Omega \subset \mathbb{R}$ we denote by ∇u the gradient, seen as a column vector, and by $\partial_i u$, $1 \leq i \leq d$, the partial derivative in direction e_i, e_i the i-th unit vector. The expression $(\nabla u)(x-y)$ means evaluating the gradient of u at $x - y$ rather than differentiating the function $x \mapsto u(x-y)$ or $y \mapsto u(x-y)$. By dx we denote the Lebesgue measure, by ε_x the point measure in a point x. Let $(E, \mathcal{B})$ be a measurable space with a topology τ. For a measure μ we denote by *topological support of* μ all points for which an open neighborhood with positive μ-measure exists. We denote by $\mathcal{L}^p(E, \mu)$, $1 \leq p \leq \infty$, the space of p-integrable functions and by $L^p(E, \mu)$ the corresponding equivalence classes. For a function space $\mathcal{L}$ we denote by $\sigma(\mathcal{L})$ the σ-algebra generated by $\mathcal{L}$, i.e., the smallest σ-algebra for which all functions in $\mathcal{L}$ are measurable. By $H^{m,p}(\Omega)$, $\Omega \subset \mathbb{R}^d$ open, $m, p \in \mathbb{N}$, $p \geq 1$, we denote the *Sobolev space* of m-times weakly differentiable L^p-functions with $L^p(\Omega, dx)$-integrable derivatives. The corresponding local spaces are marked by the subindex loc, they are introduced in Section 7.5. In a metric space $(E, \mathbf{d})$ we denote by $B_r(x)$ the ball with radius $r > 0$ around $x \in E$. By $\mathrm{dist}(x, A)$, $x \in E$, $A \subset E$, we denote the *distance* of x to A.

2 Construction of $\mathcal{L}^p$-Strong Feller Processes

In this chapter we provide a general construction scheme for $\mathcal{L}^p$-strong Feller processes on locally compact separable metric spaces. The construction result yields that starting from certain regularity conditions on the semigroup associated with a symmetric Dirichlet form, one obtains a diffusion process which solves the corresponding martingale problem for every starting point from an explicitly known set. In Theorem 2.3.10 we mention further useful properties of the process, formulated also as pointwise statements. In Section 2.4 we provide concrete examples. Our results and their proofs are based on [AKR03] and [Doh05]. We got also many ideas from [FG07], [FG08] and [Sti10]. For the construction of classical Feller processes from strongly continuous contraction semigroups on spaces of continuous functions vanishing at infinity, see e.g. [BG68, Ch. I, Theo 9.4]. There are also results on the construction of Hunt processes from resolvents of kernels, see [Sto83] and Remark 2.3.1 below.

We have published the results of this chapter already in [BGS13].

2.1 A General Construction Scheme

For readers, who are unfamiliar with the concepts of Dirichlet forms or $\mathcal{L}^p$-strong Feller processes, it might help to have a look at the examples provided in Section 2.4 first.

Throughout Section 2.2 and 2.3 we fix a metric space $(E, \mathbf{d})$, a measure μ on the Borel σ-algebra $\mathcal{B}(E)$ and a Dirichlet form $(\mathcal{E}, D(\mathcal{E}))$ on $L^2(E, \mu)$.

We assume the following conditions.

Condition 2.1.1.
(i) $(E, \mathbf{d})$ is a locally compact separable metric space.
(ii) μ is a locally finite Borel measure with full topological support.
(iii) $(\mathcal{E}, D(\mathcal{E}))$ is symmetric, regular and strongly local.

Except for the locality assumption these are the standard assumptions under which Dirichlet forms and stochastic processes are considered in [FOT11]. With locally finite Borel measure we mean that μ is defined on the Borel σ-algebra and is finite on compact sets. With full topological support we mean that for every $x \in E$ there exists a neighborhood U of x such that $\mu(U) > 0$. By the Beurling-Deny theorem there exists an associated strongly continuous contraction semigroup on $L^r(E, \mu)$ (L^r-s.c.c.s) $(T_t^r)_{t>0}$ with generator $(L_r, D(L_r))$ for every $1 \leq r < \infty$, see Theorem 7.2.3. If $r > 1$ then $(T_t^r)_{t>0}$ is the restriction of an analytic semigroup. Here associated means that for $f \in L^1(E, \mu) \cap L^\infty(E, \mu)$ it holds $T_t^2 f = T_t^r f$ for every $t \geq 0$ where $(T_t^2)_{t \geq 0}$ is the unique L^2-s.c.c.s associated with $(\mathcal{E}, D(\mathcal{E}))$.

We assume the following stronger conditions that are needed to get refined pointwise results.

Condition 2.1.2.
There exists a Borel set $E_1 \subset E$ with $\text{cap}_{\mathcal{E}}(E \setminus E_1) = 0$ and $1 < p < \infty$ such that

(i) $D(L_p) \hookrightarrow C^0(E_1)$ and the embedding is locally continuous, i.e., for $x \in E_1$ there exists an E_1-neighborhood U and a constant $C_1 = C_1(U) < \infty$ such that

$$\sup_{y \in U} |\tilde{u}(y)| \leq C_1 \|u\|_{D(L_p)} \quad \text{for all } u \in D(L_p). \tag{2.1}$$

Here $\tilde{u}$ denotes the continuous version of u.

(ii) For each point $x \in E_1$ there exists a sequence of functions $(u_n)_{n \in \mathbb{N}}$ in $D(L_p)$ such that

a) Either $\{u_n^2 \mid n \in \mathbb{N}\} \subset D(L_p)$ or $0 \leq u_n \leq 1$ and $u_n(x) = 1$ for all $n \in \mathbb{N}$.

b) The sequence $(u_n)_{n \in \mathbb{N}}$ is point separating in x.

Here $C^0(S)$ denotes the space of all continuous functions on a topological space S. By $\|\cdot\|_{D(L_p)}$ we denote the graph norm of $(L_p, D(L_p))$. Point separating in x means that for every $y \neq x$, $y \in E$, there exists u_n such that $u_n(y) = 0$ and $u_n(x) = 1$. We adjoin to E an extra point Δ which is not contained in E. We endow $E^\Delta := E \cup \{\Delta\}$ with the topology of the Alexandrov one-point compactification of E. The open neighborhoods of Δ are given by the complements of compact subsets of E.

Under Condition 2.1.1 and Condition 2.1.2 we obtain the following theorem.

Theorem 2.1.3. *There exists a diffusion process (i.e., a strong Markov process having continuous sample paths on the time interval $[0, \infty)$)* $\mathbf{M} = (\Omega, \mathcal{F}, (\mathcal{F}_t)_{t \geq 0}, (\mathbf{X}_t)_{t \geq 0}, (\mathbb{P}_x)_{x \in E \cup \{\Delta\}})$ *with state space E and cemetery Δ, the Alexandrov point of E. The process leaves $E_1 \cup \{\Delta\}$ $\mathbb{P}_x$-a.s., $x \in E_1 \cup \{\Delta\}$, invariant. The transition semigroup $(P_t)_{t \geq 0}$ is associated with $(T_t^2)_{t \geq 0}$ and is $\mathcal{L}^p$-strong Feller, i.e., $P_t \mathcal{L}^p(E, \mu) \subset C^0(E_1)$ for $t > 0$. The process has continuous paths on $[0, \infty)$ and it solves the martingale problem associated with $(L_p, D(L_p))$ for starting points in E_1, i.e.,*

$$M_t^{[u]} := \widetilde{u}(\mathbf{X}_t) - \widetilde{u}(x) - \int_0^t L_p u(\mathbf{X}_s) \, ds, \ t \geq 0,$$

is an $(\mathcal{F}_t)$-martingale under $\mathbb{P}_x$ for all $u \in D(L_p)$ and $x \in E_1$. As filtration $(\mathcal{F}_t)_{t \geq 0}$ we take the natural filtration, defined in (7.8) below.

Here $(P_t)_{t \geq 0}$ being associated with $(T_t^2)_{t \geq 0}$ means that $P_t f$ is a μ-version of $T_t^2 f$ for $f \in \mathcal{L}^1(E, \mu) \cap \mathcal{B}_b(E)$ (the space of Borel-measurable bounded functions). By $\mathcal{L}^p(E, \mu)$ we denote the space of all p-integrable functions on (E, μ).

Remark 2.1.4. The continuity holds with respect to topology of the Alexandrov one-point compactification of E^Δ. This means that the process has continuous paths in E and reaches Δ only by leaving continuously every compact set of E. The integral in $M_t^{[u]}$ exists and is independent of the μ-version of $L_p u$. This will be seen in the proof below.

The theorem is proven in Section 2.3, see page 31 and Theorem 2.3.11 below. Further useful properties of the constructed process are proven in Theorem 2.3.10 below.

Under additional conditions, the corresponding resolvent of kernels $(R_\lambda)_{\lambda > 0}$ are even *strong Feller*, i.e., $R_\lambda \mathcal{B}_b(E) \subset C^0(E_1)$. More precisely, we have the following theorem.

Theorem 2.1.5. *Assume the following conditions.*

(i) *For every $x \in E_1$ there exists a neighborhood $U \subset E_1$ such that for the closure in E it holds $\overline{U} \subset E_1$ and $\overline{U}$ is compact.*

(ii) *For every sequence $(u_n)_{n \in \mathbb{N}}$ in $D(L_p)$ such that $((1 - L_p)u_n)_{n \in \mathbb{N}}$ is uniformly bounded in the $\| \cdot \|_{L^\infty}$-norm it holds that $(u_n)_{n \in \mathbb{N}}$ is equicontinuous.*

Then $(R_\lambda)_{\lambda>0}$ is strong Feller.

For the proof see Section 2.2 (page 21).

Remark 2.1.6. In [AKR03] it is shown that strong Feller property of $(R_\lambda)_{\lambda>0}$ and conservativity of $(\mathcal{E}, D(\mathcal{E}))$ imply that $(P_t)_{t>0}$ is strong Feller. The proof generalizes to the case considered here.

Having strong Feller properties of the resolvent family at hand, we can provide a conservativity criterion for the process **M**.

Corollary 2.1.7. *Assume that $(\mathcal{E}, D(\mathcal{E}))$ is conservative and $(R_\lambda)_{\lambda>0}$ is strong Feller, then* **M** *from Theorem 2.1.3 is conservative for every starting point $x \in E_1$.*

See p. 34 for the proof.

For constructions in the later chapters it is convenient to consider the so-called restriction of the process from Theorem 2.1.3 to $E_1 \cup \{\Delta\}$. We obtain the following corollary.

Corollary 2.1.8. *Let* $\mathbf{M} = (\Omega, \mathcal{F}, (\mathcal{F}_t)_{t\geq0}, (\mathbf{X}_t)_{t\geq0}, (\mathbb{P}_x)_{x\in E\cup\{\Delta\}})$ *be the diffusion process constructed in Theorem 2.1.3. Let the restricted process* $\mathbf{M}^1 := (\Omega^1, \mathcal{F}^1, (\mathcal{F}_t^1)_{t\geq0}, (\mathbf{X}_t^1)_{t\geq0}, (\mathbb{P}_x^1)_{x\in E_1\cup\{\Delta\}})$ *be defined as in Definition 7.3.18 with $\widetilde{E}_1 := E_1 \cup \{\Delta\}$. Then* $\mathbf{M}^1$ *is a $\mathcal{L}^p$-strong Feller diffusion process with state space E_1 and cemetery Δ. The transition semigroup $(P_t^{E_1\cup\{\Delta\}})_{t\geq0}$ is absolutely continuous on E_1.*

See p. 35 for the proof.

Remark 2.1.9. The filtration $(\mathcal{F}_t^1)_{t\geq0}$ and $\mathcal{F}^1$ are important for the construction of additive functionals in Chapter 6. There some subsets of Ω have full $\mathbb{P}_x$-measure for $x \in E_1 \cup \{\Delta\}$ only.

2.2 Construction of $\mathcal{L}^p$-strong Feller Kernels

We start with the construction of a semigroup of kernels $(P_t)_{t>0}$ and resolvent of kernels $(R_\lambda)_{\lambda>0}$ which yield a μ-version of $(T_t^p)_{t>0}$ and $(G_\lambda^p)_{\lambda>0}$. For this we assume Condition 2.1.1 and Condition 2.1.2 and fix a $1 < p < \infty$ as in Condition 2.1.2. We adapt the structure of [AKR03, Sec. 3] and modify the statements and proofs there in order to cover the abstract setting.

Remark 2.2.1. The restriction of μ to $\mathcal{B}(E_1)$ is also strictly positive on non-empty open sets. Indeed, let $\widetilde{U} \subset E_1$ be non-empty and open w.r.t. the trace topology. Then there exists $U \subset E$ open such that $\widetilde{U} = U \cap E_1$. So $U \setminus \widetilde{U} \subset E \setminus E_1$, the latter is of capacity zero and hence has also μ measure equal to zero. So $\mu(\widetilde{U}) = \mu(U) > 0$. In particular, if $\tilde{u}$ is continuous on E_1 and equal to zero μ-almost everywhere in E_1, then $\tilde{u}$ is equal to zero on E_1. This implies that if u $\mathcal{B}(E)$-measurable has a continuous version on E_1, then this version is unique on E_1.

For the associated L^p-resolvent $(G_\lambda^p)_{\lambda>0}$ it holds $D(L_p) = G_\lambda^p L^p(E, \mu)$. So for $f \in L^p(E, \mu)$, $G_\lambda^p f$ has a unique continuous version denoted by $\widetilde{G_\lambda^p f}$. The boundedness of $G_\lambda : L^p(E, \mu) \to D(L_p)$ together with (2.1) yields a constant $C_2 = C_2(\lambda, U) < \infty$, U as in Condition 2.1.2, such that

$$\sup_{y \in U} |\widetilde{G_\lambda^p f}(y)| \leq C_1 \|G_\lambda^p f\|_{D(L_p)} \leq C_2 \|f\|_{L^p(E,\mu)} \quad \text{for all } f \in L^p(E, \mu). \tag{2.2}$$

Since $1 < p < \infty$, the L^p-semigroup $(T_t^p)_{t>0}$ is the restriction of an analytic semigroup. Thus $T_t^p L^p(E, \mu) \subset D(L_p)$ for $t > 0$. So for $u \in L^p(E, \mu)$, $t > 0$, $T_t^p f$ has a unique continuous version, denoted by $\widetilde{T_t^p f}$. From Lemma 7.2.2(ii) we get that there exists a constant $C_3 < \infty$ such that

$$\|T_t^p f\|_{D(L_p)} \leq \left(1 + \frac{C_3}{t}\right) \|f\|_{L^p(E,\mu)}, \quad t > 0, \text{ for all } f \in L^p(E, \mu).$$

Combining with (2.1) we obtain for $t > 0$ a constant $C_4 = C_4(t, U) < \infty$ such that

$$\sup_{y \in U} |\widetilde{T_t^p f}(y)| \leq C_1 \|T_t^p f\|_{D(L_p)} \leq C_4 \|f\|_{L^p(E,\mu)} \quad \text{for all } f \in L^p(E, \mu). \tag{2.3}$$

In particular, L^p-convergence of a sequence $(f_n)_{n \in \mathbb{N}}$ to f implies pointwise convergence of $(\widetilde{T_t^p f_n}(x))_{n \in \mathbb{N}}$ to $\widetilde{T_t^p f}(x)$, $t > 0$, and pointwise convergence of $(\widetilde{G_\lambda^p f_n}(x))_{n \in \mathbb{N}}$ to $\widetilde{G_\lambda^p f}(x)$, $\lambda > 0$, for $x \in E_1$.

If $u \in D(L_p)$ then $\|T_t u\|_{D(L_p)} \leq \|u\|_{D(L_p)}$, $t \geq 0$, and hence

$$\sup_{y \in U} |\widetilde{T_t^p u}(y)| \leq C_1 \|u\|_{D(L_p)}, \quad t \geq 0. \tag{2.4}$$

Most of the time we omit the upper index p when applying the L^p-semigroup or resolvent to a function $f \in L^p(E, \mu)$.

The following well-known lemma is useful for monotone class arguments.

Lemma 2.2.2. *Let $(E, \mathbf{d})$ be a locally compact separable metric space. Then there exists a sequence of compact sets $(K_n)_{n \in \mathbb{N}}$, $K_n \subset \overset{\circ}{K}_{n+1}$ (the interior of K_{n+1}), with $E = \bigcup_{n \in \mathbb{N}} K_n$. Furthermore, if μ is a locally finite measure on $\mathcal{B}(E)$, then $\mathcal{B}(E)$ is generated by the open sets of finite measure.*

For the proof of the existence of such a covering, see e.g. Lemma 7.1.1 or [CB06, Cor. 2.77]. The second statement follows then directly.

For many constructions we need the Functional Monotone Class Theorem, see e.g. [BG68, Ch. 0, Theo. 2.3]. See also [Wer11, Lem. VII.1.5] for a similar statement. Denote by $\mathcal{B}_b(E, \mathcal{B})$ the space of all bounded $\mathcal{B}$-measurable real-valued functions on the measurable space $(E, \mathcal{B})$.

Theorem 2.2.3. *Let $(E, \mathcal{B})$ be a measurable space and $\mathcal{L} \subset \mathcal{B}$ be an intersection-stable generator of $\mathcal{B}$. Let $\mathcal{H} \subset \mathcal{B}_b(E, \mathcal{B})$ be a subset having the following three properties:*

(i) $\mathcal{H}$ is a vector space over $\mathbb{R}$.

(ii) $\mathcal{H}$ contains 1_E as well as the indicator function 1_F for $F \in \mathcal{L}$.

(iii) If $f_n \in \mathcal{H}$, $f_n \geq 0$, $n \in \mathbb{N}$, such that $f_n \uparrow f$ as $n \to \infty$ and f is bounded, then $f \in \mathcal{H}$.

Then $\mathcal{H} = \mathcal{B}_b(E, \mathcal{B})$.

If $\mathcal{B} = \mathcal{B}(E)$ is the Borel σ-algebra, we just write $\mathcal{B}_b(E)$ instead of $\mathcal{B}_b(E, \mathcal{B}(E))$ for the space of bounded Borel-measurable functions. The following corollary is useful to apply the functional monotone class theorem.

Corollary 2.2.4. *Assume that $(E, \mathbf{d})$ is a metric space and $\mathcal{H} \subset \mathcal{B}_b(E)$ fulfills 2.2.3(i) and 2.2.3(iii). If $C_b(E) \subset \mathcal{H}$, then $\mathcal{H} = \mathcal{B}_b(E)$.*

Proof. Define $\mathcal{L}$ to be the family of open sets in $(E, \mathbf{d})$. Note that this is an intersection-stable generator of $\mathcal{B}(E)$. For every open set U, there exists a sequence of bounded continuous functions $(f_n)_{n \in \mathbb{N}}$ with $f_n \uparrow 1_U$. By 2.2.3(iii) also $1_U \in \mathcal{H}$. So $\mathcal{H}$ fulfills 2.2.3(ii). Since $\mathcal{L}$ is an intersection-stable generator of $\mathcal{B}(E)$, we get $\mathcal{H} = \mathcal{B}_b(E)$. $\qquad\square$

However, in many applications we have to restrict to continuous bounded functions that are sufficiently integrable. Therefore, the following corollary is even more suited to our application.

Corollary 2.2.5. *Assume that $(E, \mathbf{d})$ is a locally compact separable metric space, μ a locally finite measure on $\mathcal{B}(E)$. Then the conclusion of Theorem 2.2.3 holds for the measurable space $(E, \mathcal{B}(E))$ if instead of (ii) we assume either*

(ii') For U open, $\mu(U) < \infty$, it holds $1_U \in \mathcal{H}$, or

(ii'') For U open, $\mu(U) < \infty$, there exist $f_n \in \mathcal{H}$, $n \in \mathbb{N}$, such that $f_n \uparrow 1_U$.

In particular, if $C_b(E) \cap L^p(E, \mu) \subset \mathcal{H}$ and $\mathcal{H}$ has 2.2.3(i) and 2.2.3(iii), then $\mathcal{H} = \mathcal{B}_b(E)$.

Proof. (ii') implies 2.2.3(ii): We have $E = \bigcup_{n\in\mathbb{N}} \overset{\circ}{K}_n$ with K_n as in Lemma 2.2.2. Thus the open sets of finite measure are an intersection stable-generator. Hence we can choose as $\mathcal{L}$ the system of all open sets with finite measure. Moreover, $1_{\overset{\circ}{K}_n} \uparrow 1_E$. So together with 2.2.3(iii) it follows $1_E \in \mathcal{H}$. Altogether, (ii') implies 2.2.3(ii).

(ii'') implies (ii'): Follows directly by 2.2.3(iii).

For the proof of the last claim, recall that for every open set U, there exists a sequence of continuous functions $(f_n)_{n\in\mathbb{N}}$ with $f_n \uparrow 1_U$. So $f_n \in C_b(E)$ and if $\mu(U) < \infty$ then also $f_n \in L^p(E, \mu)$. So if $C_b(E) \cap L^p(E, \mu) \subset \mathcal{H}$, then $\mathcal{H}$ fulfills (ii''). $\qquad\square$

Now we construct the semigroup of kernels $(P_t)_{t\geq 0}$ on $E_1 \times \mathcal{B}(E)$. As in [AKR03, Lem. 3.1] we may apply the Daniell-Stone theorem to get the following result.

Lemma 2.2.6. *Let $t > 0$, $x \in E_1$. Then the map*

$$\mathcal{L}^p(E, \mu) \ni f \mapsto \widetilde{T_t f}(x) \in \mathbb{R} \tag{2.5}$$

is a Daniell integral, cf. [Bau78, Def. 39.1], and there exists a unique positive measure $P_t(x, dy)$ on $\mathcal{B}(E)$ such that

$$\widetilde{T_t f}(x) = \int_E f(y) P_t(x, dy). \tag{2.6}$$

Proof. By positivity of $T_t f$ and continuity on E_1 we have that $\widetilde{T_t f}(x) \geq 0$ for every $x \in E_1$ if $f \geq 0$. Using linearity of T_t and continuity of $\widetilde{T_t f}$ we get that the mapping (2.5) is also linear. Moreover, if $f_n \downarrow 0$ μ-a.e. this convergence also holds in $L^p(E, \mu)$ by Lebesgue's dominated convergence. So (2.3) yields $\widetilde{T_t f_n}(x) \xrightarrow{n \to \infty} 0$.

By positivity this convergence is also monotone. Thus the map is a Daniell integral. By [Bau78, Satz 39.4] there exists a positive measure denoted by $P_t(x, dy)$ on $\sigma(\mathcal{L}^p)$ (the σ-algebra generated by $\mathcal{L}^p$) such that (2.6) holds. Note that for every set $M \in \mathcal{B}(E)$ of finite measure it holds $1_M \in \mathcal{L}^p(E, \mu)$. By Lemma 2.2.2 $\mathcal{B}(E)$ is generated by the open sets of finite measure. Hence $\mathcal{B}(E) = \sigma(\mathcal{L}^p(E, \mu))$. Moreover, the measure is unique since the open sets of finite measure are an intersection-stable generator.   □

Remark 2.2.7. Note that the map (2.5) is formulated on the $\mathcal{L}^p$-functions rather than on the μ-equivalence classes from $L^p(E, \mu)$. However, the operator T_t respects μ-equivalence classes. So two different representatives of an element in $L^p(E, \mu)$ lead to the same equivalence class $T_t f$ and to the same unique continuous version $\widetilde{T_t f}$. So the mapping (2.5) is also well-defined as a mapping $L^p(E, \mu) \to \mathbb{R}$.

The kernels $(P_t)_{t>0}$ naturally induce linear operators acting on function spaces. These operators we define next and denote them also by $(P_t)_{t \geq 0}$.

Definition 2.2.8. For $t > 0$, $x \in E_1$ and $f \in \mathcal{L}^1(E, P_t(x, dy)) \cup \mathcal{B}^+(E)$ we define

$$P_t f(x) := \int_E f(y) P_t(x, dy)$$

and $P_0 f(x) := f(x)$.

We generalize [AKR03, Prop. 3.2] to obtain the following important properties of the semigroup of kernels $(P_t)_{t \geq 0}$.

Theorem 2.2.9.

(i) Let $x \in E_1$. It holds $P_t 1_E(x) \leq 1$. There exists a $\mathcal{B}(E_1 \times E)$-measurable map $(x, y) \mapsto p_t(x, y)$, $t > 0$, such that $P_t(x, dy) = p_t(x, y) d\mu(y)$.

Furthermore, $P_t(x, E \setminus E_1) = 0$, so the kernels defined in Definition 2.2.8 can be considered as kernels on E_1, denoted by the same symbol below. Moreover, $\mathcal{L}^p(E, \mu) \subset \mathcal{L}^1(E, P_t(x, dy))$ and $P_t f(x) = \widetilde{T_t^p f}(x)$ for all $x \in E_1$, $f \in \mathcal{L}^p(E, \mu)$, i.e., $P_t f$ is the unique continuous version of $T_t^p f$.

(ii) $(P_t)_{t>0}$ is a semigroup of kernels on E_1 which is $\mathcal{L}^p$-strong Feller, i.e., $P_t f \in C^0(E_1)$ for all $t > 0$, $f \in \mathcal{L}^p(E, \mu)$. Moreover, $P_{t+s}f = P_t P_s f$ for $f \in \mathcal{L}^p(E, \mu)$ and $t, s \geq 0$.

(iii) For $f \in L^p(E, \mu)$ and $s > 0$

$$\lim_{t \to 0} P_{t+s}f(x) = P_s f(x) \quad \text{for all } x \in E_1.$$

For $f \in D(L_p)$ we have, denoting by $\widetilde{f}$ the continuous version of f on E_1,

$$\lim_{t \to 0} P_t f(x) = \widetilde{f}(x) \quad \text{for all } x \in E_1.$$

(iv) $(P_t)_{t>0}$ is a measurable semigroup on E_1, i.e., for $f \in \mathcal{B}_b(E)$ the map $(t, x) \mapsto P_t f(x)$ is $\mathcal{B}([0, \infty) \times E_1)$-measurable. This holds also for $f \in \mathcal{L}^p(E, \mu)$.

Proof. (i): Let $K_n \subset E$, $n \in \mathbb{N}$, be the sequence of Lemma 2.2.2 with $\mu(K_n) < \infty$ and $E = \bigcup_{n \in \mathbb{N}} K_n$. Then $T_t^p 1_{K_n} \leq 1$ μ-a.e. on E. By continuity we have $P_t 1_{K_n}(x) = \widetilde{T_t^p 1_{K_n}}(x) \leq 1$ for every $x \in E_1$. By monotone convergence it holds $P_t 1_E(x) = \sup_{n \in \mathbb{N}} P_t 1_{K_n}(x) \leq 1$ for $x \in E_1$ and $t > 0$.

Let $N \in \mathcal{B}(E)$ with $\mu(N) = 0$. Then $1_N = 0$ μ-a.e., so $T_t^p 1_N = 0$ μ-a.e. So by continuity $P_t(x, N) = P_t 1_N(x) = \widetilde{T_t^p 1_N}(x) = 0$ for every $x \in E_1$. Hence $P_t(x, dy)$ is absolutely continuous w.r.t. μ.

Thus by the Radon-Nikodym theorem there exists a map $p_t(x, y)$ such that $p_t(x, y)$ is the density of $P_t(x, dy)$ for all $x \in E_1$ and $t > 0$.

To prove measurability note that for $U \in \mathcal{B}(E)$ with $\mu(U) < \infty$ the mapping $x \to P_t(x, U) = \widetilde{T_t^p 1_U}(x)$ is measurable in x, hence by monotone approximation using Lemma 2.2.2 also for arbitrary $U \in \mathcal{B}(E)$. The existence of a $\mathcal{B}(E_1 \times E)$-measurable density for P_t follows by Theorem 7.1.7, see also [Doo53, Theo. 2.5] and [Doo53, Exa. 2.7].

Since $\text{cap}_\varepsilon(E \setminus E_1) = 0$, we have $\mu(E \setminus E_1) = 0$. So $P_t(x, E \setminus E_1) = 0$. Now let $f \in \mathcal{L}^p(E, \mu)$ and consider f^+. Then for all $x \in E_1$, $t > 0$, we have by construction of P_t

$$P_t f^+(x) = \widetilde{T_t^p f^+}(x) < \infty.$$

Thus $f^+ \in \mathcal{L}^1(E, P_t(x, dy))$. The same reasoning works for f^-, thus $f \in \mathcal{L}^1(E, P_t(x, dy))$ and $P_t f(x) = \widetilde{T_t^p f}(x)$.

(ii): Let $f \in \mathcal{L}^p(E, \mu)$. Then by the $L^p(E, \mu)$-semigroup property and continuity we have for all $x \in E_1$, $t, s \geq 0$

$$\widetilde{T_{t+s}f}(x) = \widetilde{T_t(\widetilde{T_s f})}(x).$$

Thus

$$P_{t+s}f(x) = P_t(P_s f)(x) \quad \text{for all } x \in E_1. \tag{2.7}$$

Note that this holds in particular for $f = 1_U$, U open with $\mu(U) < \infty$. To prove it for Borel bounded functions observe that the system of functions in $\mathcal{B}_b(E)$ for which property (2.7) holds is a vector space satisfying 2.2.3(iii). So by Corollary 2.2.5 it follows that (2.7) holds also for $u \in \mathcal{B}_b(E)$. The other statements are clear by construction.

(iii): First, we prove the statement for $s = 0$ and $f \in D(L_p)$. Note that the L^p-semigroup $(T_t)_{t \geq 0}$ is strongly continuous on $D(L_p)$ w.r.t. the graph norm of $(L_p, D(L_p))$, see Lemma 7.2.1. So for $f \in D(L_p)$ we get using (2.1) for $x \in E_1$

$$|\widetilde{T_t f}(x) - \widetilde{f}(x)| \leq C_1 \|T_t f - f\|_{D(L_p)} \xrightarrow{t \to 0} 0.$$

Since $P_t f(x) = \widetilde{T_t f}(x)$, we get

$$\lim_{t \to 0} P_t f(x) = \lim_{t \to 0} \widetilde{T_t f}(x) = \widetilde{f}(x) = P_0 \widetilde{f}(x) \quad \text{for all } x \in E_1.$$

For $s > 0$, $f \in L^p(E, \mu)$, we have by analyticity of $(T_t)_{t > 0}$ that $P_s f = \widetilde{T_s f} \in D(L_p)$. By the semigroup property of the kernels $(P_t)_{t \geq 0}$ we have

$$\lim_{t \to 0} P_{t+s}f(x) = \lim_{t \to 0} P_t(P_s f)(x) = P_0(P_s f)(x) = P_s f(x) \text{ for all } x \in E_1.$$

(iv): First, let $f \in \mathcal{L}^p(E, \mu)$. Define for $n \in \mathbb{N}_0$, $S_n := \{k2^{-n} \mid k \in \mathbb{N}_0\}$, $s_k^{(n)} := k2^{-n}$, $k \in \mathbb{N}_0$, $M_k^{(n)} = (s_{k-1}^{(n)}, s_k^{(n)}]$, $k \in \mathbb{N}$, and $M_0^{(n)} = \{0\}$. For $t > 0$ define $t_n := \min\{s \in S_n \mid t \leq s\}$ and for $t = 0$ define $t_n := 0$, $n \in \mathbb{N}$. Clearly $t_n \downarrow t$ as $n \to \infty$. We define for $n \in \mathbb{N}$ and $x \in E_1$

$$P^n f : [0, \infty) \ni (t, x) \mapsto P_t^n f(x) := P_{t_n} f(x).$$

Then for $A \in \mathcal{B}(\mathbb{R})$ it holds

$$(P^n f(\cdot))^{-1}(A) = \bigcup_{k \in \mathbb{N}_0} M_k^{(n)} \times (P_{s_k^{(n)}} f)^{-1}(A) \in \mathcal{B}(\mathbb{R}_0^+ \times E_1).$$

Thus $P^n f$ is measurable. Now note that for $t = 0$ we have $P^n_t f(x) = P_0 f(x)$ and for $t > 0$ we have $P^n_t f(x) = P_{t_n} f(x) \xrightarrow{n \to \infty} P_t f(x)$ for $x \in E_1$ by (iii). So $P_t f$ is measurable for $f \in \mathcal{L}^p(E, \mu)$. Then measurability for general $u \in \mathcal{B}_b(E)$ follows as in (ii) using Corollary 2.2.5.

$\square$

For the resolvent we obtain similar statements. These are obtained by generalizing [AKR03, Lem. 3.4] and [AKR03, Prop. 3.5].

Lemma 2.2.10. *Let $0 < \lambda < \infty$ and $x \in E_1$. Then the map*

$$\mathcal{L}^p(E, \mu) \ni f \mapsto \widetilde{G^p_\lambda f}(x) \in \mathbb{R}$$

is a Daniell integral, hence there exists a unique positive measure $R_\lambda(x, dy)$ on $\mathcal{B}(E)$ such that

$$\widetilde{G^p_\lambda f}(x) = \int_E f(y) R_\lambda(x, dy) \quad \text{for all } f \in \mathcal{L}^p(E, \mu).$$

Definition 2.2.11. For $\lambda > 0$, $x \in E_1$ and $f \in L^1(E, R_\lambda(x, dy)) \cup \mathcal{B}^+(E)$ we define

$$R_\lambda f(x) := \int_E f(y) R_\lambda(x, dy).$$

Theorem 2.2.12.

(i) *For $\lambda > 0$ it holds $\lambda R_\lambda 1 \leq 1$. There exists a $\mathcal{B}(E_1 \times E)$-measurable map $(x, y) \mapsto r_\lambda(x, y)$ such that $R_\lambda(x, dy) = r_\lambda(x, y) d\mu(y)$. In particular, $R_\lambda(x, E \setminus E_1) = 0$, so $(R_\lambda)_{\lambda > 0}$ can be considered as kernels on E_1, denoted by the same symbol. Moreover, $\mathcal{L}^p(E, \mu) \subset \mathcal{L}^1(E_1, R_\lambda(x, dy))$ and $R_\lambda f(x) = \widetilde{G^p_\lambda f}(x)$ for all $x \in E_1$, i.e., $R_\lambda f$ is the unique continuous version of $\widetilde{G^p_\lambda f}$. In particular, the integral $R_\lambda f$ coincides for all μ-version of f.*

(ii) *$(R_\lambda)_{\lambda > 0}$ is a resolvent of kernels on E_1 which is $\mathcal{L}^p$-strong Feller, i.e., $R_\lambda f \in C^0(E_1)$ for all $f \in \mathcal{L}^p(E, \mu)$.*

(iii) *For all $u \in D(L_p)$ and all $x \in E_1$*

$$R_\lambda u(x) = \int_0^\infty \exp(-\lambda t) P_t u(x)\, dt, \quad \lambda > 0. \tag{2.8}$$

(iv) For all $u \in D(L_p)$

$$\lim_{\lambda \to \infty} \lambda R_\lambda u(x) = \widetilde{u}(x) \quad \text{for all } x \in E_1.$$

Proof. The proofs of (i) and (ii) work analogously to those of Theorem 2.2.9. (iii): Let $\lambda > 0$ and $u \in D(L_p)$. Note that by the properties of G_λ^p and T_t^p it holds for μ-a.e. $x \in E$

$$G_\lambda u(x) = \int_0^\infty \exp(-\lambda t) T_t u \, dt \,(x) = \int_0^\infty \exp(-\lambda t) P_t u \, dt \,(x).$$

The integral is obtained as the L^p-limit and hence by dropping to a subsequence also as the μ-a.e. limit of Riemannian sums. Let $x \in E_1$. The mapping $[0, \infty) \ni t \mapsto P_t u(x) \in \mathbb{R}$ is continuous and bounded by (2.4), hence the mapping $[0, \infty) \ni t \mapsto \exp(-\lambda t) P_t u(x) \in \mathbb{R}$ is Lebesgue-integrable and the integral is obtained as the limit of Riemannian sums. Thus

$$\int_0^\infty \exp(-\lambda t) P_t u \, dt \,(x) = \int_0^\infty \exp(-\lambda t) P_t u(x) \, dt \quad \text{for } \mu\text{-a.e. } x.$$

Thus we get for almost all $x \in E_1$

$$R_\lambda u(x) = G_\lambda u(x) = \int_0^\infty \exp(-\lambda t) T_t u \, dt \,(x) = \int_0^\infty \exp(-\lambda t) P_t u(x) \, dt. \tag{2.9}$$

By (2.4) we have $P_t u(\cdot) = \widetilde{T_t u}(\cdot)$ is uniformly bounded in $t \geq 0$ and locally bounded w.r.t. $x \in E_1$. So by Lebesgue's dominated convergence the right-hand side of (2.9) is continuous in x.
Thus for all $x \in E_1$

$$R_\lambda u(x) = \int_0^\infty \exp(-\lambda t) P_t u(x) \, dt.$$

(iv): Observe that $(\lambda G_\lambda)_{\lambda > 0}$ is strongly continuous also on $(D(L_p), \|\cdot\|_{D(L_p)})$, see Lemma 7.2.1, so we get using (2.1) for $x \in E_1$

$$|\lambda R_\lambda u(x) - \widetilde{u}(x)| \leq C_1 \|\lambda G_\lambda u - u\|_{D(L_p)} \overset{\lambda \to \infty}{\longrightarrow} 0.$$

$\square$

Now we can prove Theorem 2.1.5, i.e., strong Feller properties for the resolvent under additional assumptions.

proof of Theorem 2.1.5. Let $f \in \mathcal{B}_b^+(E)$. Set $f_n := 1_{K_n} f$, $n \in \mathbb{N}$, with K_n as in Lemma 2.2.2. Then $f_n \in \mathcal{L}^p(E, \mu)$ and $f_n \uparrow f$. Define $u_n := R_1 f_n$, then $(1 - L)u_n = f_n$ is uniformly bounded in the L^∞-norm. Moreover, $(u_n)_{n\in\mathbb{N}}$ is also bounded in L^∞-norm since R_1 is sub-Markovian. So by assumption $(u_n)_{n\in\mathbb{N}}$ is equicontinuous.

For $x \in E_1$, choose $U \subset E_1$ such that $\overline{U} \subset E_1$ is compact. Then by Arzela-Ascoli the sequence $(u_n|_{\overline{U}})_{n\in\mathbb{N}}$ in $C^0(\overline{U})$ possesses a subsequence converging uniformly to a function $v \in C^0(\overline{U})$. By the properties of R_1 and $(f_n)_{n\in\mathbb{N}}$ it holds that $R_1 f_n(x) \uparrow R_1 f(x)$ for every $x \in E_1$. So $R_1 f(x) = v(x)$ for $x \in \overline{U}$. So for every $x \in E_1$ there exists a neighborhood of x such that $R_1 f$ is continuous on U. Hence $R_1 f \in C^0(E_1)$. The claim for general $f \in \mathcal{B}_b(E)$ follows by linearity now.

$\square$

We prove an enforced version of Theorem 2.2.12(iv) which we need later on for the solution of the martingale problem.

Lemma 2.2.13. *For all $x \in E_1$ and $f \in L^p(E, \mu) \cup \mathcal{B}_b(E)$ (2.8) holds.*

Proof. Let $x \in E_1$. First assume that $f \in L^p(E, \mu) \cap L^\infty(E, \mu)$, $f \geq 0$ and $\|f\|_{L^\infty} \leq C_5 < \infty$. Set $f_n := nG_n f \in D(L_p)$, $n \in \mathbb{N}$. Since nG_n is sub-Markov, we have $f_n(x) \leq C_5$ μ-a.e. and f_n converges to f in $L^p(E, \mu)$. By Theorem 2.2.12(iii) we have

$$R_\lambda f_n(x) = \int_0^\infty \exp(-\lambda t) P_t f_n(x) dt \quad \text{for all } n \in \mathbb{N}. \qquad (2.10)$$

Using (2.2) and L^p-convergence the left-hand side of (2.10) converges to $R_\lambda f(x)$. Furthermore, (2.3) implies for $t > 0$ $\lim_{n\to\infty} P_t f_n(x) = P_t f(x)$. Since P_t is sub-Markovian, we get $|\exp(-\lambda t) P_t f_n(x)| \leq \exp(-\lambda t) C_5$. Hence the right-hand side of (2.10) converges to $\int_0^\infty \exp(-\lambda t) P_t f(x) dt$ by Lebesgue's dominated convergence. Thus (2.8) holds for those f.

Now let $f \in L^p(E, \mu)$ with $f \geq 0$ and set $f_n := f \wedge n$. Then $f_n \uparrow f$ and (2.10) holds for all f_n. By monotone convergence on both sides we get the identity for f.

Now for $f \in L^p(E, \mu)$ observe that $P_t f(x) = P_t f^+(x) - P_t f^-(x)$ and therefore $|P_t f(x)| \leq P_t f^+(x) + P_t f^-(x)$. Since by the proven statement $\int_0^\infty \exp(-\lambda t) P_t f^{+/-}(x) dt < \infty$, we get that $\int_0^\infty \exp(-\lambda t) P_t f(x) dt$ exists.

Then (2.10) follows by linearity of R_λ and P_t. Thus the class of all functions satisfying (2.10) fulfills condition (i) and (iii) of Theorem 2.2.3 and condition (ii') of Corollary 2.2.5. So the statement for $f \in \mathcal{B}_b(E)$ follows using Corollary 2.2.5.   □

Based on this lemma we prove a pointwise equation relating the semigroup of kernels and the generator. This formula is essential for the solution of the martingale problem.

Lemma 2.2.14. *For $x \in E_1$, $u \in D(L_p)$ it holds for all $t > 0$*

$$P_t u(x) - \widetilde{u}(x) = \int_0^t P_s L_p u(x) ds = \int_0^t \widetilde{L_p P_s} u(x) ds \qquad (2.11)$$

and the integral is well-defined. Here $\widetilde{u}(x)$ and $\widetilde{L_p P_s} u(x)$ denote the value of the respective continuous version at x.

Proof. Let $x \in E_1$, $t > 0$ be fixed. First note that for $f \in L^p(E, \mu)$ the map $[0, t] \ni s \mapsto P_s f(x) \in \mathbb{R}$ is integrable. Indeed, consider $f \geq 0$ first. Then

$$\int_0^t P_s f(x) ds = \int_0^t \exp(s) \exp(-s) P_s f(x) ds$$

$$\leq \exp(t) \int_0^t \exp(-s) P_s f(x) ds \leq \exp(t) R_1 f(x) < \infty.$$

Note that the integral value is independent of the μ-version of f. Since $|P_s f| \leq P_s |f|$, the statement follows for general $f \in L^p(E, \mu)$.
Let $u \in D(L_p)$, set $f := (1 - L_p)u$, then $u = G_1 f$. Since $T_t G_1 f = G_1 T_t f$, we have by continuity $P_t R_1 f(x) = R_1 P_t f(x)$. Using Lemma 2.2.13 and the semigroup property we get

$$\exp(-t) P_t R_1 f(x) = R_1 \exp(-t) P_t f(x) = \int_0^\infty \exp(-(t + s)) P_{t+s} f(x) \, ds$$

$$= \int_t^\infty \exp(-s) P_s f(x) \, ds.$$

Thus

$$\exp(-t) P_t R_1 f(x) - R_1 f(x) = -\int_0^t \exp(-s) P_s f(x) ds$$

$$= -\int_0^t \exp(-s) P_s (1 - L) u(x) ds.$$

By construction $R_1 f$ is the unique continuous version of $G_1 f$, thus

$$\exp(-t)P_t u(x) - \tilde{u}(x) = -\int_0^t \exp(-s)P_s(1 - L)u(x)ds.$$

So for every $x \in E_1$ the mapping $[0, t] \ni s \mapsto \exp(-s)P_s\tilde{u}(x)$ is absolutely continuous with integrable weak derivative $\exp(-s)P_s(L - 1)u(x)$, $s > 0$. So by the product rule also the mapping $s \mapsto P_s\tilde{u}(x)$ is absolutely continuous with weak derivative $P_s Lu(x)$, $s \in [0, t]$. This proves the first equality of (2.11). For the second equality note that $L_p T_s u = T_s L_p u$ for $s > 0$ and since $T_s u \in D(L_p^2)$, we have by continuity that $\widetilde{L_p P_s}u(x) = P_s L_p u(x)$ for every $x \in E_1$. $\qquad\square$

2.3 Construction of the $\mathcal{L}^p$-strong Feller Process

In this section we use the $\mathcal{L}^p$-strong Feller kernels $(P_t)_{t>0}$ and $(R_\lambda)_{\lambda>0}$ to construct a diffusion process solving the martingale problem. Throughout this section the same assumptions as in Section 2.2 are assumed. The construction is based on techniques developed in [AKR03] and [Doh05].

Remark 2.3.1. Stoica ([Sto83]) provides an interesting construction result of Hunt processes from resolvent of kernels $(R_\lambda)_{\lambda>0}$. Besides a Feller type property for the resolvent of kernels he assumes pointwise strong continuity of $(R_\lambda)_{\lambda>0}$ and a condition generalizing the decay at infinity condition for classical Feller resolvents.

The family of resolvent of kernels $(R_\lambda)_{\lambda>0}$ we constructed in the previous section does not exactly match the assumptions made by Stoica. But it might be possible, perhaps under a suitable modification of our assumptions, to prove that our resolvent of kernels also fulfill the assumption of Stoica.

We first construct a process with dyadic time-parameter having $(P_t)_{t\in S}$ as transition semigroup. In the first step we show that the set of all paths that admit a continuous extension to $(0, \infty)$ have full $\mathbb{P}_x$-measure for $x \in E_1$, see Lemma 2.3.5 below. In [Doh05] this has been proven in the case of strong Feller kernels. This result has been modified in [AKR03] to cover also the $\mathcal{L}^p$-strong Feller case for $p < \infty$. So we follow [AKR03] here.

In the second step we prove right-continuity at $t = 0$. This follows using super-martingale convergence, see Lemma 2.3.7 below. For this lemma it is essential that the domain of the L^p-generator contains point separating functions for every point $x \in E_1$.

The process is constructed on p. 31.

For some construction below we need a probability measure on E that has the same nullsets as μ.

Lemma 2.3.2. *Let $(S, \mathcal{B}, \mu)$ be σ-finite with $\mu(S) > 0$. Then there exists a probability measure ν, i.e., $\nu(E) = 1$, absolutely continuous to μ such that $\nu(A) = 0$ implies $\mu(A) = 0$.*

Proof. Choose an increasing sequence of measurable sets $F_k \in \mathcal{B}$ with $\mu(F_k) < \infty$, $k \in \mathbb{N}$, and $\bigcup_{k \in \mathbb{N}} F_k = S$. Define $G_1 := F_1$ and inductively $G_{k+1} := F_{k+1} \setminus F_k$, $k \in \mathbb{N}$. Set $\alpha_k := 2^{-k} \frac{1}{\mu(G_k)} < \infty$ if $\mu(G_k) > 0$ and $\alpha_k = 0$ otherwise. Define $\alpha = \sum_{k=1}^\infty \alpha_k 1_{G_k}$. Then $\alpha > 0, \mu$-a.e., and $0 < \int_S \alpha d\mu < \infty$. Now normalize α and set $\nu := \alpha \mu$.

If $\nu(A) = 0$ for some $A \in \mathcal{B}$, then for G_k with $\mu(G_k) \neq 0$ it holds $\int_{G_k \cap A} \alpha \mu = 0$. Since $\alpha > 0$ μ-a.e. on G_k, it follows $\mu(G_k \cap A) = 0$. So by construction of the G_k it follows $\mu(A) = 0$. $\qquad\square$

Now we fix such a measure ν for $(E, \mathcal{B}(E), \mu)$. Thus we get a probability distribution of starting points that has the same nullsets as μ.

We extend E by the cemetery point Δ and endow $E^\Delta := E \cup \{\Delta\}$ with the topology of the Alexandrov one-point compactification, i.e., the open neighborhoods of Δ are given by the complements of compact subsets of E. Note that there exists a complete metric on E^Δ inducing this topology, see e.g. [CB06, Cor. 3.45]. Thus E^Δ is a Polish space. Next we extend the semigroup of kernels $(P_t)_{t \geq 0}$ from $E_1 \times \mathcal{B}(E)$ to $E^\Delta \times \mathcal{B}(E^\Delta)$. Here $\mathcal{B}(E^\Delta) = \mathcal{B}(E) \cup \{\mathcal{B}(E) \cup \Delta\}$.

Definition 2.3.3. Let $(P_t)_{t \geq 0}$ be the kernels on $E_1 \times \mathcal{B}(E)$ from Theorem 2.2.9. Define $P_t^\Delta : E^\Delta \times \mathcal{B}(E^\Delta)$ by: For $x \in E^\Delta$, $t > 0$, $A \in \mathcal{B}(E^\Delta)$ define

$$P_t^\Delta(x, A) := \begin{cases} P_t(x, A \cap E_1) + (1 - P_t(x, E))\varepsilon_\Delta(A) & \text{if } x \in E_1 \\ \varepsilon_x(A) & \text{else} \end{cases}$$

and for $t = 0$ define $P_t^\Delta(x, A) := \varepsilon_x(A)$, $x \in E^\Delta$, $A \in \mathcal{B}(E^\Delta)$. Here ε_x is the point measure in x.

A straightforward calculation gives that $(P_t^\Delta)_{t \geq 0}$ is a semigroup of kernels. For $x \in E_1$ the measure $P_t^\Delta(x, \cdot)$ consists of the part $P_t(x, \cdot)$ which is absolutely continuous w.r.t. μ and the singular part ε_Δ. We extend each

$\mathcal{B}(E)$-measurable f to a $\mathcal{B}(E^\Delta)$-measurable function by $f(\Delta) := 0$. For $n \in \mathbb{N}$ we set

$$S_n := \{k2^{-n} \mid k \in \mathbb{N} \cup \{0\}\} \text{ and } S := \bigcup_{m \in \mathbb{N}} S_m.$$

$(P_t^\Delta(x, dy))_{t \geq 0}$ is a semigroup of probability kernels on the polish space E^Δ, so by Kolmogorov's standard construction scheme, see e.g. [BG68, Ch. I, Theo. 2.11], there exists a family of probability measures $\mathbb{P}_x^0$, $x \in E^\Delta$, on $\Omega^0 := (E^\Delta)^S$, equipped with the product σ-field $\mathcal{F}^0$, such that

$$\mathbf{M}^0 := \left(\Omega^0, \mathcal{F}^0, (\mathcal{F}_s^0)_{s \in S}, (\mathbf{X}_s^0)_{s \in S}, (\mathbb{P}_x^0)_{x \in E^\Delta} \right)$$

is a normal Markov process having transition kernels $(P_t^\Delta(x, \cdot))_{t \in S}$. Here $\mathbf{X}_s^0 : \Omega^0 \to E^\Delta$ are the coordinate maps and $\mathcal{F}_t^0 := \sigma(\mathbf{X}_s^0 \mid s \leq t, s \in S)$.

This process is defined only for dyadic time parameters at first. Next we show that this process can be uniquely extended to a process with time parameter $t \in [0, \infty)$ having continuous paths on $(0, \infty)$.

Under Condition 2.1.2(ii) this process has even right-continuous paths at $t = 0$.

We use the result of [FOT11, Theo. 4.5.3], see also [MR92, Ch. V, Theo. 1.11].

They yield the existence of a diffusion on $[0, \infty)$ (compare Definition 7.3.12)

$$\hat{\mathbf{M}} = \left(\hat{\Omega}, \hat{\mathcal{F}}, (\hat{\mathcal{F}}_t)_{t \geq 0}, (\hat{\mathbf{X}}_t)_{t \geq 0}, (\hat{\mathbb{P}}_x)_{x \in E^\Delta} \right)$$

that is properly associated with the regular Dirichlet form $(\mathcal{E}, D(\mathcal{E}))$, i.e., $\hat{P}_t f$ is a $\mathcal{E}$-quasi-continuous version of $T_t f$ for $f \in \mathcal{L}^2(E, \mu) \cap \mathcal{B}_b(E)$ where $\hat{P}_t f(x) = \mathbb{E}_x[f(\hat{\mathbf{X}}_t)]$, see [MR92, Ch. III, Def. 2.5] or Definition 7.3.14. See also Definition 7.3.12, Theorem 7.3.15 and Proposition 7.3.17 in the appendix for a summary of the existence results.

For a Borel set K define $\sigma_K = \inf\{t > 0 \mid \hat{\mathbf{X}}_t \in K\}$, the first hitting time. Moreover, define the lifetime $\mathcal{X} := \inf\{t > 0 \mid \hat{\mathbf{X}}_t = \Delta\}$. The strong local property of $(\mathcal{E}, D(\mathcal{E}))$ implies that $\hat{\mathbf{M}}$ enters the cemetery only continuously, see [FOT11, Theo. 4.5.3]. So $(\hat{\mathbf{X}}_t)_{t \geq 0}$ is even continuous for all $t \in [0, \infty)$ and not only for $t \in [0, \mathcal{X})$. Thus we redefine $\hat{\Omega}$ such that

$$\hat{\Omega} := C^0([0, \infty), E^\Delta).$$

Note that the process provided by the general theory gives a martingale solution for $\mathcal{E}$-quasi-every starting point only.

Define

$$\hat{\boldsymbol{\Omega}}_0 := \left\{ \omega \in \hat{\boldsymbol{\Omega}} \mid \omega(t) \in E \text{ for } 0 \le t < \mathcal{X} \text{ and } \omega(t) = \Delta \text{ for } t \ge \mathcal{X} \right\}.$$

We get

$$\hat{\mathbb{P}}_x(\hat{\boldsymbol{\Omega}}_0) = 1 \text{ for } \mu - a.e. \ x \in E.$$

Let ν be the probability measure defined after Lemma 2.3.2. We may consider ν also as a probability measure on $\mathcal{B}(E^\Delta)$ $(\nu(\{\Delta\}) := 0)$. Define a path measure on $(\hat{\boldsymbol{\Omega}}, \hat{\mathcal{F}})$ by

$$\hat{\mathbb{P}}_\nu (\cdot) := \int_E \hat{\mathbb{P}}_x (\cdot) \, d\nu(x), \qquad (2.12)$$

see also (7.5) in Section 7.3 for details on this definition. Then

$$\hat{\mathbb{P}}_\nu(\hat{\boldsymbol{\Omega}}_0) = 1,$$

i.e., with $\hat{\mathbb{P}}_\nu$-probability 1 we observe continuous paths which reach Δ only continuously. Define the map $G : \hat{\boldsymbol{\Omega}} \to \boldsymbol{\Omega}^0$ by

$$\hat{\boldsymbol{\Omega}} \ni \omega = (\omega(t))_{t \in [0,\infty)} \mapsto G(\omega) := (\omega(s))_{s \in S} \in \boldsymbol{\Omega}^0.$$

Since every continuous function is uniquely determined by its values on a dense set, G is injective. Moreover, for $\hat{\mathcal{F}}^0 := \sigma(\hat{\mathbf{X}}_s \mid s \in S)$ it holds $\hat{\boldsymbol{\Omega}}_0 \in \hat{\mathcal{F}}^0$ and G is $\hat{\mathcal{F}}^0 / \mathcal{F}^0$-measurable.

With the following lemma we can connect the measures $\mathbb{P}_\nu^0$ and $\hat{\mathbb{P}}_\nu$.

Lemma 2.3.4. *Define $\hat{\mathbb{P}}_\nu'$ as the image measure of the restriction of $\hat{\mathbb{P}}_\nu$ to $\hat{\mathcal{F}}^0$ under G, i.e.,*

$$\hat{\mathbb{P}}_\nu' := \hat{\mathbb{P}}_\nu |_{\hat{\mathcal{F}}^0} \circ G^{-1}.$$

Then

$$\hat{\mathbb{P}}_\nu' = \mathbb{P}_\nu^0$$

where $\mathbb{P}_\nu^0$ is defined on $(\boldsymbol{\Omega}^0, \mathcal{F}^0)$ analogously as in (2.12) with $\hat{\mathbb{P}}_x$ replaced by $\mathbb{P}_x^0$.

Proof. Define

$$\mathcal{A} := \left\{ A = \{\omega \in \mathbf{\Omega}^0 \mid (\omega(t_1), ..., \omega(t_n)) \in A_1 \times ... \times A_n\} \right.$$

$$\text{for some } 0 \leq t_1 \leq ... \leq t_n < \infty, t_i \in S, 1 \leq i \leq n,$$

$$\left. A_{t_1}, ..., A_{t_n} \in \mathcal{B}(E^\Delta), n \in \mathbb{N} \right\}.$$

Note that both $(P_t)_{t \geq 0}$ and $(\hat{P}_t)_{t \geq 0}$ yield a μ-version (hence also a ν-version) of $(T_t^2)_{t \geq 0}$. Both are extended to functions on E^Δ as in Definition 2.3.3. So for every $f \in \mathcal{B}_b(E^\Delta)$ we have

$$P_t^\Delta f(x) = \hat{P}_t^\Delta f(x) \quad \text{for } t > 0 \text{ and } \nu - \text{a.e. } x.$$

Let $A \in \mathcal{A}$, $0 \leq t_1 \leq ... \leq t_n$ and $(A_1, ..., A_n)$ as in the definition of $\mathcal{A}$. Using the Markov property of $\mathbf{M}^0$ and $\hat{\mathbf{M}}$ we get

$$\hat{\mathbb{P}}_\nu'(A) = \hat{\mathbb{E}}_\nu[1_{A_1}(\hat{\mathbf{X}}_{t_1})...1_{A_n}(\hat{\mathbf{X}}_{t_n})]$$

$$= \int_{E^\Delta} \hat{P}_{t_1}^\Delta (1_{A_1} \hat{P}_{t_2-t_1}^\Delta (1_{A_2}... \hat{P}_{t_{n-1}-t_{n-2}}^\Delta (1_{A_{n-1}} \hat{P}_{t_n-t_{n-1}}^\Delta 1_{A_n})...)) (x) \, d\nu(x)$$

$$= \int_{E^\Delta} P_{t_1}^\Delta (1_{A_1} P_{t_2-t_1}^\Delta (1_{A_2}... P_{t_{n-1}-t_{n-2}}^\Delta (1_{A_{n-1}} P_{t_n-t_{n-1}}^\Delta 1_{A_n})...)) (x) \, d\nu(x)$$

$$= \mathbb{P}_\nu^0(A).$$

So $\hat{\mathbb{P}}_\nu'$ and $\mathbb{P}_\nu^0$ coincide on $\mathcal{A}$. This is an intersection-stable generator of $\mathcal{F}^0$, so they coincide on $\mathcal{F}^0$, too. $\square$

As in [AKR03, Lem. 4.3] we get using Lemma 2.3.4 the following lemma.

Lemma 2.3.5. $G(\hat{\mathbf{\Omega}}_0) \in \mathcal{F}^0$ *and* $\mathbb{P}_y^0(G(\hat{\mathbf{\Omega}}_0)) = 1$ *for* μ-*a.e.* $y \in E$.

Consider the time shift operator $\theta_s : \mathbf{\Omega}^0 \to \mathbf{\Omega}^0$, $\theta_s(\omega) = \omega(\cdot + s)$, $s \geq 0$, and define

$$\mathbf{\Omega}_1^0 := \bigcap_{\substack{s > 0 \\ s \in S}} \theta_s^{-1}(G(\hat{\mathbf{\Omega}}_0)),$$

i.e., all paths with time parameter in S that come from a path on $[0, \infty)$ which is continuous in $(0, \infty)$. Then using Lemma 2.3.5 we get.

Lemma 2.3.6.

$$\mathbb{P}_x^0(\mathbf{\Omega}_1^0) = 1 \quad \text{for } x \in E_1 \cup \{\Delta\}.$$

Proof. Let $s > 0$, $s \in S$, $x \in E_1$. We prove $\mathbb{P}_x^0(\theta_s^{-1}(G(\hat{\mathbf{\Omega}}_0))) = 1$.
Using the Markov property of $\mathbf{M}^0$ in $*$ (actually Lemma 7.3.7(ii)) we get

$$\mathbb{P}_x^0(\theta_s^{-1}(G(\hat{\mathbf{\Omega}}_0))) = \mathbb{P}_x^0((\mathbf{X}_{t+s}^0)_{t \in S} \in G(\hat{\mathbf{\Omega}}_0))$$

$$= \mathbb{E}_x\left[\mathbb{E}_x\left[1_{\{(\mathbf{X}_{t+s}^0)_{t \in S} \in G(\hat{\mathbf{\Omega}}_0)\}} | \mathcal{F}_s^0\right]\right] \overset{*}{=} \mathbb{E}_x\left[\mathbb{P}_{\mathbf{X}_s^0}^0(G(\hat{\mathbf{\Omega}}_0))\right]$$

$$= \int_E p_s(x, y)\mathbb{P}_y^0(G(\hat{\mathbf{\Omega}}_0))d\mu(y) + (1 - P_s(x, E))\mathbb{P}_\Delta^0(G(\hat{\mathbf{\Omega}}_0)).$$

By definition of $\mathbb{P}_\Delta^0$ and $\hat{\mathbf{\Omega}}_0$ we have $\mathbb{P}_\Delta^0(G(\hat{\mathbf{\Omega}}_0)) = 1$ and by Lemma 2.3.5
we have for μ-a.e. $y \in E$, $\mathbb{P}_y^0(G(\hat{\mathbf{\Omega}}_0)) = 1$. Thus

$$\mathbb{P}_x^0(\theta_s^{-1}(G(\hat{\mathbf{\Omega}}_0))) = 1.$$

But then we also have

$$\mathbb{P}_x^0\left(\bigcap_{\substack{s>0 \\ s \in S}} \theta_s^{-1}(G(\hat{\mathbf{\Omega}}_0))\right) = 1.$$

$\square$

To get right-continuity of the paths at $t = 0$ we use the point separating
Condition 2.1.2(ii). We apply the same argument as in [AKR03, Lem. 4.6]
and [Doh05, Lem. 3.2].

Lemma 2.3.7. *For $x \in E_1$ it holds*

$$\lim_{s \downarrow 0, s \in S} \mathbf{X}_s^0 = x \quad \mathbb{P}_x^0\text{-a.s.}$$

Proof. Let $x \in E_1$, $u \in D(L_p)$. Denote the continuous version of u also by
u. Set $f := (1 - L_p)u$. Then

$$u = (1-L_p)^{-1}(1-L_p)u = (1-L_p)^{-1}((1-L_p)u)^+ - (1-L_p)^{-1}((1-L_p)u)^-$$

$$= (1 - L_p)^{-1}f^+ - (1 - L_p)^{-1}f^-.$$

Since f^+ and f^- are positive functions, it follows using Lemma 2.2.12(iii)
and the Markov property that $(\exp(-s)R_1 f^{+/-}(\mathbf{X}_s^0))_{s \in S}$ are positive super-
martingales w.r.t.

$(\mathcal{F}_s^0)_{s \geq 0}$ (similar calculation as in the proof of Lemma 2.2.14). So by the martingale convergence theorem $\mathbb{P}_x^0$-a.s.

$$\lim_{s \downarrow 0, \, s \in S} (\exp(-s) R_1 f^{+/-}(\mathbf{X}_s^0)) \text{ exists in } \mathbb{R},$$

but then also $\mathbb{P}_x^0$-a.s.

$$\lim_{s \downarrow 0, \, s \in S} u(\mathbf{X}_s^0) \text{ exists in } \mathbb{R}.$$

Assume that $0 \leq u \leq 1$ and $u(x) = 1$. Then using Theorem 2.2.9.(iii) we get

$$\mathbb{E}_x \left[|u(\mathbf{X}_s^0) - u(x)| \right] = 1 - \mathbb{E}_x \left[u(\mathbf{X}_s^0) \right] = 1 - P_s u(x) \xrightarrow{s \to 0} 1 - u(x) = 0.$$

Now assume that $u^2 \in D(L_p)$ instead. Then we get

$$\mathbb{E}_x \left[(u(\mathbf{X}_s^0) - u(x))^2 \right] = \mathbb{E}_x \left[u(\mathbf{X}_s^0)^2 - 2u(\mathbf{X}_s^0)u(x) + u(x)^2 \right]$$
$$= P_s u^2(x) - 2u(x) P_s u(x) + u^2(x) \xrightarrow{s \to 0} 0.$$

So in both cases $\lim_{s \downarrow 0, \, s \in S} u(\mathbf{X}_s^0) = u(x)$ $\mathbb{P}_x^0$-a.s. Choose now the sequence $(u_k)_{k \in \mathbb{N}}$ of point separating functions for $x \in E_1$ according to Condition 2.1.2.

Then $\mathbb{P}_x^0$-a.s.

$$\lim_{s \downarrow 0, \, s \in S} u_k(\mathbf{X}_s^0) = u_k(x) \text{ for all } k \in \mathbb{N}. \tag{2.13}$$

Now let $\omega \in \mathbf{\Omega}^0$ such that (2.13) holds. If there exists a sequence $(s_l)_{l \in \mathbb{N}}$ with $s_l \xrightarrow{l \to \infty} 0$ and $\lim_{l \to \infty} \mathbf{X}_{s_l}^0(\omega) = \Delta$, we get a contradiction to (2.13) since $u_k(\Delta) = 0$. Let $(K_n)_{n \in \mathbb{N}}$ be a sequence of compact sets covering E. If we assume that $\mathbf{X}_{s_l}^0(\omega)$ leaves every compact set, then we get again a subsequence converging to Δ what is not possible. Hence there exists a compact set K_{n_0} such that $\mathbf{X}_{s_l}^0(\omega) \in K_{n_0}$ for $l \in \mathbb{N}$. Then $(\mathbf{X}_{s_l}^0(\omega))_{l \in \mathbb{N}}$ has a convergent subsequence. By (2.13) the limit of this subsequence is x. Thus every subsequence has a subsubsequence converging to x. So with the standard contradiction argument we get $\mathbb{P}_x^0$-a.s

$$\lim_{s \downarrow 0, \, s \in S} \mathbf{X}_s^0 = x.$$

$\square$

The following lemma allows us to get back from paths on dyadic time points to paths with time-parameter in $[0, \infty)$.

Lemma 2.3.8. *Let*

$$\mathbf{\Omega}_0^0 := \mathbf{\Omega}_1^0 \cap \left(\{ \omega \in \mathbf{\Omega}^0 \mid \lim_{s \downarrow 0, s \in S} \omega(s) = \omega(0) \in E \} \cup \{ \omega \mid \omega(s) = \Delta \, for \, s \in S \} \right).$$

Then $G : \hat{\mathbf{\Omega}}_0 \to \mathbf{\Omega}_0^0$ is bijective with inverse $H : \mathbf{\Omega}_0^0 \to \hat{\mathbf{\Omega}}_0$, given by $\mathbf{\Omega}_0^0 \ni \omega^0 \mapsto \lim_{s \downarrow t, s \in S} \omega^0(s) =: \omega(t), t \geq 0$.

Proof. First of all, we have to show $G(\hat{\mathbf{\Omega}}_0) \subset \mathbf{\Omega}_0^0$. Let $\omega \in \hat{\mathbf{\Omega}}_0$. Then $\omega(t) \in E^\Delta$ for all $t > 0$ and $\omega(0) \in E^\Delta$. Furthermore, $\omega : [0, \infty) \to E^\Delta$ is continuous. In particular, $\omega|_S$ is right-continuous at 0 with limit in E^Δ. If $\omega(0) = \Delta$ then $\omega(t) = \Delta$ for all $t \geq 0$.

So it is left to show that $G(\hat{\mathbf{\Omega}}_0) \subset \mathbf{\Omega}_1^0$. Clearly, $\hat{\mathbf{\Omega}}_0$ is shift-invariant. Let $\omega \in \hat{\mathbf{\Omega}}_0$ then $G(\theta_s \omega) = \theta_s(G(\omega))$ for $s \in S$. So $G(\omega) \in \theta_s^{-1}(G(\hat{\mathbf{\Omega}}_0))$ for all $s \in S$, hence $G(\omega) \in \mathbf{\Omega}_0^0$. Thus $G(\hat{\mathbf{\Omega}}_0) \subset \mathbf{\Omega}_0^0$.

By continuity of the paths in $\hat{\mathbf{\Omega}}_0$ we get that G is injective.

Finally, we prove that G is surjective onto $\mathbf{\Omega}_0^0$. Let $\omega^0 \in \mathbf{\Omega}_0^0$. Define

$$\omega(t) := \lim_{s \downarrow t, s \in S} \omega^0(s), t \geq 0.$$

Clearly, for $t = 0$ the limit exists. We show that the limit exists also for $t > 0$. Let $n \in \mathbb{N}$ arbitrary but fixed. Assume that $S \ni s \geq \frac{1}{n}$. Since $\omega^0 \in \mathbf{\Omega}_1^0$, there exists $\omega_n \in \hat{\mathbf{\Omega}}_0$ such that $\theta_{1/n} \omega^0 = \omega_n|_S$. So $\omega^0(s) = \omega^0(s - \frac{1}{n} + \frac{1}{n}) = \omega_n(s - \frac{1}{n})$. Thus for $t \geq \frac{1}{n}$

$$\omega(t) = \lim_{s \downarrow t, s \in S} \omega^0(s) = \lim_{s \downarrow t, s \in S} \omega_n \left(s - \frac{1}{n} \right) = \omega_n \left(t - \frac{1}{n} \right). \tag{2.14}$$

Note that the value $\omega(t)$ does not depend on the choice of ω_n. Indeed, assume there is another $\omega_n' \in \hat{\mathbf{\Omega}}_0$ such that $\theta_{1/n} \omega^0 = \omega_n'|_S$. Then $\omega_n'|_S = \omega_n|_S$ and by continuity $\omega_n' = \omega_n$. Furthermore, for $s \in S$ with $s \geq \frac{1}{n}$ we have $\omega(s) = \omega_n(s - \frac{1}{n}) = \omega^0(s)$. So $G(\omega) = \omega^0$ and $\omega(t) \in E^\Delta$ for all $t \geq 0$.

We show that $\omega \in \hat{\mathbf{\Omega}}_0$. For every $n \in \mathbb{N}$ we have by (2.14) that ω is continuous on $[\frac{1}{n}, \infty)$ with values in $E \cup \Delta$. Thus ω is continuous on $(0, \infty)$.

We already know that ω is right-continuous at $t = 0$ if restricted to dyadic time-parameters. If $\omega(0) = \Delta$, then by choice of $\mathbf{\Omega}_0^0$ we have $\omega^0(s) = \Delta$ for all $s \in S$ and thus also $\omega(t) = \Delta$ for all $t \geq 0$.

So assume $\omega(0) \neq \Delta$. Since ω^0 is right-continuous at $t = 0$, there exists $\delta_1 > 0$ such that $\omega^0(s) \in E$ for $0 \leq s < \delta_1$. Let $\varepsilon > 0$, choose $0 < \delta_2 < \delta_1$

such that $\mathbf{d}(\omega(s), \omega(0)) = \mathbf{d}(\omega^0(s), \omega^0(0)) < \frac{\varepsilon}{2}$ for all $s < \delta_2$ and $s \in S$. Let $t \in (0, \delta_2)$ arbitrary and choose $n \in \mathbb{N}$ with $\frac{1}{n} < t$. Choose $\omega_n \in \mathbf{\Omega}_1^0$ with $\omega_n(\cdot) = \omega^0(\frac{1}{n} + \cdot)$ on S. Then $\omega(t) = \omega_n(t - \frac{1}{n})$ for $t \geq \frac{1}{n}$. By continuity we can find $\delta_3 > 0$ such that for $\frac{1}{n} \leq t', s' < \delta_2$ with $|t' - s'| < \delta_3$ it holds $\mathbf{d}(\omega(t'), \omega(s')) < \frac{\varepsilon}{2}$.

For our given t choose $s \in S$ with $\frac{1}{n} \leq s < \delta_2$ and $|t - s| < \delta_3$. Then

$$\mathbf{d}(\omega(t), \omega(0)) \leq \mathbf{d}(\omega(t), \omega(s)) + \mathbf{d}(\omega^0(s), \omega^0(0)) < \varepsilon.$$

So altogether, ω is continuous on $[0, \infty)$.

It is left to prove that $\omega(t) = \Delta$ for all $t \geq \mathcal{X}(\omega)$. If $\omega(0) = \Delta$ then $\omega^0(0) = \Delta$. From definition of $\mathbf{\Omega}_0^0$ we get $\omega^0(s) = \Delta$ for all $s \in S$ and hence $\omega(t) = \Delta$ for all $t \in \mathbb{R}_0^+$.

Assume $\mathcal{X}(\omega) > 0$. Choose $n \in \mathbb{N}$ with $\frac{1}{n} \leq \mathcal{X}(\omega)$, ω_n as above. Then $\mathcal{X}(\omega) = \mathcal{X}(\omega_n) + \frac{1}{n}$. For $t \geq \mathcal{X}(\omega)$ it holds $t - \frac{1}{n} \geq \mathcal{X}(\omega_n)$. Thus $\omega(t) = \omega_n(t - \frac{1}{n}) = \Delta$ for those t.

Altogether, $\omega \in \hat{\mathbf{\Omega}}_0$ and $G(\omega) = \omega^0$. So $G : \hat{\mathbf{\Omega}}_0 \to \mathbf{\Omega}_0^0$ is surjective.

□

After these preparations we can construct the $\mathcal{L}^p$-strong Feller process.

Construction of the process of Theorem 2.1.3. Let $\mathbf{\Omega} := \hat{\mathbf{\Omega}}_0$, $\mathbf{X}_t : \mathbf{\Omega} \to E^\Delta$, $\mathbf{X}_t(\omega) := \omega(t)$, $t \geq 0$, $\mathcal{F}_t' := \sigma(\mathbf{X}_s \,|\, 0 \leq s \leq t)$, $\mathcal{F}' := \sigma(\mathbf{X}_s \,|\, 0 \leq s < \infty)$, $\theta_s(\omega) := \omega(s + \cdot)$, $0 \leq s < \infty$. With H as in Lemma 2.3.8 define

$$\mathbb{P}_x := \mathbb{P}_x^0\big(H^{-1}(\cdot) \cap \mathbf{\Omega}_0^0\big) = \mathbb{P}_x^0\big(G(\cdot) \cap \mathbf{\Omega}_0^0\big).$$

Then we have for $x \in E^\Delta$

$$\mathbb{P}_x(\mathbf{\Omega}) = \mathbb{P}_x^0 \circ G(\hat{\mathbf{\Omega}}_0) = \mathbb{P}_x^0(\mathbf{\Omega}_0^0).$$

For $x \in E^\Delta \setminus E_1$ we have $\mathbb{P}_x^0(\mathbf{\Omega}_0^0) = 1$ since the process stays in the starting point x. For $x \in E_1$ we get $\mathbb{P}_x^0(\mathbf{\Omega}_0^0) = 1$ by Lemma 2.3.6 and Lemma 2.3.7.

Thus we get a stochastic process $\mathbf{M}' = (\mathbf{\Omega}, \mathcal{F}', (\mathcal{F}_t')_{t \geq 0}, (\mathbf{X}_t)_{t \geq 0}, (\mathbb{P}_x)_{x \in E^\Delta})$ with continuous paths and shift operator $(\theta_t)_{t \geq 0}$.

We show that the *law* $\mathcal{L}(\mathbf{X}_t)$, $t \geq 0$, is given by $P_t^\Delta(x, \cdot)$ under $\mathbb{P}_x$, $x \in E_1^\Delta$. For $t \in S$ we have for $A \in \mathcal{B}(E^\Delta)$

$$\mathbb{P}_x\big(\mathbf{X}_t \in A\big) = \mathbb{P}_x^0\big(G(\{\omega(t) \in A\})\big) = \mathbb{P}_x^0\big(\{\omega|_S(t) \in A\}\big)$$
$$= \mathbb{P}_x^0\big(\mathbf{X}_t^0 \in A\big) = P_t^\Delta 1_A(x) \text{ for every } x \in E^\Delta. \quad (2.15)$$

So $\mathcal{L}(\mathbf{X}_t) = P_t^\Delta(x, \cdot)$ for every $x \in E^\Delta$ and $t \in S$.

So $\mathcal{L}(\mathbf{X}_t) = P_t^\Delta(x, \cdot)$ for all $t \in S$ and $x \in E^\Delta$. For $x = E \setminus E_1 \cup \{\Delta\}$ this also true for all $t \geq 0$ since the process is then constantly equal x.

Let $x \in E_1$. Let $u \in C_b(E) \cap \mathcal{L}^p(E, \mu)$. Then by (2.15) we get $\mathbb{E}_x[u(\mathbf{X}_s)] = P_s u(x)$ for $s \in S$. Let $t > 0$ and choose a dyadic sequence $(s_n)_{n \in \mathbb{N}}$ converging monotonically to t. Using Lebesgue's dominated convergence in the first equality and Theorem 2.2.9(iii) in the third one we get

$$\mathbb{E}_x[u(\mathbf{X}_t)] = \lim_{n \to \infty} \mathbb{E}_x[u(\mathbf{X}_{s_n})] = \lim_{n \to \infty} P_{s_n} u(x) = P_t u(x).$$

Corollary 2.2.5 implies that

$$\mathbb{E}_x[u(\mathbf{X}_t)] = P_t u(x)$$

holds for $u \in \mathcal{B}_b(E)$. Furthermore, note that $\mathbb{P}_x(\mathbf{X}_t = \Delta) = 1 - \mathbb{P}_x(\mathbf{X}_t \in E) = 1 - P_t 1_E(x)$. Thus $\mathcal{L}_x(\mathbf{X}_t) = P_t^\Delta(x, \cdot)$ for $t \geq 0$ and $x \in E_1$.

Define $\mathcal{F}_t$, $0 \leq t < \infty$, by

$$\mathcal{F}_t := \bigcap_{\nu \in \mathcal{P}(E^\Delta)} (\mathcal{F}_t')^{\mathbb{P}_\nu}$$

and $\mathcal{F} := \bigcap_{\nu \in \mathcal{P}(E^\Delta)} (\mathcal{F}')^{\mathbb{P}_\nu}$. Then $(\mathcal{F}_t)_{t \geq 0}$ is the so-called natural filtration, see also (7.8). Note that the path measures $(\mathbb{P}_x)_{x \in E^\Delta}$ naturally extend to $\mathcal{F}$.

Define

$$\mathbf{M} = (\mathbf{\Omega}, \mathcal{F}, (\mathcal{F}_t)_{t \geq 0}, (\mathbf{X}_t)_{t \geq 0}, (\mathbb{P}_x)_{x \in E^\Delta}).$$

The path regularity properties are clear. So it is left to show the (strong) Markov property. First we show the measurability condition (iii) in Definition 7.3.2.

Let $\mathcal{A}$ as in the proof of Lemma 2.3.4. Let $A \in \mathcal{A}$, $0 \leq t_1 \leq \ldots \leq t_n \in S$ and $(A_1, \ldots, A_n)$ as in the definition of $\mathcal{A}$.

Then for every $x \in E^\Delta$ we get

$$\mathbb{P}_x(A) = \mathbb{P}_x^0(G(\{\mathbf{X}_{t_1} \in A_1, \ldots \mathbf{X}_{t_n} \in A_n\}))$$
$$= \mathbb{P}_x^0(\{\mathbf{X}_{t_1}^0 \in A_1, \ldots \mathbf{X}_{t_n}^0 \in A_n\})$$
$$= P_{t_1}^\Delta (1_{A_1} P_{t_2 - t_1}^\Delta 1_{A_2} \ldots P_{t_n - t_{n-1}}^\Delta 1_{A_n})(x).$$

Since $P_t^\Delta u$ is $\mathcal{B}(E^\Delta)$-measurable for $u \in \mathcal{B}_b(E^\Delta)$, we get that the expression on the right-hand side is $\mathcal{B}(E^\Delta)$-measurable. By a monotone class argument we get that $\mathbb{P}_x(A)$ is $\mathcal{B}(E^\Delta)$-measurable for $A \in \sigma(\mathbf{X}_t \,|\, t \in S)$. Since $\mathbf{X}_t(\omega)$ is continuous for $\omega \in \Omega$, we have $\mathcal{F}' = \sigma(\mathbf{X}_t \,|\, 0 \leq t < \infty) = \sigma(\mathbf{X}_t \,|\, t \in S)$. Thus measurability holds for general $A \in \mathcal{F}'$. Applying Lemma 7.3.4(ii) to $\mathcal{F}$ we conclude that $x \mapsto \mathbb{P}_x(\Gamma)$ is $\mathcal{B}^*(E^\Delta)$-measurable for $\Gamma \in \mathcal{F}$.

We prove that $(\mathbf{X}_t)_{t \geq 0}$ has the Markov property w.r.t. $(\mathcal{F}_t)_{t \geq 0}$, i.e., Definition 7.3.2(iv) is fulfilled. The proof consists of several steps. Note that for $x \in E^\Delta \setminus E_1$ the Markov property is trivial. So we restrict to the case $x \in E_1$. First we prove that $(\mathbf{X}_t)_{t \in S}$ is Markov w.r.t. $(\hat{\mathcal{F}}_t^0)_{t \in S}$, $\hat{\mathcal{F}}_t^0 := \sigma(\mathbf{X}_s \,|\, s \in S, s \leq t)$. Define $\mathcal{A}_t$, $t \geq 0$, similarly as $\mathcal{A}$ but with the restriction that $t_1 \leq ... \leq t_n \leq t$ for the time-parameters occurring in the definition of $\mathcal{A}$. Let s and t in S, $A \in \mathcal{A}_t$. The relation between $\mathbb{P}_x$ and $\mathbb{P}_x^0$ together with the Markov property of $\mathbf{M}^0$ yields

$$\mathbb{E}_x\big[f(\mathbf{X}_{s+t})1_A\big] = \mathbb{E}_x\big[f(\mathbf{X}_{s+t}^0)1_{G(A)}\big] = \mathbb{E}_x\big[\mathbb{E}_{\mathbf{X}_t^0}[f(\mathbf{X}_s^0)]1_{G(A)}\big]$$
$$= \mathbb{E}_x\big[P_s^\Delta f(\mathbf{X}_t^0)1_{G(A)}\big] = \mathbb{E}_x\big[P_s^\Delta f(\mathbf{X}_t)1_A\big]$$
$$\text{for all } f \in C_b(E) \cap L^p(E,\mu), \; x \in E_1. \qquad (2.16)$$

For $s \notin S$ choose a sequence $(s_n)_{n \in \mathbb{N}}$ converging monotonically to s. Plugging s_n into (2.16) and letting n to ∞ we get the equality also for s itself. Using a monotone class argument we get the equality for all $A \in \hat{\mathcal{F}}_t^0$.

Let $t \in \mathbb{R}^+$, $s \in \mathbb{R}_0^+$, $A \in \hat{\mathcal{F}}_{t+}^0$. Choose a dyadic sequence $(t_n)_{n \in \mathbb{N}}$ converging monotonically to t. Since $A \in \hat{\mathcal{F}}_{t_n}^0$ for $n \in \mathbb{N}$, we get

$$\mathbb{E}_x\big[f(\mathbf{X}_{s+t_n})1_A\big] = \mathbb{E}_x\big[P_s^\Delta f(\mathbf{X}_{t_n})1_A\big] \quad \text{for } f \in C_b(E) \cap L^p(E,\mu), \; x \in E_1.$$
$$(2.17)$$

To get convergence we need the following observation: Note that at this point we do not know whether the process stays in $E_1 \cup \{\Delta\}$. However, since $P_r(x, E \setminus E_1) = 0$ for $x \in E_1$ and $r > 0$ we know

$$\mathbb{P}_x\Big(\{\mathbf{X}_t \in E \setminus E_1\} \cup \bigcup_{n \in \mathbb{N}} \{\mathbf{X}_{t_n} \in E \setminus E_1\}\Big) = 0.$$

Furthermore, if $\mathbf{X}_t \in E_1$, then by right-continuity (on E^Δ) there exists $N_0 \in \mathbb{N}$ such that for $n > N_0$ it holds $\mathbf{X}_{t_n} \in E$ $\mathbb{P}_x$-a.s. for $x \in E_1$. So altogether, for

$$G := \big\{\mathbf{X}_t = \Delta, \mathbf{X}_{t_n} = \Delta \text{ for } n \in \mathbb{N}\big\}$$
$$\cup \big\{\mathbf{X}_t \in E_1, \exists N \in \mathbb{N} : \mathbf{X}_{t_n} \in E_1 \text{ for } n > N\big\}$$

we have

$$\mathbb{P}_x(G) = 1 \quad \text{for } x \in E_1.$$

Since $P_s^\Delta f(\Delta) = 0$ and $P_s^\Delta f$ is continuous on E_1, we have for $\omega \in G$ $P_s^\Delta f(\mathbf{X}_{t_n}) \overset{n \to \infty}{\longrightarrow} P_s^\Delta f(\mathbf{X}_t)$. So we may let $n \to \infty$ in (2.17) and get:

$$\mathbb{E}_x\big[f(\mathbf{X}_{s+t})1_A\big] = \mathbb{E}_x\big[P_s^\Delta f(\mathbf{X}_t)1_A\big]$$
$$= \mathbb{E}_x\big[\mathbb{E}_{\mathbf{X}_t}[f(\mathbf{X}_s)]1_A\big] \quad \text{for all } f \in C_b(E) \cap L^p(E, \mu)$$

and $x \in E_1$. Using Lemma 7.3.7 we get the Markov property w.r.t. $(\hat{\mathcal{F}}_{t+}^0)_{t \geq 0}$. Since $\mathbf{X}_t$ is $\hat{\mathcal{F}}_{t+}^0$-adapted, we have $\mathcal{F}_t' \subset \mathcal{F}_{t+}' \subset \hat{\mathcal{F}}_{t+}^0$, $t \geq 0$, and thus $(\mathbf{X}_t)_{t \geq 0}$ is also Markov w.r.t. $(\mathcal{F}_t')_{t \geq 0}$ and $(\mathcal{F}_{t+}')_{t \geq 0}$, see also Lemma 7.3.6(i). From Lemma 7.3.6(iii) we get that $(\mathbf{X}_t)_{t \geq 0}$ is also Markov with respect to $(\mathcal{F}_t)_{t \geq 0}$. Lemma 7.3.6(iv) yields that $(\mathcal{F}_t)_{t \geq 0}$ is right-continuous.

Before continuing with the strong Markov property, we have to prove that $(\mathbf{X}_t)_{t \geq 0}$ stays in $E_1 \cup \{\Delta\}$ for starting points in $E_1 \cup \{\Delta\}$. By assumption $\text{cap}_\mathcal{E}(E \setminus E_1) = 0$, thus we get this from Lemma 2.3.10(ii) below.

We prove the strong Markov property, Definition 7.3.8(iii), and want to apply Theorem 7.3.10. Set $\mathbf{L} = C_b(E) \cap L^p(E, \mu)$. Assumption (ii) of the theorem is clearly fulfilled, see e.g. the proof of the last claim in Corollary 2.2.5. We consider the potential operators of $(\mathbf{X}_t)_{t \geq 0}$, defined in (7.10), see also Lemma 2.3.9 below.

Let us prove assumption (i), i.e., $\mathbb{P}_x$-a.s. right-continuity of $t \mapsto U^\lambda f(\mathbf{X}_t)$ on $[0, \mathcal{X})$, $f \in \mathbf{L}$, $0 < \lambda < \infty$. For $x \in E^\Delta \setminus E_1$, the process $(\mathbf{X}_t)_{t \geq 0}$ is constantly equal to x, hence the mapping is right-continuous. For $x \in E_1$ we have $U^\lambda f(x) = R_\lambda f(x)$ for $0 < \lambda < \infty$ by Lemma 2.3.9 below. Since $R_\lambda f$ is continuous on E_1 and $\mathbf{X}_t \in E_1$, $t < \mathcal{X}$, $\mathbb{P}_x$-a.s. for $x \in E_1$, the mapping $U_\alpha f(\mathbf{X}_t)$ is also $\mathbb{P}_x$-a.s. right-continuous in $t \in [0, \mathcal{X})$ for $x \in E_1$. So by applying Theorem 7.3.10 we get that $(\mathbf{X}_t)_{t \geq 0}$ is strong Markov. That the process solves the corresponding martingale problem is shown below, see Theorem 2.3.11. $\qquad \square$

We can prove the conservativity criterion for $\mathbf{M}$, i.e., Corollary 2.1.7.

proof of Corollary 2.1.7. Assume that $(\mathcal{E}, D(\mathcal{E}))$ is conservative and $(R_\lambda)_{\lambda > 0}$ is strong Feller. Then we conclude as in [AKR03, Prop. 3.8] that $P_t 1_E(x) = 1$ for every $t \geq 0$ and $x \in E_1$. This implies that the constructed process $\mathbf{M}$ is conservative. $\qquad \square$

Let us now prove the statement on the restriction of the process, i.e., Corollary 2.1.8.

proof of Corollary 2.1.8. That $\mathbf{M}^1$ is again a Hunt process is well-known, see e.g. Theorem 7.3.19. The path regularity properties are clear from the definition of the process. Denote the expectation for $\mathbf{M}^1$ by $\mathbb{E}^1$ and the semigroup by $(P_t^1)_{t \geq 0}$. For $f \in \mathcal{L}^p(E_1, \mu)$, define $\tilde{f} := 1_{E_1} f \in \mathcal{L}^p(E, \mu)$. We have for $x \in E_1$ and $t > 0$

$$P_t^1 f(x) = \mathbb{E}_x^1[f(\mathbf{X}_t^1)] = \mathbb{E}_x[\tilde{f}(\mathbf{X}_t)] = P_t \tilde{f}(x) = \int_{E_1} f(y)\, p_t(x,y)\, d\mu(y).$$

Thus $(P_t^1)_{t>0}$ is $\mathcal{L}^p(E_1, \mu)$-strong Feller and absolutely continuous on E_1. $\square$

In the next lemma we consider the potential operator $(U^\lambda)_{\lambda > 0}$ as defined in Definition 7.10.

Lemma 2.3.9. *Let $\mathbf{M}$ be the Markov process from Theorem 2.1.3. Then the potential operators $(U^\lambda)_{\lambda > 0}$ induce mappings $U^\lambda : \mathcal{L}^p(E, \mu) \to \mathcal{B}(E_1)$. For $u \in \mathcal{L}^p(E, \mu) \cup \mathcal{B}_b(E)$ it holds*

$$U^\lambda u(x) = \mathbb{E}_x \left[\int_0^\infty \exp(-\lambda s) u(\mathbf{X}_s) ds \right] = R_\lambda u(x) \quad \text{for all } x \in E_1, \lambda > 0.$$

(2.18)

Proof. Let $0 < \lambda < \infty$. First of all, note that the mapping $\Omega \times [0, \infty) \ni (\omega, s) \mapsto \exp(-\lambda s) u(\mathbf{X}_s(\omega))$ is $\mathcal{F}' \otimes \mathcal{B}([0, \infty))$-measurable for a continuous functions u. For $u \in \mathcal{B}_b(E)$ we can conclude measurabililty of this mapping by Corollary 2.2.5. For general positive measurable functions the measurability follows by monotone approximation.

Let $u \in \mathcal{L}^p(E, \mu)$ and $u \geq 0$. Let $x \in E_1$. From Lemma 2.2.13 and the fact that $\mathcal{L}_x(\mathbf{X}_t) = P_t(x, \cdot)$ we get

$$\int_0^\infty \int_\Omega \exp(-\lambda s) u(\mathbf{X}_s(\omega)) d\mathbb{P}_x(\omega) ds = \int_0^\infty \exp(-\lambda s) P_s u(x) ds$$
$$= R_\lambda u(x) < \infty.$$

So by Fubini-Tonelli it follows

$$\mathbb{E}_x \left[\int_0^\infty \exp(-\lambda s) u(\mathbf{X}_s) ds \right] = \int_\Omega \int_0^\infty \exp(-\lambda s) u(\mathbf{X}_s(\omega))\, ds\, d\mathbb{P}_x(\omega)$$
$$= \int_0^\infty \int_\Omega \exp(-\lambda s) u(\mathbf{X}_s(\omega))\, d\mathbb{P}_x(\omega)\, ds = R_\lambda u(x).$$

So for $x \in E_1$, $U^\lambda u(x)$ is well-defined and $U^\lambda u$ is a measurable positive function on E_1 and (2.18) holds.

The statement for general $u \in \mathcal{L}^p(E, \mu)$ follows by applying the argumentation to $|u|$, then to u^+ and u^- and then using linearity. For $u \in \mathcal{B}_b(E)$ (2.18) follows using Corollary 2.2.5. $\qquad\square$

Next we state some properties of the process $\mathbf{M}$ which can be transferred from the process $\hat{\mathbf{M}}$ to pointwise statements for $\mathbf{M}$ on E_1.

Theorem 2.3.10. *Let* $\mathbf{M} = (\Omega, \mathcal{F}, (\mathcal{F}_t)_{t \geq 0}, (\mathbf{X}_t)_{t \geq 0}, (\mathbb{P}_x)_{x \in E_1 \cup \{\Delta\}})$ *be the Markov process from Theorem 2.1.3. Then the following properties hold:*

(i) *If* $U \subset E_1$ *has* μ*-measure zero, then* $dx(\{s \in \mathbb{R}_0^+ \mid \mathbf{X}_s \in U\}) = 0$, dx *the Lebesgue measure on* $\mathbb{R}_0^+$, $\mathbb{P}_x$*-a.s. for* $x \in E_1$.

(ii) *If* $U \in \mathcal{B}(E)$ *has the property* $cap_\mathcal{E}(U) = 0$, *then* $\mathbb{P}_x(\sigma_U < \infty) = 0$ *for* $x \in E_1$.

(iii) *Let* $(K_n)_{n \in \mathbb{N}}$ *be an increasing sequence of sets in* E_1 *with*

$$\lim_{n \to \infty} cap_\mathcal{E}(K \setminus K_n) = 0 \quad \text{for every compact} \, K \subset E.$$

Set $\widetilde{E_1} := \bigcup_{n \in \mathbb{N}} K_n \subset E_1$. *Assume that* $\bigcup_{n \in \mathbb{N}} \overset{\circ}{K_n} = \bigcup_{n \in \mathbb{N}} K_n$. *Then it holds*

$$\mathbb{P}_x \left(\lim_{n \to \infty} \sigma_{K_n^c} \geq \mathcal{X} \right) = 1 \quad \text{for all } x \in \widetilde{E_1}.$$

Proof. (i): If $\mu(U) = 0$ then $G_1 1_U(x) = 0$ for μ-a.e. $x \in E$. By the $\mathcal{L}^p$-strong Feller property it holds $R_1 1_U(x) = 0$ for $x \in E_1$. Thus by Lemma 2.3.9

$$\mathbb{E}_x \left[\int_0^\infty \exp(-t) 1_U(\mathbf{X}_t) dt \right] = R_1 1_U(x) = 0.$$

So $\mathbb{P}_x$-a.s. it holds $\exp(-t) 1_U(\mathbf{X}_t) = 0$ for dx-a.e. t, hence $1_U(\mathbf{X}_t) = 0$ for dx-a.e. t.

(ii): Since U has capacity zero, it is also exceptional, in particular $\mathbb{P}_\mu(\{\sigma_U < \infty\}) = 0$, see Theorem 7.6.8(ii) and Definition 7.6.5(i). Using the Markov property of $\mathbf{M}$ and absolute continuity of $(P_t)_{t \geq 0}$ we get $\mathbb{P}_x(\{\sigma_U \circ \theta_s < \infty\}) = 0$ for all $x \in E_1$.

Thus for all $s > 0$ it holds $\mathbb{P}_x(\exists t > s$ such that $X_t \in U) = 0$. So, $\mathbb{P}_x(\exists 0 < t$ such that $X_t \in U) = 0$ for $x \in E_1$. But this just means that $\sigma_U = \infty$ $\mathbb{P}_x$-a.s. for every $x \in E_1$.

(iii): From Lemma 7.6.9 we get

$$\mathbb{P}_x \left(\lim_{n \to \infty} \sigma_{K_n^c} \geq \mathcal{X} \right) = 1 \quad \text{for } \mu - \text{a.e. } x \in E.$$

Using the Markov property and absolute continuity of $(P_t)_{t \geq 0}$ we get

$$\mathbb{P}_x \left(\left\{ \lim_{n \to \infty} \sigma_{K_n^c}(\theta_s \cdot) \geq \mathcal{X}(\theta_s \cdot) \right\} \right) = 1 \quad \text{for } s > 0 \text{ and } x \in E_1,$$

hence

$$\mathbb{P}_x \left(\bigcap_{\substack{s > 0 \\ s \in S}} \left\{ \lim_{n \to \infty} \sigma_{K_n^c}(\theta_s \cdot) \geq \mathcal{X}(\theta_s \cdot) \right\} \right) = 1 \quad \text{for every } x \in E_1.$$

Let $x \in \widetilde{E_1}$. Then $x \in \overset{\circ}{K_{\widetilde{n}}}$ for some $\widetilde{n} \in \mathbb{N}$. Define

$$\Omega^x := \left\{ \omega \in \Omega \mid X_t(\omega) \text{ is continuous on } [0, \infty) \text{ and } \mathbf{X}_0(\omega) = x \right\}$$
$$\cap \bigcap_{\substack{s > 0 \\ s \in S}} \left\{ \lim_{n \to \infty} \sigma_{K_n^c}(\theta_s \cdot) \geq \mathcal{X}(\theta_s \cdot) \right\}.$$

Choose $\omega \in \Omega^x$. By right-continuity of $\mathbf{X}_t(\omega)$ we can find $0 < s' < \infty$ such that $\mathbf{X}_\cdot(\omega)|_{[0,s']} \in \overset{\circ}{K_{\widetilde{n}}}$. So $\sigma_{K_n^c}(\omega) \geq s'$ for $n \geq \widetilde{n}$ and $\mathcal{X}(\omega) \geq s'$. This implies

$$\sigma_{K_n^c}(\omega) = s' + \sigma_{K_n^c}(\theta_{s'}\omega)$$

for $n \geq \widetilde{n}$ and

$$\mathcal{X}(\omega) = s' + \mathcal{X}(\theta_{s'}\omega).$$

So by choice of ω we have

$$\lim_{n \to \infty} \sigma_{K_n^c}(\omega) = \lim_{n \to \infty} \sigma_{K_n^c}(\theta_{s'}\omega) + s' \geq \mathcal{X}(\theta_{s'}\omega) + s' = \mathcal{X}(\omega).$$

So $\lim_{n \to \infty} \sigma_{K_n^c} \geq \mathcal{X}$ on Ω^x. Since $\mathbb{P}_x(\Omega^x) = 1$, we get the claim. $\qquad \square$

We prove that the process solves the martingale problem for functions $(L_p, D(L_p))$.

Theorem 2.3.11. *The diffusion* $\mathbf{M} = (\Omega, \mathcal{F}, (\mathcal{F}_t)_{t\geq 0}, (\mathbf{X}_t)_{t\geq 0}, (\mathbb{P}_x)_{x \in E_1 \cup \{\Delta\}})$ *of Theorem 2.1.3 solves the martingale problem for* $(L_p, D(L_p))$, *i.e.,*

$$M_t^{[u]} := \widetilde{u}(\mathbf{X}_t) - \widetilde{u}(x) - \int_0^t L_p u(\mathbf{X}_s) \, ds, \ t \geq 0,$$

is an $(\mathcal{F}_t)$*-martingale under* $\mathbb{P}_x$ *for all* $u \in D(L_p)$, $x \in E_1$.

Proof. Let $x \in E_1$, $u \in D(L_p)$. Then by Lemma 2.2.14 it holds

$$P_t u(x) - \widetilde{u}(x) = \int_0^t P_s \, L_p u \, (x) ds$$

and the integral is well-defined. Since both u and $L_p u$ are extended by 0 to E^Δ, we have that this equality also holds with P_t and P_s replaced by P_t^Δ and P_s^Δ, respectively. Since $\int_0^t \mathbb{E}_x[(L_p u)^{+/-}(\mathbf{X}_s)]ds = \int_0^t P_s(L_p u)^{+/-}(x)ds < \infty$, the integral $\int_0^t L_p u(\mathbf{X}_s)ds$ exists $\mathbb{P}_x$-a.s. Note that by Theorem 2.3.10(i) the integral is independent of the μ-version of $L_p u$. Now the statement follows using the Markov property of $(\mathbf{X}_t)_{t\geq 0}$ and that $(P_t^\Delta)_{t\geq 0}$ is the transition semigroup of $(\mathbf{X}_t)_{t\geq 0}$. $\square$

2.4 Some Examples

We close this section with examples. These examples are based on [AKR03], [BGS13] and [Sti10]. We show that the results presented there can be deduced from our construction scheme together with the elliptic regularity results which are used in the cited works.

First we consider the distorted Brownian motion on $\mathbb{R}^d$, $d \in \mathbb{N}$. This is done in [AKR03]. Let $\varrho : \mathbb{R}^d \to \mathbb{R}_0^+$ be Borel measurable. Define $\mu := \varrho \, dx$, the measure with density ϱ with respect to the Lebesgue measure. Assume the following conditions.

Condition 2.4.1.

(i) $\sqrt{\varrho} \in H_{\mathrm{loc}}^{1,2}(\mathbb{R}^d)$ and $\varrho > 0$ dx-a.e.

(ii) $\frac{|\nabla \varrho|}{\varrho} \in L_{\mathrm{loc}}^p(\mathbb{R}^d, \mu)$ for some $p > d$.

Observe that the integrability condition in (ii) is formulated with respect to the measure μ not to the Lebesgue measure. This allows potentials with strong singularities. Define

$$\mathcal{E}(u,v) := \int_{\mathbb{R}^d} (\nabla u, \nabla v) \, d\mu, \ u,v \in C_c^\infty(\mathbb{R}^d).$$

Due to Condition 2.4.1(i) the Dirichlet form admits a partial integration formula and hence it is closable. The closure $(\mathcal{E}, D(\mathcal{E}))$ is a regular strongly local symmetric Dirichlet form on $L^2(\mathbb{R}^d, \mu)$. Furthermore, the set $\{\varrho = 0\}$ is of zero capacity with respect to $(\mathcal{E}, D(\mathcal{E}))$. Together with Condition 2.4.1(ii) one obtains that ϱ has a continuous version, see [AKR03, Cor. 2.2]. We denote this continuous version by ϱ as well. We choose as $E := \mathbb{R}^d$ and as $E_1 := \{\varrho > 0\}$ which is open due to continuity. One easily sees that Condition 2.1.1 is fulfilled.

We have to check Condition 2.1.2. First of all we claim that $C_c^\infty(\mathbb{R}^d) \subset D(L_p)$ for p as in Condition 2.4.1(ii). Using partial integration we get (according to [AKR03, (2.2)]

$$\mathcal{E}(u,v) = -\int_{\mathbb{R}^d} \left(\Delta u + \left(\frac{\nabla \varrho}{\varrho}, \nabla u \right) \right) v \, d\mu \quad \text{for } u,v \in C_c^\infty(\mathbb{R}^d).$$

Together with Condition 2.4.1(ii) this yields $C_c^\infty(\mathbb{R}^d) \subset D(L_p)$ and $L_p u = \Delta u + (\frac{\nabla \varrho}{\varrho}, \nabla u)$. Thus $D(L_p)$ is point separating on $\mathbb{R}^d$.

Next we check the regularity condition, i.e., Condition 2.1.2(i). We apply the regularity result for the resolvent of [AKR03] which is obtained using the elliptic regularity results of [BKR97] and [BKR01]. So [AKR03, Cor. 2.3] states that for $f \in L^r(\mathbb{R}^d, \mu)$, $p \le r < \infty$, it holds

$$\varrho G_\lambda f \in H_{\text{loc}}^{1,p}(\mathbb{R}^d)$$

and for any open ball $B \subset \overline{B} \subset \{\varrho > 0\}$ there exists $c_{B,\lambda} \in (0,\infty)$, independent of f, such that

$$\|\varrho G_\lambda f\|_{H^{1,p}(\mathbb{R}^d)} \le c_{B,\lambda}(\|G_\lambda f\|_{L^1(B,\mu)} + \|f\|_{L^p(B,\mu)}). \tag{2.19}$$

Since $p > d$, we get by Sobolev embedding that $\varrho G_\lambda f$ is continuous on $\{\varrho > 0\}$. So also $G_\lambda f$ is continuous on $\{\varrho > 0\}$. The L^p-resolvent is surjective on $D(L_p)$, hence we get the required embedding $D(L_p) \hookrightarrow C^0(\{\varrho > 0\})$. Estimate (2.19) ensures local continuity of the embedding. So Condition 2.1.2 is fulfilled. A proper modification of (2.19) yields that the constructed resolvent kernels are even strong Feller, see Theorem 2.1.5. From Theorem 2.1.3 we obtain the following result:

Theorem 2.4.2. *There exists a diffusion process (i.e., a strong Markov process having continuous sample paths on the time interval $[0, \infty)$)* $\mathbf{M} = (\Omega, \mathcal{F}, (\mathcal{F}_t)_{t \geq 0}, (\mathbf{X}_t)_{t \geq 0}, (\mathbb{P}_x)_{x \in E_1 \cup \{\Delta\}})$ *with state space $\mathbb{R}^d$ and cemetery Δ, the Alexandrov point of $\mathbb{R}^d$. The process leaves $\{\varrho > 0\} \cup \{\Delta\}$ invariant. The transition semigroup $(P_t)_{t \geq 0}$ is associated with $(T_t^2)_{t \geq 0}$ and is $\mathcal{L}^p$-strong Feller, i.e., $P_t \mathcal{L}^p(E, \mu) \subset C^0(\{\varrho > 0\})$ for $t > 0$. The process has continuous paths on $[0, \infty)$ and it solves the martingale problem associated with $(L_p, D(L_p))$ for starting points in $\{\varrho > 0\}$, i.e.,*

$$M_t^{[u]} := \widetilde{u}(\mathbf{X}_t) - \widetilde{u}(x) - \int_0^t \Delta u(\mathbf{X}_s) + \left(\frac{\nabla \varrho}{\varrho}(\mathbf{X}_s), \nabla u(\mathbf{X}_s) \right) \, ds, \ t \geq 0,$$

is an $(\mathcal{F}_t)$-martingale under $\mathbb{P}_x$ for all $u \in D(L_p)$ and $x \in E_1$. In particular the process solves the martingale problem for functions in $C_c^\infty(\mathbb{R}^d)$. If $(\mathcal{E}, D(\mathcal{E}))$ is conservative, then $\mathbf{M}$ is conservative under $\mathbb{P}_x$ for every starting point $x \in E_1$.

This theorem reproduces [AKR03, Theo. 1.1]. The authors prove that the transition semigroup (and resolvent) is even $\mathcal{L}^r$-strong Feller on $\{\varrho > 0\}$ for $r \in [p, \infty)$.

These results are applied to construct stochastic dynamics for finite particle systems with singular interaction and diffusions in random media, see [AKR03, Sec. 6]. Using different techniques, these results have been further developed to yield even strong solutions, see [KR05].

Next we present the results obtained in [BGS13] and [Sti10]. There a gradient-Dirichlet form with variable coefficient matrix A and density ϱ on a general open set is considered. They construct a diffusion process on this set with absorbing boundary condition. This result is obtained in [Sti10] without applying the general construction scheme we presented here. In [BGS13] the general construction scheme is presented and the result of [Sti10] is then reproduced. We follow [BGS13]. For a detailed workout the reader should consult [Sti10].

So let us fix an open set $\Omega \subset \mathbb{R}^d$, a matrix-valued mapping $A = (a_{ij})_{i,j=1}^d :$ $\Omega \to \mathbb{R}^{d \times d}$, a Borel measure μ with density $\varrho : \Omega \to [0, \infty)$, i.e., $\mu := \varrho \, dx$. Fix $p \in \mathbb{N}$. Assume the following conditions.

Condition 2.4.3. Assume $p > d$.

(i) $\sqrt{\varrho} \in H^{1,2}_{\text{loc}}(\Omega)$, $\varrho > 0$ dx-a.e.

(ii) $\frac{\nabla \varrho}{\varrho} \in L^p_{\text{loc}}(\Omega, \mu)$.

(iii) A is measurable, symmetric and locally strictly elliptic on Ω dx-a.e. Furthermore, $a_{ij} \in H_{loc}^{1,\infty}(\Omega)$, $1 \le i, j \le d$.

In the case $A = 1$, the identity matrix, $\Omega = \mathbb{R}^d$ we are back in the setting of [AKR03].

Consider the (pre-) gradient-Dirichlet form

$$\mathcal{E}(u,v) := \int_\Omega (A\nabla u, \nabla v) \, d\mu = \int_\Omega \sum_{i,j=1}^{d} a_{ij} \, \partial_i u \, \partial_j v \, d\mu, \; u, v \in \mathcal{D} := C_c^\infty(\Omega).$$

The form is closable to a regular, strongly local and symmetric Dirichlet form. Furthermore, $\{\varrho > 0\}$ is of zero capacity, see [BGS13, Prop. 4.2] and [BGS13, Prop. 4.4]. Using partial integration we get $C_c^\infty(\Omega) \subset D(L_p)$ and

$$L_p u = \sum_{i,j=1}^{d} \partial_i(a_{ij} \, \partial_j u) + \partial_i (\ln \varrho) \, a_{ij} \, \partial_j u \quad \text{for } u \in C_c^\infty(\Omega).$$

This can be rewritten as

$$L_p u = \sum_{i,j=1}^{d} a_{ij} \, \partial_i \partial_j u + \sum_{i=1}^{d} \sum_{j=1}^{d} (\partial_j a_{ij} + \partial_j(\ln \varrho) a_{ij}) \, \partial_i u.$$

See [BGS13, Prop. 4.3]. In [BGS13, Sec. 5] an elliptic regularity result from [BKR01] is generalized. This result is then applied to conclude $D(L_p) \subset C^0(\{\varrho > 0\})$, the embedding being locally continuous. So again from Theorem 2.1.3 we get a $\mathcal{L}^p$-strong Feller process solving the martingale problem for $u \in C_c^\infty(\Omega)$ and $x \in \{\varrho > 0\}$.

Note that after compactifying Ω to $\Omega \cup \{\Delta\}$ the euclidean boundary of Ω is identified with Δ. Thus if the process hits the euclidean boundary, it is transfered to Δ, i.e., absorbed.

3 Elliptic Regularity up to the Boundary

The aim of this chapter is to prove a Sobolev space regularity result for weak solutions of elliptic equations. This result will be used for the construction of $\mathcal{L}^p$-strong Feller processes in the next chapter. It will be also used for the construction of the boundary local time.

3.1 Elliptic Regularity up to the Boundary

We partially generalize a regularity result of Morrey to the case of local assumptions on the coefficients and data. Morrey's result applies for Ω being a relatively compact set and coefficients fulfilling certain integrability conditions and bounds on an open set Γ with $\overline{\Omega} \subset \Gamma$. In particular, it is assumed that the coefficient matrix A is uniformly elliptic.

We only assume conditions that hold *locally* in $\overline{\Omega}$. In particular, we need only *local* uniform ellipticity of the matrix. This is important for our application since there the matrix is ϱA and the density ϱ might be zero at some points. On the other hand, our result is more specific because we do not consider the case of lower order singular coefficients, as it was done by Morrey. Nevertheless, to some extent we can also handle singular drifts since the drift term of the generator allows singularities at the zero set of the density, see (4.5) and Lemma 4.1.13 in the next chapter.

We give a detailed version of the proof. Morrey provided only a short sketch of the proof of his result, see [Mor66, Theo. 5.5.4'] and [Mor66, Theo. 5.5.5']. Our proof uses methods and estimates from Shaposhnikov, see [Sha06]. Shaposhnikov gives a detailed proof of Morrey's a-priori estimate there. It is shown that solutions in $H^{1,q}(\Omega)$, $2 \leq q < \infty$, satisfy a (global) norm estimate similar to (3.2) below. We emphasize that the a-priori estimate and the existence of an $H^{1,2}(\Omega)$ solution alone are not sufficient to conclude the $H^{1,q}$-regularity, we prove, of this solution.

If Ω is relatively compact and has C^1-smooth boundary, then our regularity result implies global regularity of the solution, provided ϱ is strictly positive on $\overline{\Omega}$. Thus we partially reproduce Morrey's result. But in all details.

Let us mention other elliptic regularity results. Fukushima and Tomisaki consider elliptic equation with bounded and uniformly elliptic matrix coefficient on a domain D with Lipschitz boundary and Hölder cusps. They prove continuity (on $\overline{D}$) of solutions with right-hand side in $L^p(D, dx)$ for certain p depending on the boundary smoothness. See [FT95] and [FT96].

Miyazaki provides several Sobolev space regularity results for L^p-resolvents, however on $\mathbb{R}^d$ or on domains with Dirichlet boundary condition, see e.g. [Miy03] or [Miy06].

Bogachev, Krylov and Röckner (see [BKR97] and [BKR01]) prove regularity results for measures which solve elliptic (or parabolic) equations in distributional form. Although we do not apply these results here directly, we got many ideas from these articles, in particular the iteration sequence used for the proof in the interior case. We have published the results stated in this chapter in [BG13].

Theorem 3.1.1. *Let $\Omega \subset \mathbb{R}^d$, $d \in \mathbb{N}$ and $d \geq 2$, be open. Let $2 \leq p < \infty$. Let $p < q \leq \frac{dp}{d-p}$ for $p < d$, or $p < q < \infty$ for $p \geq d$. Let $x \in \overline{\Omega}$ and $r > 0$ such that $B_r(x) \subset \Omega$ if $x \in \Omega$ and $B_r(x) \cap \partial\Omega$ is C^1-smooth if $x \in \partial\Omega$. Assume that $A = (a_{ij})_{1 \leq i,j \leq d}$ is a matrix-valued mapping of symmetric, strictly elliptic matrices in $\overline{B_r(x)} \cap \overline{\Omega}$ and continuous at x. Let $c : \overline{\Omega} \to \mathbb{R}$ be continuous. Then there exists $0 < r' < r$ such that the following conclusion holds:*

Assume that $u \in H^{1,2}(B_r(x) \cap \Omega)$ and there exist $f \in L^p(B_r(x) \cap \Omega, dx)$, $e = (e_i)_{1 \leq i \leq d}$, $e_i \in L^q(B_r(x) \cap \Omega, dx)$ such that u solves

$$\int_{B_r(x) \cap \Omega} (A\nabla u, \nabla v)\, dx + \int_{B_r(x) \cap \Omega} c\, u\, v\, dx$$
$$= \int_{B_r(x) \cap \Omega} f\, v\, dx + \int_{B_r(x) \cap \Omega} (e, \nabla v)\, dx \quad \text{for all } v \in C_c^1(B_r(x) \cap \overline{\Omega}). \tag{3.1}$$

Then $u \in H^{1,q}(B_{r'}(x) \cap \Omega)$ and there exists a constant $C_1 < \infty$ independent of u, e and f such that

$$\|u\|_{H^{1,q}(B_{r'}(x) \cap \Omega)} \leq C_1(\|f\|_{L^p(B_r(x) \cap \Omega, dx)} + \|e\|_{(L^q)^d(B_r(x) \cap \Omega, dx)}$$
$$+ \|u\|_{H^{1,2}(B_r(x) \cap \Omega)}). \tag{3.2}$$

Remark 3.1.2. To prove the regularity of the resolvent, we choose $c := \lambda \varrho$ and as matrix ϱA, see Theorem 4.2.5 in Chapter 4.

Note that in the case $x \in \Omega$ the test functions have compact support in Ω while in the boundary case the support can contain a compact part of the smooth boundary. We prove the theorem in several steps. First we state some preparatory definitions and lemmata.

Definition 3.1.3. We define Green's function by

$$K(y) = -\frac{1}{d(d-2)\omega_d} \frac{1}{|y|^{d-2}}$$

for $d > 2$ and

$$K(y) = \frac{1}{2\pi}\ln(|y|)$$

for $d = 2$. Here $|\cdot|$ denotes the Euclidean norm on $\mathbb{R}^d$ and ω_d the surface measure of the unit sphere in $\mathbb{R}^d$.

See also [Sha06, Def. 1] or [Mor66, Sec. 2.4.1].
We need the following results on functions in Sobolev spaces.

Lemma 3.1.4.

(i) Let $\Omega \subset \mathbb{R}^d$ be a bounded domain, $u \in H_0^{1,1}(\Omega)$. Then for almost all $y \in \Omega$ it holds

$$u(y) = \int_\Omega ((\nabla K)(y-x), \nabla u(x))\, dx = \sum_{i=1}^d \frac{1}{\omega_d d} \int_\Omega \frac{(y_i - x_i)\partial_i u(x)}{|x-y|^d}\, dx.$$

(ii) Let $1 \le p_0 < d$, $p_1 \le \frac{dp_0}{d-p_0}$, $\Omega \subset \mathbb{R}^d$ a domain satisfying the cone property. Then $H^{1,p_0}(\Omega) \hookrightarrow L^{p_1}(\Omega, dx)$ and the embedding is continuous.

For the proof of (i) see [Mor66, Theo. 3.7.2], for the proof of (ii) see [AD75, Ch. V, Lem. 5.14]. Here $H_0^{1,1}(\Omega)$ denotes the closure of $C_c^\infty(\Omega)$ in $H^{1,1}(\Omega)$. Both Lemma 3.1.4(ii) and Theorem 7.5.12 in the appendix are called *Sobolev embedding Theorem*.

Lemma 3.1.5.   *(i) Let $0 < R < \infty$ ($R < \frac{1}{2}$ if $d = 2$) be fixed, define for $0 < r < R$ the linear operator P_r by*

$$f \mapsto -\int_{B_r(0)} K(\cdot - x) f(x)\, dx$$

where f is a scalar-valued measurable function such that the Lebesgue integral exists. Then for $1 < \hat{p} < d$, $1 < \hat{q} \leq \frac{d\hat{p}}{d-\hat{p}}$ P_r defines a bounded linear operator from $L^{\hat{p}}(B_r(0), dx) \to H^{1,\hat{q}}(B_r(0))$ and there is a constant $C_2 = C_2(d, \hat{q}, \hat{p}, R) < \infty$ such that

$$\|P_r f\|_{H^{1,\hat{q}}(B_r(0))} \leq C_2 \|f\|_{L^{\hat{p}}(B_r(0), dx)}.$$

(ii) Let $0 < R < \infty$. Define for $0 < r < R$ the linear operator Q_r by

$$e \mapsto -\int_{B_r(0)} \big((\nabla K)(\cdot - x), e(x)\big)\, dx$$

where $e : B_r(0) \to \mathbb{R}^d$, $e = (e_1, ..., e_d)$ is a measurable vector field such that the Lebesgue integral exists. Then for $1 < \hat{q} < \infty$, Q_r defines a bounded linear operator from $(L^{\hat{q}})^d(B_r(0), dx) \to H^{1,\hat{q}}(B_r(0))$ and there exists a constant $C_3 = C_3(d, \hat{q}, R)$ (independent of r!) such that

$$\|Q_r e\|_{H^{1,\hat{q}}(B_r(0))} \leq C_3 \|e\|_{(L^{\hat{q}})^d(B_r(0), dx)}. \tag{3.3}$$

Here the $(L^s)^d$-norm, $1 \leq s < \infty$, is defined with the s-norm on $\mathbb{R}^d$, i.e.,

$$\|e\|_{(L^s)^d(B_r(0), dx)} := \left(\sum_{j=1}^{d} \|e_j\|_{L^s(B_r(0), dx)}^s \right)^{1/s}$$

$$= \left(\sum_{j=1}^{d} \int_{B_r(0)} |e_j(x)|^s dx \right)^{1/s}.$$

As $H^{1,s}(B_r(0))$-norm we take

$$\|u\|_{H^{1,s}(B_r(0))} := \left(\|u\|_{L^s(B_r(0), dx)}^s + \|\nabla u\|_{(L^s)^d(B_r(0), dx)}^s \right)^{1/s}.$$

(iii) Let $0 < R < \infty$ ($R < \frac{1}{2}$ if $d = 2$), $1 < \hat{q} < \infty$, A a matrix-valued function that is continuous in 0 with $A(0) = 1$. Then there exists an $0 < \widetilde{R} < R$ depending on R, $\hat{q}$ such that for $R_0 \leq \widetilde{R}$ the operator $T_{R_0} : H^{1,\hat{q}}(B_{R_0}(0)) \to H^{1,\hat{q}}(B_{R_0}(0))$

$$v \mapsto v + \int_{B_{R_0}(0)} \left((\nabla K)\left(\cdot - x \right), (A(x) - 1)\nabla v(x) \right) dx$$

is a continuously invertible bijective operator on $H^{1,\hat{q}}(B_{R_0}(0))$.

Proof. The first two statements can be concluded from [Sha06, Lem. 1]. However, an estimate of the $L^{\hat{q}}(B_r(0))$ norm of $P_r f$ was given only in the case $R < \frac{1}{2}$. The norm estimate for $d > 2$ and general $R > 0$ follows from [GT77, Lem. 7.12] with $\mu = \frac{2}{d}$.

For the third statement note that for $R > 0$, $v \in H^{1,\hat{q}}(B_R(0))$ it holds

$$T_R v = v - Q_R\big((A - 1)\nabla v\big).$$

This defines a bounded linear operator on $H^{1,\hat{q}}(B_R(0))$. Choose $0 < \widetilde{R} < R$ so small that with the constant $C_3 = C_3(d, \hat{q}, R)$ of (3.3) it holds

$$C_3 \left(\sum_{i=1}^{d} \max_{1 \leq j \leq d} \sup_{x \in B_{\widetilde{R}}(0)} |a_{ij}(x) - \delta_{ij}|^{\hat{q}} \right)^{1/\hat{q}} < \frac{1}{2}.$$

Since for $0 < R_0 < R$ the operator norm of Q_{R_0} is bounded by the constant C_3, we get for $R_0 \leq \widetilde{R}$

$$\left\| Q_{R_0}\big((A - 1)\nabla \cdot \big) \right\|_{L\left(H^{1,\hat{q}}(B_{R_0}(0))\right)} < \frac{1}{2}.$$

By Neumann series we obtain that T_{R_0} is a continuously invertible bijective operator on $H^{1,\hat{q}}(B_{R_0}(0))$ for $0 < R_0 \leq \widetilde{R}$. $\qquad\square$

We state one more lemma concerning transformations of weak solutions.

Lemma 3.1.6. *Let $0 \in M_1 \subset \mathbb{R}^d$ open, $A = (a_{ij})_{1 \leq i,j \leq d}$ a matrix-valued mapping of symmetric and strictly elliptic matrices that is continuous in 0. Let $c : \overline{M_1} \to \mathbb{R}$ be a continuous mapping. Let $u \in H^{1,p_0}(M_1)$, $f \in L^p(M_1, dx)$, $e \in (L^q)^d(M_1, dx)$, $p_0, p, q \geq 1$.*

Furthermore, let $M_2 \subset \mathbb{R}^d$ *open and* W_1, W_2 *relatively compact open neighborhoods of* 0 *and* $\psi : W_1 \to W_2$ *a* C^1*-diffeomorphism with* $\psi(W_1 \cap M_1) = W_2 \cap M_2$, $\psi(W_1 \cap \partial M_1) = W_2 \cap \partial M_2$, $\psi(0) = 0$. *Define* $\hat{u} := u \circ \psi^{-1}$, $\hat{f} := \frac{1}{|DetD\psi|} f \circ \psi^{-1}$, $\hat{e} := \frac{1}{|DetD\psi|}((D\psi)e) \circ \psi^{-1}$, $\hat{A} := ((D\psi)A(D\psi)^{\top}) \frac{1}{|DetD\psi|} \circ \psi^{-1}$ *and* $\hat{c} := \frac{1}{|DetD\psi|} c \circ \psi^{-1}$.

Then $\hat{u} \in H^{1,p_0}(W_2 \cap M_2)$, $\hat{f} \in L^p(W_2 \cap M_2, dx)$, $\hat{e} \in (L^q)^d(W_2 \cap M_2, dx)$. *Moreover,* $\hat{A}$ *is a matrix-valued mapping of symmetric and strictly elliptic matrix that is continuous in* 0. *Furthermore, if* u *solves*

$$\int_{W_1 \cap M_1} (A\nabla u, \nabla v) \, dx + \int_{W_1 \cap M_1} c \, u \, v \, dx$$

$$= \int_{W_1 \cap M_1} f \, v \, dx + \int_{W_1 \cap M_1} (e, \nabla v) \, dx \quad \text{for all } v \in C_c^1(W_1 \cap \overline{M_1})$$

then $\hat{u}$ *solves*

$$\int_{W_2 \cap M_2} (\hat{A}\nabla \hat{u}, \nabla \hat{v}) \, dy + \int_{W_2 \cap M_2} \hat{c} \, \hat{u} \, \hat{v} \, dy$$

$$= \int_{W_2 \cap M_2} \hat{f} \, \hat{v} \, dy + \int_{W_2 \cap M_2} (\hat{e}, \nabla \hat{v}) \, dy \quad \text{for all } \hat{v} \in C_c^1(W_2 \cap \overline{M_2}).$$

Here $D\psi$, respectively $Det\, D\psi$, denotes the Jacobi matrix, respectively the determinant of the Jacobi matrix, of ψ.

Remark 3.1.7. In the sequel we apply this lemma for M_1, M_2 open subsets of $\mathbb{R}^d$ and $M_1 \subset \Omega$ an open set, $M_2 = \mathbb{R}^d \cap \{x_d > 0\}$. In the latter case ψ is the diffeomorphism flattening the boundary of Ω locally.

Using these lemmata we prove the regularity result. We start with the interior point case. We provide a local representation for weak solutions in terms of potentials. This representation has been already derived in [Sha06, Lem. 2], but we include it here to keep our exposition self-contained.

Proof of Theorem 3.1.1, interior point case.
After translation we may assume $x = 0$. The proof is divided into several steps. First we consider the case $A(0) = \mathbf{1}$. We prove that for $0 < R < r$ there exists an $0 < R' < R$ such that for $1 < \hat{p} < d$ and $\hat{p} \le p$, $\hat{p} < \hat{q} \le q \wedge \frac{d\hat{p}}{d-\hat{p}}$ it holds for u fulfilling (3.1) that $u \in H^{1,\hat{p}}(B_R(0))$ implies $u \in H^{1,\hat{q}}(B_{R'}(0))$. The statement follows then by induction. In the last step we reduce the case of general A by transformation to the case $A(0) = \mathbf{1}$.

So let us prove the first step. We assume $A(0) = 1$ and $1 < \hat{p} < d$, $\hat{p} \leq p$, $\hat{p} < \hat{q} \leq q \wedge \frac{d\hat{p}}{d-\hat{p}}$. From Lemma 3.1.5(iii) we conclude that there exists an $R_0 > 0$ with $R_0 < R$ such that the operator T_{R_0} is bijective both on $H^{1,\hat{p}}(B_{R_0}(0))$ and $H^{1,\hat{q}}(B_{R_0}(0))$. Now choose $R' < R_0$ and a smooth cutoff ϕ for $B_{R'}(0)$ in $B_{R_0}(0)$.

Then for $v \in C^1(B_R(0))$ it holds $\phi v \in C_c^1(B_R(0)) \subset C_c^1(B_r(0))$ so using (3.1) we get

$$
\int_{B_R(0)} \big(A\nabla(\phi u), \nabla v\big)\, dx = \int_{B_R(0)} \big(A(\phi \nabla u + u \nabla \phi), \nabla v\big)\, dx
$$

$$
= \int_{B_R(0)} \big(A\nabla u, \nabla(\phi v)\big)\, dx - \int_{B_R(0)} v\big(A\nabla u, \nabla\phi\big)\, dx + \int_{B_R(0)} \big(uA\nabla\phi, \nabla v\big)\, dx
$$

$$
= \int_{B_R(0)} \big(f\phi v + (e, \nabla(\phi v))\big)\, dx - \int_{B_R(0)} c\,\phi u\, v\, dx - \int_{B_R(0)} v\,\big(A\nabla u, \nabla\phi\big)\, dx
$$

$$
+ \int_{B_R(0)} \big(uA\nabla\phi, \nabla v\big)\, dx
$$

$$
= \int_{B_R(0)} \big(\phi f + (e, \nabla\phi) - (A\nabla u, \nabla\phi) - c\phi u\big)\, v\, dx + \int_{B_R(0)} \big(\phi e + uA\nabla\phi, \nabla v\big)\, dx.
$$

Set $\hat{f} := \phi f + (e, \nabla\phi) - (A\nabla u, \nabla\phi) - c\phi u$, $\hat{e} := \phi e + uA\nabla\phi$. Then $\hat{f} \in L^{\hat{p}}(B_R(0), dx)$. By Sobolev embedding (Lemma 3.1.4(ii)) we get that $\hat{e} \in (L^{\hat{q}})^d(B_R(0), dx)$. Now for every $\psi \in C_c^\infty(B_R(0))$, we have that for $v := \int_{B_R(0)} K(\cdot - y)\psi(y)dy \in C^\infty(B_R(0))$ the above equality holds.

Using Fubini and the Fundamental Lemma of Calculus of Variations we get for almost all $y \in B_R(0)$:

$$
-\int_{B_R(0)} \big((\nabla K)(y-x), A\nabla(\phi u)(x)\big)\, dx
$$

$$
= \int_{B_R(0)} K(y-x)\,\hat{f}(x)\, dx - \int_{B_R(0)} \big((\nabla K)(y-x), \hat{e}(x)\big)\, dx.
$$

Here we used $K(y - x) = K(x - y)$ and $(\nabla K)(y - x) = -(\nabla K)(x - y)$. Using Lemma 3.1.4(i) we get for almost all $y \in B_R(0)$:

$$
\begin{aligned}
\phi u(y) &= \int_{B_R(0)} \left((\nabla K)(y - x), \nabla(\phi u)(x) \right) dx \\
&= -\int_{B_R(0)} \left((\nabla K)(y - x), (A - 1)\nabla(\phi u)(x) \right) dx \\
&\quad + \int_{B_R(0)} \left((\nabla K)(y - x), A\nabla(\phi u)(x) \right) dx \\
&= -\int_{B_R(0)} \left((\nabla K)(y - x), (A - 1)\nabla(\phi u)(x) \right) dx \\
&\quad - \int_{B_R(0)} K(y - x)\, \hat{f}(x)\, dx \\
&\quad + \int_{B_R(0)} \left((\nabla K)(y - x), \hat{e}(x) \right) dx.
\end{aligned}
$$

Note that by the support property of ϕ all integrals can be replaced by integrals over $B_{R_0}(0)$. Using the potentials defined in Lemma 3.1.5 we get

$$T_{R_0}\phi u = P_{R_0}\hat{f} - Q_{R_0}\hat{e}. \tag{3.4}$$

By the mapping properties of P_{R_0} and Q_{R_0} we have that the right-hand side is in $H^{1,\hat{q}}(B_{R_0}(0))$. Since T_{R_0} is bijective on $H^{1,\hat{q}}(B_{R_0}(0))$ there exists a unique element $w \in H^{1,\hat{q}}(B_{R_0}(0)) \subset H^{1,\hat{p}}(B_{R_0}(0))$ with $T_{R_0}w = P_{R_0}\hat{f} - Q_{R_0}\hat{e}$. By injectivity of T_{R_0} on $H^{1,\hat{p}}(B_{R_0}(0))$ and (3.4) we have $\phi u = w$. So $\phi u \in H^{1,\hat{q}}(B_{R_0}(0))$, hence $u \in H^{1,\hat{q}}(B_{R'}(0))$.

The $L^{\hat{p}}(B_{R_0}(0), dx)$-norm of $\hat{f}$ can be estimated by the $L^p(B_R(0), dx)$-norm of f, the $L^q(B_R(0), dx)$-norm of e and the $H^{1,\hat{p}}(B_R(0))$-norm of u. Analogously, the $L^{\hat{q}}(B_{R_0}(0))$-norm of $\hat{e}$ can be estimated by the $H^{1,\hat{p}}(B_{R_0}(0))$-norm of u and the $L^q(B_R(0), dx)$-norm of e. Since T_{R_0} is continuously invertible on $H^{1,\hat{q}}(B_{R_0}(0))$, the norm estimate follows. The operator norms of P_{R_0} and Q_{R_0} depend on d, $\hat{q}$, $\hat{p}$ and R, so the constant in the norm estimate depends only on d, $\hat{q}$, $\hat{p}$, R, A and ϕ.

Now assume that $d > 2$, $p < d$ and $q \le \frac{dp}{d-p}$. Then for $2 \le s < d$ it holds

$$\frac{\frac{ds}{d-s}}{s} = \frac{d}{d - s} \ge \frac{d}{d - 2} > 1.$$

Since $q < \infty$, there exists $\tilde{q} < d$ such that $q = \frac{d\tilde{q}}{d-\tilde{q}}$ (i.e., $\tilde{q} = \frac{dq}{d+q}$). Thus there exists a finite increasing sequence of q_k, $k \le K$, $K \in \mathbb{N}$, with $q_1 = 2$

and $q_{K-1} = \tilde{q}$, $q_K = q$ and $q_{k+1} \leq \frac{dq_k}{d-q_k}$ for $1 \leq k \leq K - 1$. If there exists $k < K$ with $q_k \geq p$, then choose k' to be the smallest index with this property and replace $q_{k'}$ with p, $q_{k'+1}$ with q and set $K := k' + 1$. Note that in this case $q_{k'} = p < d$ and $q_{k'+1} = q \leq \frac{dp}{d-p} = \frac{dq_{k'}}{d-q_{k'}}$. So by applying iteratively the proof step from above starting with $q_1 = 2$ and $(\hat{q}, \hat{p}) := (q_{k+1}, q_k)$, $0 \leq k < K$, we get an $R'_K > 0$ such that $u \in H^{1,q}(B_{R'_K}(0))$. Set $r' := R'_K$. Estimate (3.2) follows by iteratively applying the derived estimates.

Assume $d > 2$, $p \geq d$ and $p < q < \infty$ arbitrary. Then we can find $\hat{p}$ with $2 \leq \hat{p} < d$ and $q \leq \frac{d\hat{p}}{d-\hat{p}}$. From $\hat{p} < p$ we get $f \in L^{\hat{p}}(B_r(x) \cap \Omega, dx)$. So we can apply the already proven statements with p replaced by $\hat{p}$ to conclude $u \in H^{1,q}(B_{r'}(0))$ for some $0 < r' < r$.

If $d = 2$, $p \geq 2$, then for $2 < q < \infty$ there exists $1 < q_0 < 2$ such that $q_1 := \frac{dq_0}{d-q_0} = q$. Hence also in this case the result follows with $(\hat{q}, \hat{p}) = (q_1, q_0)$.

Now for the case with general A, we find a transformation S with $SA(0)S^\top = 1$, by first diagonalizing $A(0)$ using an unitary matrix and then by multiplying from both sides with a properly chosen diagonal matrix. Since S is an isomorphism, $S(B_r(0))$ is open and there exists an $R_2 > 0$ with $B_{R_2}(0) \subset S(B_r(0))$. Set $W_1 := S^{-1}(B_{R_2}(0))$, $W_2 := B_{R_2}(0)$, define $\psi : W_1 \to W_2$ by $\psi(x) = Sx$. This defines a C^1-isomorphism.

Now let $u \in H^{1,2}(B_r(0))$, $f \in L^p(B_r(0), dx)$, $e = (e_i)_{1 \leq i \leq d}$, $e_i \in L^q(B_r(0), dx)$ fulfilling (3.1). Then also $u \in H^{1,2}(W_1)$, $f \in L^p(W_1, dx)$, $e \in (L^q)^d(W_1, dx)$. Define $\hat{u}$, $\hat{f}$, $\hat{e}$, $\hat{A}$ and $\hat{c}$ as in Lemma 3.1.6 where we choose $M_1 = W_1$ and $M_2 = W_2$. Then $\hat{u} \in H^{1,2}(W_2)$, $\hat{f} \in L^p(W_2, dx)$ and $\hat{e} \in (L^q)^d(W_2, dx)$. From Lemma 3.1.6 we get

$$\int_{W_2} \left(\hat{A}\nabla\hat{u}, \nabla\hat{v} \right) dy + \int_{W_2} \hat{c}\,\hat{u}\,\hat{v}\,dy$$

$$= \int_{W_2} \hat{f}\,\hat{v}\,dy + \int_{W_2} \left(\hat{e}, \nabla\hat{v} \right) dy \quad \text{for all } \hat{v} \in C^1_c(W_2).$$

Here $|\mathrm{Det}D\psi^{-1}| = \frac{1}{|\mathrm{Det}S|}$ is constant, thus we can drop this factor in $\hat{e}$, $\hat{f}$, $\hat{A}$ and $\hat{c}$. So $\hat{A}(0) = 1$, hence by the already proven facts $\hat{u} \in H^{1,q}(B_{\tilde{R}}(0))$ for some $\tilde{R} < R_2$. Then there exists an r' with $B_{r'}(0) \subset \psi^{-1}(B_{\tilde{R}}(0))$. Consequently, $u \in H^{1,q}(B_{r'}(0))$. Furthermore, the $H^{1,q}(B_{r'}(0))$-norm of u can be estimated by the $H^{1,q}(B_{\tilde{R}}(0))$-norm of $\hat{u}$. This norm can be estimated by the $H^{1,2}(B_{R_2}(0))$-norm of $\hat{u}$, the $L^p(B_{R_2}(0), dx)$-norm of $\hat{f}$ and the $(L^q)^d(B_{R_2}(0), dx)$-norm of $\hat{e}$. Since these norms can be estimated by the corresponding norms of e and f on $B_r(0)$, the norm estimate follows. $\square$

Theorem 3.1.1 at the boundary . We consider the case of a half-ball around
0, i.e., $\Omega = B_r(0) \cap \{x_d > 0\}$ for some $0 < r$ and A a matrix-valued
mapping on $B_r(0) \cap \{x_d \geq 0\}$ with $a_{id}(0) = a_{di}(0) = 0$ for $i \neq d$. Here
$B_r^+(0) = B_r(0) \cap \{x_d > 0\}$, $B_r^-(0) \cap \{x_d < 0\}$. Define $\hat{A} = (\hat{a}_{ij})_{1 \leq i,j \leq d}$ on
$B_r(0)$ by

$$\hat{a}_{ij}(x', x_d) = \begin{cases} a_{ij}(x', x_d) & \text{if } x_d \geq 0 \\ (-1)^{\delta_{id}}(-1)^{\delta_{jd}} a_{ij}(x', -x_d) & \text{else.} \end{cases}$$

Then $\hat{A}$ is strictly elliptic and by the additional assumption on $A(0)$ also
continuous in 0. Choose $r' < r$ according to Theorem 3.1.1 for the interior
point case with matrix $\hat{A}$. Denote this r' by R'.

Let $u \in H^{1,2}(B_r^+(0))$, $f \in L^p(B_r^+(0), dx)$, $e \in (L^q)^d(B_r^+(0), dx)$ fulfilling
(3.1). Define $\hat{u}$, $\hat{c}$ and $\hat{f}$ by even reflection that is

$$\hat{u}(x', x_d) = \begin{cases} u(x', x_d) & \text{if } x_d \geq 0 \\ u(x', -x_d) & \text{else,} \end{cases}$$

the same for $\hat{f}$ and $\hat{c}$. The vector field $\hat{e}$ is defined by even reflection for
the first $d - 1$ coordinates and by odd reflection for e_d, i.e., $e_d(x', x_d) =$
$-e_d(x', -x_d)$ and $e_i(x', -x_d) = e_i(x', x_d)$ for $i < d$. Then $\hat{u} \in H^{1,2}(B_r(0))$,
$\hat{f} \in L^p(B_r(0), dx)$ and $\hat{e} \in (L^q)^d(B_r(0), dx)$. Now for $v \in C_c^1(B_r(0))$ we
have

$$\int_{B_r(0)} (\hat{A}\nabla\hat{u}, \nabla v)\, dx + \int_{B_r(0)} \hat{c}\,\hat{u}\, v\, dx$$

$$= \int_{B_r^+(0)} (A\nabla\hat{u}, \nabla v)\, dx + \int_{B_r^+(0)} \hat{c}\,\hat{u}\, v\, dx + \int_{B_r^-(0)} (\hat{A}\nabla\hat{u}, \nabla v)\, dx$$

$$+ \int_{B_r^-(0)} \hat{c}\,\hat{u}\, v\, dx$$

$$= \int_{B_r^+(0)} \hat{f}(x)\, v(x)\, dx + \int_{B_r^+(0)} (\hat{e}, \nabla v)\, dx + \int_{B_r^-(0)} (\hat{A}\nabla\hat{u}, \nabla v)\, dx$$

$$+ \int_{B_r^-(0)} \hat{c}\,\hat{u}\, v\, dx.$$

Define $\hat{v}(x', x_d) := v(x', -x_d)$. Then using the chain rule and transformation rule we get

$$\int_{B_r^-(0)} (\hat{A}\nabla\hat{u}, \nabla v)\, dx + \int_{B_r^-(0)} \hat{c}\,\hat{u}\,v\, dx$$

$$= \int_{B_r^-(0)} \sum_{i,j=1}^{d} (-1)^{\delta_{id}} (-1)^{\delta_{jd}} a_{ij}(x', -x_d)\, (-1)^{\delta_{jd}} (\partial_j u)(x', -x_d) \times$$

$$\times (\partial_i v)(x', x_d)\, dx + \int_{B_r^-(0)} \hat{c}\,\hat{u}\,v\, dx$$

$$\overset{*}{=} \int_{B_r^+(0)} (A\nabla u, \nabla\hat{v})\, dx + \int_{B_r^+(0)} c\,u\,\hat{v}\, dx$$

$$= \int_{B_r^+(0)} f(x)\hat{v}(x) + \int_{B_r^+(0)} (e, \nabla\hat{v})\, dx$$

$$= \int_{B_r^-(0)} \hat{f}(x)\,v(x)\, dx + \int_{B_r^-(0)} (\hat{e}, \nabla v)\, dx.$$

In $*$ and the last equality we used $(\partial_i \hat{v})(x', x_d) = (-1)^{\delta_{id}}(\partial_i v)(x', -x_d)$. Thus $\hat{u}$ fulfills (3.1).

So by the choice of R' we have $\hat{u} \in H^{1,q}(B_{R'}(0))$ and thus $u \in H^{1,q}(B_{R'}^+(0))$. The norm estimate is now a consequence of the norm estimate for the interior applied to $\hat{u}$ on $B_{R'}(0)$.

Now we prove the general case. Let $\Omega \subset \mathbb{R}^d$, $x \in \partial\Omega$, $r > 0$ such that $B_r(x) \cap \partial\Omega$ is C^1-smooth. W.l.o.g. we assume $x = 0$. By definition of C^1-smoothness there exist open neighborhoods W of 0, V of 0 and a C^1-diffeomorphism $\psi : W \to V$, such that $\psi(W \cap \Omega) = V \cap \mathbb{R}_+^d$ and $\psi(W \cap \partial\Omega) = V \cap \{x_d = 0\}$. Define $A' = (a'_{ij})_{1 \leq i,j \leq d}$ by $a'_{ij}(y) := (D\psi A D\psi^\top)_{ij}(\psi^{-1}(y))$. Define $B = (b_{ij})_{1 \leq i,j \leq d}$ by $b_{ii} = 1$, $1 \leq i \leq d$, $b_{id}(0) = -\frac{a'_{id}}{a'_{dd}}(0)$, $1 \leq i < d$, the other entries 0. Then

$$(BA'(0)B^\top)_{id} = (BA'(0)B^\top)_{di} = 0 \quad \text{for } i \neq d.$$

It holds $B\,\mathbb{R}^d \cap \{x_d > 0\} = \mathbb{R}^d \cap \{x_d > 0\}$ and $B\,\mathbb{R}^d \cap \{x_d = 0\} = \mathbb{R}^d \cap \{x_d = 0\}$, because $b_{di} = 0$ for $i \neq d$. Set $\psi_1 = B \circ \psi$. Then $D\psi_1 = B(D\psi)$ and $\psi_1 : W \to \psi_1(W)$ is a C^1-diffeomorphism. Thus $\psi_1(W)$ is open and there exists a $B_R(0) \subset \psi_1(W)$. Set $W_1 := \psi_1^{-1}(B_R(0))$, $W_2 := B_R(0)$, $M_1 := \Omega$, $M_2 := \mathbb{R}^d \cap \{x_d > 0\}$.

Define $\hat{u}$, $\hat{f}$, $\hat{e}$, $\hat{A}$ and $\hat{c}$ as in Lemma 3.1.6 using as diffeomorphism ψ_1. Then $\hat{u} \in H^{1,q_0}(B_R^+(0))$, $\hat{f} \in L^p(B_R^+(0), dx)$, $\hat{e} \in (L^q)^d(B_R^+(0), dx)$. By Lemma 3.1.6 it holds

$$
\int_{B_R^+(0)} \left(\hat{A}\nabla\hat{u}, \nabla\hat{v}\right) dy + \int_{B_R^+(0)} \hat{c}\,\hat{u}\,\hat{v}\,dy
$$

$$
= \int_{B_R^+(0)} \hat{f}\,\hat{v}\,dy + \int_{B_R^+(0)} (\hat{e}, \nabla\hat{v})\,dy \quad \text{for all } \hat{v} \in C_c^1\big(B_R(0) \cap \{x_d \geq 0\}\big).
$$

Note that $\hat{A}(0)$ fulfills the additional assumption $\hat{a}_{id}(0) = \hat{a}_{di}(0) = 0$ for $i \neq d$. So by the statements proven above we get $\hat{u} \in H^{1,q}(B_{R'}^+(0))$ for some $R' < R$. Set $U := \psi_1^{-1}(B_{R'}(0))$ and choose $r' > 0$ such that $B_{r'}(0) \subset U$. Then $u \in H^{1,q}(B_{r'}(0) \cap \Omega)$. The norm estimate is a consequence of the norm estimate for $\hat{u}$. $\square$

4 Construction of Elliptic Diffusions

In this chapter we apply the construction result of Chapter 2 and the elliptic regularity result of Chapter 3 to construct elliptic diffusions as $\mathcal{L}^p$-strong Feller process associated with gradient Dirichlet forms. This is done in the next two sections. We apply the regularity result from Chapter 3 to prove local Hölder continuity of the L^p-resolvent for specified p. We postpone the proof to Section 4.2. It is applied only in Theorem 4.1.9. Furthermore, we can identify a class of C^2-functions with Neumann-type boundary condition as elements in the domain of the L^p-generator. The constructed $\mathcal{L}^p$-strong Feller process solves the martingale problem for the L^p-generator for every starting point where the drift is not singular and that is either in the interior of the domain or at a C^2-smooth boundary point.

Let us now describe results on construction of (strong) Feller processes from Dirichlet forms obtained by other authors and compare them with ours. Albeverio, Kondratiev and Röckner (see [AKR03]) construct distorted Brownian motion with singular drift on $\mathbb{R}^d$, $d \in \mathbb{N}$. See also Section 2.4. Fattler and Grothaus (see [FG07], [FG08] and [Fat08]) generalize these methods to construct Brownian motion with singular drift in the interior and reflecting boundary behavior on domains with certain smoothness assumptions. In both cases drifts with very strong (repulsive) singularities are allowed, in particular potentials of Lennard-Jones type can be treated. As already mentioned in the introduction we follow the methods used in these articles, i.e., we construct elliptic diffusion with singular drift as $\mathcal{L}^p$-strong Feller processes associated with Dirichlet forms.

Bass and Hsu construct reflected Brownian motion on Lipschitz domains as a strong Feller process associated with a gradient Dirichlet form, see [BH91].

Fukushima and Tomisaki construct elliptic diffusions with reflecting boundary behavior on Lipschitz domains with Hölder cusps, see [FT95] and [FT96]. They consider a gradient-Dirichlet form with matrix but without density. The matrix is assumed to be uniformly elliptic and bounded. The process is obtained even as a classical Feller process. However, their assumptions exclude the case of singular drift terms.

Ambrosio, Savaré and Zambotti (see [ASZ09]) consider gradient Dirichlet forms with logarithmic concave reference measure on convex sets of Hilbert spaces. Among other results, an associated semigroup of transition kernels having the strong Feller property is constructed. The corresponding process has, however, right-continuous paths for quasi-every starting point only. These results are applied in other articles to describe dynamics of wetting models, see also Section 5.2 and Section 6.5.

Sturm provides an entirely different approach for constructing diffusions on general locally compact metric spaces with given reference measure, see [Stu98a] and [Stu98b]. Instead of a gradient Dirichlet form, he considers a family of Dirichlet forms consisting of difference operators. Then convergence of these forms to a limit Dirichlet form is proven. Under an additional property, namely the measure contraction property, a Feller process associated with the limiting Dirichlet form is constructed. We have published the results stated in this chapter in [BG13].

4.1 Gradient Dirichlet Forms and Construction of Associated Diffusions

Let $\Omega \subset \mathbb{R}^d$ be an open set, $d \in \mathbb{N}$, $d \geq 2$, with $\partial\Omega$ locally Lipschitz smooth (see Definition 7.5.2) and Lebesgue measure zero. For a matrix-valued measurable mapping $A = (a_{ij})_{1 \leq i,j \leq d}$ of symmetric strictly elliptic matrices and measurable density $\varrho \geq 0$ we consider the pre-Dirichlet form

$$\mathcal{E}(u,v) = \int_\Omega (A\nabla u, \nabla v)\, d\mu,$$

$$u, v \in \mathcal{D} := \left\{ u \in C_c(\overline{\Omega}) \,\middle|\, u \in H^{1,1}_{\mathrm{loc}}(\Omega), \mathcal{E}(u,u) < \infty \right\}, \quad (4.1)$$

in the Hilbert space $L^2(\overline{\Omega}, \mu)$ where $\mu := \varrho dx$, dx the Lebesgue measure on $\mathbb{R}^d$. See Definition 7.5.21 for the precise definition of local Sobolev spaces. We assume at least the following continuity conditions on the density and matrix.

Condition 4.1.1. For each $x \in \overline{\Omega}$ the matrix $A(x)$ is symmetric and strictly elliptic, i.e., there exists an $\gamma(x) > 0$ such that

$$\gamma(x)(\xi, \xi) \leq (A(x)\xi, \xi) \quad \text{for all } \xi \in \mathbb{R}^d.$$

Condition 4.1.2. It holds $A \in C^0(\overline{\Omega}; \mathbb{R}^{d \times d})$, $\varrho \in C^0(\overline{\Omega})$ and $\varrho > 0$, $d.x.$-a.e.

This condition implies that $\operatorname{supp}[\mu] = \overline{\Omega}$, $\operatorname{supp}[\mu]$ denoting the topological support of the measure μ. The continuity of A implies that the constant $\gamma(x)$ can be chosen independently of $x \in \Omega_0$ for $\Omega_0 \subset\subset \overline{\Omega}$, i.e., A is uniformly elliptic on Ω_0 and locally uniformly elliptic on $\overline{\Omega}$. For two subsets $A, B \subset \overline{\Omega}$ we write $A \subset\subset B$ if the closure $\overline{A}$ of A in $\overline{\Omega}$ is compact and $\overline{A} \subset B$. In this case we say: A is *compactly contained* in B. (As topology on $\overline{\Omega}$ we use the trace topology of the topology on $\mathbb{R}^d$ induced by any norm on $\mathbb{R}^d$).

Remark 4.1.3. If B is open in $\overline{\Omega}$, then $A \subset\subset B$ implies $\operatorname{dist}(A, \overline{\Omega} \setminus B) > 0$. Moreover, one finds a set A', which is open in $\overline{\Omega}$, such that $A \subset\subset A' \subset\subset B$.

These conditions already imply that the pre-Dirichlet form is closable. We get the following theorem.

Theorem 4.1.4. *Assume Conditions 4.1.1 and 4.1.2. Then the form $(\mathcal{E}, \mathcal{D})$ is closable with closure denoted by $(\mathcal{E}, D(\mathcal{E}))$. The closure is a strongly local, regular Dirichlet form.*

Remark 4.1.5. Closability still holds if one assumes instead of Condition 4.1.2 that A is uniformly elliptic on $\overline{\Omega}$, i.e., γ from Condition 4.1.1 can be chosen independently of $x \in \overline{\Omega}$, and that ϱ fulfills the Hamza condition. In this case the proof of the closability works as in [FG08, Prop. 2.6]. However, for the elliptic regularity result we anyway need the stronger Condition 4.1.2. Note that the Hamza condition holds if ϱ is continuous.

Proof. Closability and Dirichlet property of $(\mathcal{E}, \mathcal{D})$ follow as in [MR92, Ch. II].

For regularity we have to show that $\mathcal{D}$ is dense in $C_c(\overline{\Omega})$ with respect to the sup-norm and dense in $D(\mathcal{E})$ with respect to the $\mathcal{E}_1$-norm. The latter is fulfilled by construction of $D(\mathcal{E})$. To prove that $\mathcal{D}$ is dense in $C_c(\overline{\Omega})$ note that $\mathcal{D}$ is closed under addition and multiplication, the latter follows by the boundedness and support property of functions in $\mathcal{D}$. Moreover, for every $x_0 \in \overline{\Omega}$ it holds that for $n \in \mathbb{N}$ the mapping $\widetilde{u}_{x_0}^{(n)} : \overline{\Omega} \to \mathbb{R}$, defined by

$$x \mapsto (1/2 \wedge (1 - n|x - x_0|)) \vee 0$$

is in $\mathcal{D}$. Hence $\mathcal{D}$ is point separating and contains for every x_0 a function that does not vanish at x_0. So by the extended Stone-Weierstrass theorem, see [Sim63, Ch.7, Sec. 38], it follows that $\mathcal{D}$ is dense with respect to the sup-norm in $C_c(\overline{\Omega})$. From this it follows that $\mathcal{D}$ is dense in $C_c(\overline{\Omega})$ with respect to the $L^2(\overline{\Omega}, \mu)$-norm. Since μ is regular, see e.g. [Rud70, Theo. 2.18], we have by [Rud70, Theo. 3.14] that $C_c(\overline{\Omega})$ is dense in $L^2(\overline{\Omega}, \mu)$. So $(\mathcal{E}, \mathcal{D})$ is densely defined. It is easy to see that $(\mathcal{E}, \mathcal{D})$ is strongly local. By the properties of $\mathcal{D}$ it follows by [FOT11, Theo. 3.1.2] and [FOT11, Exer. 3.1.1] that also $(\mathcal{E}, D(\mathcal{E}))$ is strongly local. $\qquad\square$

We refer to [FG08] and [Sti10, Sec. 5] for further details.

As remarked on page 10 there exists an associated strongly continuous contraction semigroup on $L^r(\overline{\Omega}, \mu)$ (L^r-s.c.c.s.) $(T_t^r)_{t>0}$ with generator $(L_r, D(L_r))$ for every $1 \le r < \infty$. The corresponding resolvent we denote by $(G_\lambda^r)_{\lambda>0}$.

Next we fix the regularity and differentiability conditions on ϱ.

Condition 4.1.6. For the density it holds $\sqrt{\varrho} \in H_{\mathrm{loc}}^{1,2}(\overline{\Omega})$. There exists $p \ge 2$ with $p > \frac{d}{2}$ such that

$$\frac{|\nabla\varrho|}{\varrho} \in L_{\mathrm{loc}}^p(\overline{\Omega} \cap \{\varrho > 0\}, \mu). \tag{4.2}$$

The condition $p > \frac{d}{2}$ implies that for $p < d$ and $q := \frac{dp}{d-p}$ it holds $q > d$. This is used for applying the regularity result from the previous chapter to get regularity of functions in $D(L_p)$.

Remark 4.1.7. In analogy to [AKR03] and [FG07] we may assume $\frac{|\nabla\varrho|}{\varrho} \in L_{\mathrm{loc}}^p(\overline{\Omega}, \mu)$, $p > d$, instead of (4.2). If Ω has locally Lipschitz boundary, then this stronger condition implies that $\varrho \in H_{\mathrm{loc}}^{1,p}(\overline{\Omega})$ and hence has a continuous version on $\overline{\Omega}$. This follows using Sobolev embedding and Hölder inequality, similar as in the reduction step in the proof of [BKR01, Theo. 2.8].

So the continuity assumption for the density is no restriction and our assumptions are even weaker.

We obtain the following lemma.

Lemma 4.1.8. *Under Condition 4.1.1, Condition 4.1.2 and Condition 4.1.6 we get*

$$cap_{\mathcal{E}}(\{\varrho = 0\}) = 0. \tag{4.3}$$

The proof is similar to the proof in [BGS13, Prop. 4.4] or [Fuk85]. Here $cap_{\mathcal{E}}$ denotes the capacity associated with the Dirichlet form $(\mathcal{E}, D(\mathcal{E}))$. Lemma 4.1.8 ensures that the constructed process never hits the set $\{\varrho = 0\}$ if started from $\{\varrho > 0\}$. Since we are only able to prove L^p-strong Feller property outside this set, this is crucial for the pointwise construction of the process $(\mathbf{X}_t)_{t \geq 0}$. From the regularity results in the previous chapter we get the following regularity for functions in $D(L_p)$. Denote by Γ_1 the C^1-smooth part of the boundary.

Theorem 4.1.9. *Assume Conditions 4.1.1 and 4.1.2. Let $p \geq 2$ with $p > \frac{d}{2}$. Define $E_1 := (\Omega \cup \Gamma_1) \cap \{\varrho > 0\}$. Then $D(L_p) \hookrightarrow C^0(E_1)$ and the embedding is locally continuous, i.e., for $x \in E_1$ there exists an open neighborhood $U_1 \subset E_1$ and a constant $\hat{C}_1 < \infty$ such that*

$$sup_{y \in U_1} |\widetilde{u}(y)| \leq \hat{C}_1 \|u\|_{D(L_p)}.$$

Let $(u_n)_{n \in \mathbb{N}}$ be a sequence in $D(L_p)$ such that $((1 - L_p)u_n)_{n \in \mathbb{N}}$ is uniformly bounded in $L^\infty(\overline{\Omega}, \mu)$-norm. Then $(u_n)_{n \in \mathbb{N}}$ is equicontinuous.

See Theorem 4.2.5(iii) and (iv) in the next section. The regularity assumptions are one part of Condition 2.1.2. To ensure right-continuity of the paths at $t = 0$ we need that the domain of the L^p-generator contains for every $x \in E_1$ a countable family of functions that is point separating in x, see Condition 2.1.2(ii). To prove this point separating property we have to pose stronger regularity conditions.

Condition 4.1.10. There exists a subset $\Gamma_2 \subset \partial\Omega$, open in $\partial\Omega$, such that the boundary is locally C^2-smooth at every $x \in \Gamma_2$ and $\mathrm{cap}_{\mathcal{E}}(\partial\Omega \setminus \Gamma_2) = 0$. For the matrix A it holds $A \in C^1(\overline{\Omega}; \mathbb{R}^{d \times d})$.

Remark 4.1.11. The whole C^2-smooth boundary part $\widetilde{\Gamma}_2$ is open in $\partial\Omega$. Indeed, let $x \in \widetilde{\Gamma}_2$. Then there exists an open neighborhood $U \subset \mathbb{R}^d$ of x such that $\partial\Omega \cap U$ is C^2-smooth. So we can take a strictly smaller open neighborhood V contained in U such that each boundary point $x' \in \partial\Omega \cap V$ has still an open neighborhood V' contained in U with boundary part $\partial\Omega \cap V'$ C^2-smooth. If $\Gamma_2 \subset \partial\Omega$ is open in $\partial\Omega$, then $(\Omega \cup \Gamma_2) \cap \{\varrho > 0\}$ is open in $\overline{\Omega}$.

Remark 4.1.12. We need the stronger smoothness assumptions in Condition 4.1.10 to construct a family of point separating functions in the domain of the L^p-generator. This is used in our construction scheme from Chapter 2 to ensure right-continuity of the sample paths at $t = 0$. In [FT95] and [FT96] diffusion processes are constructed as classical Feller processes associated with the Dirichlet form considered therein. A crucial ingredient therein is a density result of the range of the L^p-resolvent in the space of continuous functions vanishing at infinity for certain p. This result implies in particular the point separating property of the domain of the L^p-generator. Since the result is based on global a-priori estimates, global assumptions on the coefficients are used for the proof, what excludes singular drift terms. In our case there seems to be not a similar argument. So we need stronger regularity conditions on the matrix and boundary smoothness to construct the required family of point separating functions by hand and cannot use construction results for classical Feller processes.

For Γ_2 as in Condition 4.1.10 we define

$$\mathcal{D}_{\mathrm{Neu}} := \left\{ u \in C_c^2((\Omega \cup \Gamma_2) \cap \{\varrho > 0\}) \,\middle|\, (\eta, A\nabla u) = 0 \text{ on } \Gamma_2 \right\}. \qquad (4.4)$$

Here $\eta(x)$ denotes the outer unit normal vector at $x \in \Gamma_2 \subset \partial\Omega$ orthogonal to $\partial\Omega$. With $C^2((\Omega \cup \Gamma_2) \cap \{\varrho > 0\})$ we denote the space of all C^2-smooth functions on $\Omega \cap \{\varrho > 0\}$, such that the functions and their derivatives admit continuous extensions to the boundary part $\Gamma_2 \cap \{\varrho > 0\}$. The subindex c marks that the functions have compact support and that the support is contained in $(\Omega \cup \Gamma_2) \cap \{\varrho > 0\}$, i.e., $\mathrm{supp}[u] \subset\subset (\Omega \cup \Gamma_2) \cap \{\varrho > 0\}$ for $u \in C_c^2((\Omega \cup \Gamma_2) \cap \{\varrho > 0\})$. For a function $u \in \mathcal{D}_{\mathrm{Neu}}$ define

$$\hat{L}u := \Delta_A u + (\nabla A, \nabla u) + \frac{1}{\varrho}(A\nabla\varrho, \nabla u)$$

$$= \sum_{i,j=1}^d a_{ij}\partial_i\partial_j u + \sum_{j=1}^d \left(\sum_{i=1}^d \partial_i a_{ij} + \sum_{i=1}^d \frac{1}{\varrho} a_{ji}\partial_i\varrho \right) \partial_j u. \qquad (4.5)$$

Here $\Delta_A u := \sum_{i,j=1}^d a_{ij}\partial_i\partial_j u$ and ∇A is defined by $(\nabla A)_j = \sum_{i=1}^d \partial_i a_{ij}$, $1 \le j \le d$.

Lemma 4.1.13. *Assume Condition 4.1.1, Condition 4.1.2, Condition 4.1.6 and Condition 4.1.10. Let p be as in Condition 4.1.6. Then $\mathcal{D}_{\mathrm{Neu}} \subset D(L_p)$ and $L_p u = \hat{L}u$ for $u \in \mathcal{D}_{\mathrm{Neu}}$, $\hat{L}$ as in (4.5). Furthermore, $\mathcal{D}_{\mathrm{Neu}}$ is point separating.*

Proof. Let $u \in C_c^2((\Omega \cup \Gamma_2) \cap \{\varrho > 0\})$. Let $K := \mathrm{supp}[u]$. Choose an open U_1 in $\overline{\Omega}$ with $U_1 \subset\subset (\Omega \cup \Gamma_2) \cap \{\varrho > 0\}$ and $K \subset\subset U_1$. Set $K_1 := \overline{U_1}$. Note that by Remark 4.1.3 such a set exists. Writing $(\nabla\varrho)^p = (\frac{\nabla\varrho}{\varrho})^p \varrho^{p-1}\varrho$ we get from the continuity of ϱ and (4.2) in Condition (4.1.6) that $\varrho \in L^p(U_1)$ and $\nabla\varrho \in (L^p)^d(U_1)$.

Let $v \in \mathcal{D}$. Then by Lemma 4.2.2(ii), see below, it holds $v \in H^{1,2}(U_1)$. Since $p \ge 2$, we have $\varrho v(A\nabla u)_i \in H^{1,1}(U_1)$, $1 \le i \le d$. Let $g := \varrho v A\nabla u$. The boundary condition of u imply $(\eta, g) = \varrho v(\eta, A\nabla u) = 0$ on Γ_2.

By divergence theorem, see Theorem 7.5.18, we get

$$\mathcal{E}(u,v) = \int_\Omega \varrho(A\nabla u, \nabla v)\, dx$$

$$= -\int_\Omega \nabla \cdot (\varrho A\nabla u)\, v\, dx + \int_{\Gamma_2} \varrho v(\eta, A\nabla u)\, d\sigma = -\int_\Omega \nabla \cdot (\varrho A\nabla u)\, v\, dx$$

$$= -\int_\Omega \frac{1}{\varrho}\nabla \cdot (\varrho A\nabla u)\, v\, \varrho dx = (-\hat{L}u, v)_{L^2(\overline{\Omega},\mu)}.$$

Here σ denotes the surface measure on $\partial\Omega$, see Definition 7.5.15, $\nabla\cdot$ the divergence operator, i.e., $\nabla \cdot g = \sum_{i=1}^d \partial_i g_i$ for a vector field $g = (g_1, ..., g_d)$. Observe that only boundary terms at Γ_2 appear since the support of u has positive distance to $\partial\Omega \setminus \Gamma_2$. Furthermore, $\frac{1}{\varrho}\nabla \cdot \varrho A\nabla u = \Delta_A u + (\nabla A, \nabla u) + \frac{1}{\varrho}(A\nabla\varrho, \nabla u) = \hat{L}u$. So the mapping $\mathcal{D} \ni v \mapsto \mathcal{E}(u,v) \in \mathbb{R}$ is continuous in L^2-norm. Hence the mapping is also continuous in L^2-norm on $D(\mathcal{E})$. So $u \in D(L_2)$ and $L_2 u = \hat{L}u$. Since u and $L_2 u$ are also in $L^p(\overline{\Omega}, \mu)$, it follows $u \in D(L_p)$ and $L_p u = L_2 u = \hat{L}u$, see Lemma 7.2.4.

We prove that $\mathcal{D}_{\text{Neu}}$ is point separating in E_1. For the interior points this is clear due to $C_c^\infty(\Omega \cap \{\varrho > 0\}) \subset \mathcal{D}_{\text{Neu}}$. For a boundary point $x_0 \in \Gamma_2 \cap \{\varrho > 0\}$ one constructs first a sequence of C^2-functions $(u_n)_{n\in\mathbb{N}}$ living on the boundary that separate x_0 from the boundary points. These functions are constructed using the C^2-diffeomorphism that flattens the boundary locally. The boundary values of the functions together with the Neumann boundary condition uniquely determine the normal derivative. Since the boundary values are C^2 and the normal derivative is C^1, these functions can be extended to the interior by C^2-functions satisfying the Neumann boundary condition. For details see Lemma 7.4.3 in the appendix.   □

So we can apply Theorem 2.1.3 and obtain:

Theorem 4.1.14. *Assume Conditions 4.1.1, 4.1.2, 4.1.6 and 4.1.10. Let p be as in Condition 4.1.6, Γ_2 as in Condition 4.1.10. Define $E_1 := (\Omega \cup \Gamma_2) \cap \{\varrho > 0\}$. Then there exists a diffusion process (i.e., a strong Markov process having continuous sample paths) $\mathbf{M} = (\mathbf{\Omega}, \mathcal{F}, (\mathcal{F}_t)_{t\geq 0}, (\mathbf{X}_t)_{t\geq 0}, (\mathbb{P}_x)_{x\in E\cup\{\Delta\}})$ with state space E and cemetery Δ. The process leaves $E_1 \cup \{\Delta\}$ $\mathbb{P}_x$-a.s., $x \in E_1 \cup \{\Delta\}$, invariant. The transition semigroup $(P_t)_{t\geq 0}$ is associated with $(T_t^p)_{t\geq 0}$ and is $\mathcal{L}^p$-strong Feller, i.e., $P_t\mathcal{L}^p(\overline{\Omega}, \mu) \subset C^0(E_1)$ for $t > 0$. The* *process has continuous paths on* $[0, \infty)$ *and it solves the martingale problem associated with* $(L_p, D(L_p))$, *i.e.,*

$$M_t^{[u]} := \widetilde{u}(\mathbf{X}_t) - \widetilde{u}(x) - \int_0^t L_p u(\mathbf{X}_s) \, ds, \ t \geq 0,$$

is an $(\mathcal{F}_t)$-*martingale under* $\mathbb{P}_x$ *for all* $u \in D(L_p)$, $x \in E_1$. *In particular, the process solves the martingale problem for functions in* $\mathcal{D}_{Neu}$. *Moreover, the resolvent kernels* $(R_\lambda)_{\lambda > 0}$ *are strong Feller in the classical sense, i.e.,* $R_\lambda \mathcal{B}_b(\overline{\Omega}) \subset C^0(E_1)$.

Here $(P_t)_{t \geq 0}$ being associated with $(T_t^p)_{t \geq 0}$ means that $P_t f$ is a μ-version of $T_t^p f$ for $f \in \mathcal{L}^1(\overline{\Omega}, \mu) \cap \mathcal{B}_b(\overline{\Omega})$ (the space of Borel measurable bounded functions). By $\mathcal{L}^p(\overline{\Omega}, \mu)$ we denote the space of all p-integrable functions on $(\overline{\Omega}, \mu)$.

Remark 4.1.15. The continuity holds with respect to the topology of the Alexandrov one-point compactification of $\overline{\Omega}$ to $\overline{\Omega}^\Delta$. This means that the process has continuous paths in $\overline{\Omega}$ and reaches Δ only by leaving continuously every compact set of $\overline{\Omega}$. If $\overline{\Omega}$ is compact, then Δ is an isolated point. Note that $\partial\Omega$ is not identified with the cemetery, hence there is no absorption at the boundary.

Proof. We aim to apply Theorem 2.1.3 and therefore have to check the conditions. In our case $E = \overline{\Omega}$, $\mathbf{d}$ is the restriction of the metric induced by the Euclidean scalar product on $\overline{\Omega}$, $\mu = \varrho\, dx$. Clearly $(E, \mathbf{d})$ and μ fulfill the assumptions of Theorem 2.1.3. Moreover, $E_1 \in \mathcal{B}(E)$ since Γ_2 is open with respect to the trace topology on $\partial\Omega$. By assumption on Γ_2 and (4.3) it holds $\mathrm{cap}_{\mathcal{E}}(E \setminus E_1) = 0$. From Lemma 4.1.13 we get $\mathcal{D}_{Neu} \subset D(L_p)$. These functions are point separating in E_1. Clearly for $u \in \mathcal{D}_{Neu}$ it holds $u^2 \in \mathcal{D}_{Neu}$. Theorem 4.2.5(iii) yields $D(L_p) \hookrightarrow C^0(E_1)$ and the embedding is locally continuous. So all assumptions of Theorem 2.1.3 are fulfilled and we obtain a diffusion $\mathbf{M}$ with the stated properties. That $(P_t)_{t \geq 0}$ is associated with $(T_t^p)_{t \geq 0}$ follows from the construction of the process. That the corresponding resolvent of kernels $(R_\lambda)_{\lambda > 0}$ is strongly Feller follows from Theorem 4.2.5(iv) together with Theorem 2.1.5. $\qquad\square$

Remark 4.1.16. The strong Feller property of the resolvent of kernels allows us to apply conservativity criterions for the constructed process $\mathbf{M}$, see Corollary 2.1.7. In particular, if $(\mathcal{E}, D(\mathcal{E}))$ is conservative, then $\mathbf{M}$ is conservative for every starting point in E_1.

In [FOT11, Theo. 5.7.3] a general criterion for conservativity of Dirichlet forms is given. For our Dirichlet form conservativity holds in particular if A is bounded and $\mu(\overline{\Omega}) < \infty$, see [Sti10, Prop. 5.10].

4.2 Regularity of L^p-Resolvents

In this section we prove local $H^{1,q}$-regularity of the L^p-resolvent, $p \geq 2$ and $p \leq q$ as in Theorem 4.2.5 below, associated with the Dirichlet form (4.1) and hence regularity for functions in $D(L_p)$. We apply the elliptic regularity result of Chapter 3. So we first have to prove local $H^{1,2}$-regularity of the resolvent. Throughout this chapter we only assume Condition 4.1.1 and Condition 4.1.2, i.e., only continuity assumptions are posed on the matrix and density but no differentiability assumptions. Moreover, we assume that $\Omega \subset \mathbb{R}^d$, $d \in \mathbb{N}$, is open and has locally Lipschitz smooth boundary.

Remark 4.2.1. Since $dx(\partial\Omega) = 0$ and hence also $\mu(\partial\Omega) = 0$, we can identify functions in $L^p(\Omega, \mu)$ and $L^p(\overline{\Omega}, \mu)$ with each other, so $L^p(\overline{\Omega}, \mu) = L^p(\Omega, \mu)$. However, $L^p_{\text{loc}}(\overline{\Omega}, \mu) \subsetneq L^p_{\text{loc}}(\Omega, \mu)$ in general since for functions f in the latter set it is only assumed that for each *interior* point there exists a neighborhood in which f is p-integrable.

Note that elements in $D(\mathcal{E})$ need not to be in a local Sobolev space at first. So we prove first local $H^{1,2}$-regularity for elements in $D(\mathcal{E})$. From this we get by an approximation argument also local regularity for the L^p-resolvent. Furthermore, we can identify the resolvent as a weak solution to an elliptic equation. Then we apply the regularity result of Chapter 3 to prove local Sobolev space regularity for the L^p-resolvent. Using Sobolev embedding we get local Hölder continuity for the L^p-resolvent for $2 \leq p < \infty$ with $p > \frac{d}{2}$.

Below we use the space $H^{1,2}_{\text{loc}}((\Omega \cup \Gamma) \cap \{\varrho > 0\})$ for some open $\Gamma \subset \partial\Omega$, possibly empty or the whole boundary. Since ϱ is continuous on $\overline{\Omega}$ and Ω is open in $\mathbb{R}^d$, we have that $\Omega \cap \{\varrho > 0\}$ is open in $\mathbb{R}^d$. Define $G := (\Omega \cup \Gamma) \cap \{\varrho > 0\}$. Then G fulfills the assumption in Definition 7.5.21. Indeed. For the interior we have $\overset{\circ}{G} = \Omega \cap \{\varrho > 0\}$ due to continuity of ϱ. It holds $\partial G \cap G = \Gamma \cap \{\varrho > 0\}$ and $dx(\Gamma \cap \{\varrho > 0\}) = 0$ since Γ is locally Lipschitz smooth. Furthermore, $\Gamma \cap \{\varrho > 0\} \subset \overline{\Omega \cap \{\varrho > 0\}}$. So Definition 7.5.21 and the corresponding lemmata in Section 7.5 apply to the set $\Omega \cap \{\varrho > 0\}$ and $(\Omega \cup \Gamma) \cap \{\varrho > 0\}$.

Lemma 4.2.2. *Let $(\mathcal{E}, \mathcal{D})$ be the pre-Dirichlet form in (4.1) and $(\mathcal{E}, D(\mathcal{E}))$ its closure.*

(i) For an open subset Ω_0 with $\Omega_0 \subset\subset \overline{\Omega}$, $T := \partial\Omega_0 \cap \partial\overline{\Omega}$ it holds

$$C_c^1(\Omega_0 \cup T) \hookrightarrow D(\mathcal{E}).$$

(ii) For an open subset $\Omega_0 \subset \Omega$ with $\Omega_0 \subset\subset \overline{\Omega} \cap \{\varrho > 0\}$, the restriction mapping $\iota : u \mapsto u|_{\Omega_0}$ maps from $D(\mathcal{E})$ to $H^{1,2}(\Omega_0)$ and the mapping $\iota : D(\mathcal{E}) \to H^{1,2}(\Omega_0)$ is continuous. In particular, for $u \in D(\mathcal{E})$ it holds $u \in H_{loc}^{1,2}(\overline{\Omega} \cap \{\varrho > 0\})$. Moreover, for $u \in D(\mathcal{E})$, $v \in C_c^1(\Omega_0 \cup T)$, T as in (i), we have

$$\mathcal{E}(u, v) = \int_{\Omega_0} (A\nabla u, \nabla v)\, \varrho dx. \tag{4.6}$$

Remark 4.2.3. By $C_c^1(\Omega_0 \cup T)$ we denote the space of C^1-differentiable functions on Ω_0 such that the derivatives up to first order admit a continuous extension to the boundary part T.

Proof. (i): The compact support property yields that each function $u \in C_c^1(\Omega_0 \cup T)$ can be continued by zero outside $\Omega_0 \cup T$ to a function in $C_c^1(\overline{\Omega})$. Indeed, if $x \notin \Omega_0 \cup T$, then $x \notin \mathrm{supp}[u]$ and hence there exists a neighborhood (in the topology on $\overline{\Omega}$) of x that does not intersect with the support of u. Clearly, $C_c^1(\overline{\Omega}) \subset D(\mathcal{E})$.

(ii): Let $\Omega_0 \subset\subset \overline{\Omega} \cap \{\varrho > 0\}$, T as in (i). Since ϱ is continuous, $\varrho > 0$ and $\overline{\Omega}_0$ is compact, there exists $\varrho_0 > 0$ such that $\varrho \geq \varrho_0 > 0$ on Ω_0. Using the continuity of A we get that A is uniformly elliptic on $\overline{\Omega}_0$ with ellipticity constant γ. Thus we get

$$\int_{\Omega_0} u^2\, dx + \int_{\Omega_0} (\nabla u, \nabla u)\, dx \leq \frac{1}{\varrho_0}\left(\int_{\Omega_0} u^2\, \varrho dx + \int_{\Omega_0} (\nabla u, \nabla u)\, \varrho dx \right)$$

$$\leq \frac{1}{\varrho_0}\left(\int_{\Omega_0} u^2\, \varrho dx + \int_{\Omega_0} \frac{1}{\gamma}(A\nabla u, \nabla u)\, \varrho dx \right) \leq K_1 \mathcal{E}_1(u, u),$$

for all $u \in \mathcal{D}$ and some $K_1 < \infty$. Thus the mapping $\iota : \mathcal{D} \to H^{1,2}(\Omega_0)$ is continuous. Since $\mathcal{D}$ is dense with respect to the $\mathcal{E}_1$-norm in $D(\mathcal{E})$, this mapping extends to $D(\mathcal{E})$.

We prove (4.6). If $u \in \mathcal{D}$ this equality holds by definition. Observe that the right-hand side of (4.6) is continuous in u in the $H^{1,2}(\Omega_0)$-norm and thus also in the $\mathcal{E}_1$-norm. So again by a density argument the equality holds for all $u \in D(\mathcal{E})$.

Note that for every $x \in \overline{\Omega} \cap \{\varrho > 0\}$ we can find an $R_0 > 0$ such that $\Omega_0 := B_{R_0}(x) \cap \Omega \subset\subset \overline{\Omega} \cap \{\varrho > 0\}$. So $u \in H^{1,2}(B_{R_0}(x) \cap \Omega)$ for $u \in D(\mathcal{E})$. Hence by Lemma 7.5.25 $u \in H_{loc}^{1,2}(\overline{\Omega} \cap \{\varrho > 0\})$. $\square$

Recall that the norms $\|\cdot\|_{\mathcal{E}_\lambda}$, $\lambda > 0$, defined by $\|u\|_{\mathcal{E}_\lambda} := \mathcal{E}_\lambda(u,u)^{1/2}$, are equivalent to each other.

Proposition 4.2.4. *Let* $2 \le p < \infty$, $\lambda > 0$. *Let* $x \in \overline{\Omega} \cap \{\varrho > 0\}$, $R_0 > 0$ *such that* $B_{R_0}(x) \cap \overline{\Omega} \subset\subset \overline{\Omega} \cap \{\varrho > 0\}$. *Then for all* $f \in L^p(\overline{\Omega}, \mu)$ *it holds for the* L^p-*resolvent* $G_\lambda^p f \in H^{1,2}(B_{R_0}(x) \cap \Omega)$ *and*

$$\lambda \int_{B_{R_0}(x) \cap \Omega} G_\lambda^p f \, v \, d\mu + \int_{B_{R_0}(x) \cap \Omega} (A\nabla G_\lambda^p f, \nabla v) \, d\mu$$

$$= \int_{B_{R_0}(x) \cap \Omega} f \, v \, d\mu \quad \text{for all } v \in C_c^1(B_{R_0}(x) \cap \overline{\Omega}). \quad (4.7)$$

For $R > R_0$ *such that* $B_R(x) \cap \overline{\Omega} \subset\subset \overline{\Omega} \cap \{\varrho > 0\}$ *there exists a constant* $C_1 < \infty$ *such that*

$$\|G_\lambda^p f\|_{H^{1,2}(B_{R_0}(x) \cap \Omega)} \le C_1(\|G_\lambda^p f\|_{L^p(B_R(x) \cap \Omega, dx)} + \|f\|_{L^p(B_R(x) \cap \Omega, dx)}).$$

$$(4.8)$$

Moreover, there exists a constant $C_2 < \infty$ *such that*

$$\|G_\lambda^p f\|_{H^{1,2}(B_{R_0}(x) \cap \Omega)} \le C_2\|f\|_{L^p(\overline{\Omega}, \mu)}. \quad (4.9)$$

In particular $G_\lambda^p f \in H^{1,2}_{loc}(\overline{\Omega} \cap \{\varrho > 0\})$.

Proof. We prove (4.8) first. Let $x \in \overline{\Omega} \cap \{\varrho > 0\}$, $R_0 > 0$ such that $B_{R_0}(x) \cap \overline{\Omega} \subset\subset \overline{\Omega} \cap \{\varrho > 0\}$. Note that since $\overline{\Omega} \cap \{\varrho > 0\}$ is open in $\overline{\Omega}$, we can always find an $R > R_0$ such that $B_R(x) \cap \overline{\Omega} \subset\subset \overline{\Omega} \cap \{\varrho > 0\}$. So if R is not given in advance, fix such an R now and set $\widetilde{\Omega} := B_R(x) \cap \Omega$. Choose a C^∞-cutoff function ϕ with compact support in $B_R(x)$ such that ϕ is constantly one in $B_{R'}(x)$ for some $R_0 < R' < R$.

Now let $f \in L^2(\overline{\Omega}, \mu)$, then $G_\lambda^2 f \in D(\mathcal{E})$ and $\phi G_\lambda^2 f \in D(\mathcal{E})$. The defining expression (4.1) of $\mathcal{E}$ on $\mathcal{D}$ extends to $u = v = \phi G_\lambda f$. To see this approximate $G_\lambda f$ by a sequence $(u_n)_{n \in \mathbb{N}}$ in $\mathcal{D}$. Then $(\phi u_n)_{n \in \mathbb{N}}$ converges both in $D(\mathcal{E})$ and in $H^{1,2}(\widetilde{\Omega})$ by Lemma 4.2.2(ii).

We have

$$\mathcal{E}_\lambda(\phi G_\lambda f, \phi G_\lambda f) = \lambda \int_{\widetilde{\Omega}} \phi G_\lambda f \, \phi G_\lambda f \, \varrho dx + \int_{\widetilde{\Omega}} (A\nabla(\phi G_\lambda f), \nabla(\phi G_\lambda f)) \, \varrho dx$$

$$= \lambda \int_{\widetilde{\Omega}} G_\lambda f \, \phi^2 G_\lambda f \varrho dx + \int_{\widetilde{\Omega}} \phi(A\nabla G_\lambda f, \nabla(\phi G_\lambda f)) \, \varrho dx$$

$$+ \int_{\widetilde{\Omega}} G_\lambda f \, (A\nabla\phi, \nabla(\phi G_\lambda f)) \, \varrho dx$$

$$= \mathcal{E}_\lambda(G_\lambda f, \phi^2 G_\lambda f) - \int_{\widetilde{\Omega}} (A\nabla G_\lambda f, \nabla\phi) \, \phi G_\lambda f \, \varrho dx$$

$$+ \int_{\widetilde{\Omega}} G_\lambda f \, (A\nabla\phi, \nabla(\phi G_\lambda f)) \, \varrho dx$$

$$= \int_{\widetilde{\Omega}} f \, \phi^2 G_\lambda f \varrho dx - \int_{\widetilde{\Omega}} (A\nabla G_\lambda f, \nabla\phi) \, \phi G_\lambda f \varrho dx$$

$$+ \int_{\widetilde{\Omega}} G_\lambda f \, (A\nabla\phi, \nabla(\phi G_\lambda f)) \varrho dx. \quad (4.10)$$

With a constant $K_1 < \infty$ depending on the bounds of A, ϱ and ϕ in $\widetilde{\Omega}$ we get

$$\left| \int_{\widetilde{\Omega}} (A\nabla G_\lambda f, \nabla\phi) \, \phi G_\lambda f \, \varrho dx \right| \leq K_1 \sum_{j=1}^{d} \int_{\widetilde{\Omega}} |\phi \partial_j G_\lambda f| |G_\lambda f| dx.$$

Using that $\phi \partial_j G_\lambda f = \partial_j(\phi G_\lambda f) - G_\lambda f \partial_j \phi$ we can estimate this further with $K_2 < \infty$:

$$\leq K_1 \sum_{j=1}^{d} \left(\int_{\widetilde{\Omega}} |\partial_j(\phi G_\lambda f)| |G_\lambda f| dx + \int_{\widetilde{\Omega}} |G_\lambda f \partial_j \phi| |G_\lambda f| dx \right)$$

$$\leq K_2 \left(\|\phi G_\lambda f\|_{H^{1,2}(\widetilde{\Omega})} \|G_\lambda f\|_{L^2(\widetilde{\Omega}, dx)} + \|G_\lambda f\|_{L^2(\widetilde{\Omega}, dx)}^2 \right)$$

$$\leq \frac{\varepsilon}{2} \|\phi G_\lambda f\|_{H^{1,2}(\widetilde{\Omega})}^2 + \left(\frac{K_2^2}{2\varepsilon} + K_2 \right) \|G_\lambda f\|_{L^2(\widetilde{\Omega}, dx)}^2. \quad (4.11)$$

The last inequality holds for all $\varepsilon > 0$ and the constant K_2 is independent of ε. Similarly the third term of (4.10) can be estimated.

By Lemma 4.2.2(ii) for $\widetilde{\Omega}$ we get a constant $K_3 < \infty$ such that

$$\|\phi G_\lambda f\|_{H^{1,2}(\widetilde{\Omega})}^2 \leq K_3 \mathcal{E}_\lambda(\phi G_\lambda f, \phi G_\lambda f).$$

Together with (4.10) and (4.11) we get with a constant $K_4 < \infty$

$$\|\phi G_\lambda f\|^2_{H^{1,2}(\widetilde{\Omega})} \leq K_4 \left(\|f\|^2_{L^2(\widetilde{\Omega},dx)} + \left(1 + \frac{1}{\varepsilon}\right) \|G_\lambda f\|^2_{L^2(\widetilde{\Omega},dx)} \right.$$

$$\left. +\varepsilon \|\phi G_\lambda f\|^2_{H^{1,2}(\widetilde{\Omega})} \right).$$

Choosing $\varepsilon = \frac{1}{2K_4}$ and using the fact that ϕ is constantly one in $B_{R'}(x) \cap \Omega$ we get a constant $K_5 < \infty$ and the estimate

$$\|G_\lambda f\|_{H^{1,2}(B_{R_0}(x)\cap\Omega)} \leq \|\phi G_\lambda f\|_{H^{1,2}(\widetilde{\Omega})} \leq K_5(\|G_\lambda f\|_{L^2(\widetilde{\Omega},dx)} + \|f\|_{L^2(\widetilde{\Omega},dx)}). \tag{4.12}$$

Assume now that $f \in L^1(\overline{\Omega}, \mu) \cap L^\infty(\overline{\Omega}, \mu)$ then in particular $f \in L^2(\overline{\Omega}, \mu) \cap L^p(\overline{\Omega}, \mu)$. By construction the corresponding L^2- and L^p-resolvent coincide and estimate (4.12) applies. Since $dx(\widetilde{\Omega}) < \infty$, the right-hand side of (4.12) can be estimated by the corresponding $L^p(\widetilde{\Omega}, dx)$-norms. The density is bounded from below on $\widetilde{\Omega}$, so the right-hand side can be then estimated further by the corresponding $L^p(\widetilde{\Omega}, \mu)$-norms which can be estimated by the corresponding $L^p(\overline{\Omega}, \mu)$-norm. By the contraction properties of the L^p-resolvent on $L^p(\overline{\Omega}, \mu)$ we end up with the estimate

$$\|G_\lambda f\|_{H^{1,2}(B_{R_0}(x)\cap\Omega)} \leq C_1(\|G_\lambda f\|_{L^p(\widetilde{\Omega},dx)} + \|f\|_{L^p(\widetilde{\Omega},dx)})$$

$$\leq K_6(\|G_\lambda f\|_{L^p(\widetilde{\Omega},\mu)} + \|f\|_{L^p(\widetilde{\Omega},\mu)}) \leq C_2\|f\|_{L^p(\overline{\Omega},\mu)},$$

with constants $C_2, K_6 < \infty$. Since every function f in $L^p(\overline{\Omega}, \mu)$ can be approximated in $L^p(\overline{\Omega}, \mu)$-norm by functions $(f_k)_{k\in\mathbb{N}}$ in $L^1(\overline{\Omega}, \mu) \cap L^\infty(\overline{\Omega}, \mu)$, it follows by applying the above estimate to the differences $(f_k - f_l)$ that $G_\lambda^p f \in H^{1,2}(B_{R_0}(x) \cap \Omega)$ and estimates (4.8) and (4.9) hold.

It is left to prove (4.7). For f in $L^1(\overline{\Omega}, \mu) \cap L^\infty(\overline{\Omega}, \mu) \subset L^2(\overline{\Omega}, \mu)$ the equation holds by (4.6) and the relation between G_λ^2 and $\mathcal{E}_\lambda$. By approximation this equality extends to all $f \in L^p(\overline{\Omega}, \mu)$.

Finally we get that $G_\lambda^p f \in H_{\text{loc}}^{1,2}(\overline{\Omega} \cap \{\varrho > 0\})$ with the same argument as in the proof of Lemma 4.2.2. $\qquad\square$

The next step is to prove additional regularity of the L^p-resolvent. Note that by (4.7) we can identify the L^p-resolvent as a weak solution of an elliptic partial differential equation in $B_{R_0}(x) \cap \overline{\Omega}$ with coefficient matrix ϱA and right-hand side $\widetilde{f} = \varrho f - \lambda \varrho G_\lambda f$. So we can apply the results of Chapter 3. Denote by Γ_1 the C^1-smooth part of $\partial\Omega$.

Theorem 4.2.5. *Let* $2 \leq p < \infty$ *and* $0 < \lambda < \infty$. *Denote by* $U_{r'} :=$ $B_{r'}(x) \cap (\Omega \cup \Gamma_1) \cap \{\varrho > 0\}$, $U_r := B_r(x) \cap (\Omega \cup \Gamma_1) \cap \{\varrho > 0\}$ *for* $r, r' > 0$, $x \in (\Omega \cup \Gamma_1) \cap \{\varrho > 0\}$.

(i) *Let* $p < q \leq \frac{dp}{d-p}$ *if* $p < d$ *and* $p < q < \infty$ *arbitrary else. For* $x \in (\Omega \cup \Gamma_1) \cap \{\varrho > 0\}$ *there exists* $0 < r' < r$ *such that: For* $f \in L^p(\overline{\Omega}, \mu)$, *it holds* $G_\lambda^p f \in H^{1,q}(B_{r'}(0) \cap \Omega)$. *Furthermore, there exist constants* $C_3, C_3' < \infty$ *such that*

$$\|G_\lambda^p f\|_{H^{1,q}(B_{r'}(0)\cap\Omega)} \leq C_3(\|f\|_{L^p(U_r, dx)} + \|G_\lambda^p f\|_{L^p(U_r, dx)})$$
$$\leq C_3'(\|f\|_{L^p(\overline{\Omega}, d\mu)} + \|G_\lambda^p f\|_{L^p(\overline{\Omega}, d\mu)}). \quad (4.13)$$

(ii) *If* $p > \frac{d}{2}$ *and* $2 \leq p < \infty$, *there exists for* $x \in (\Omega \cup \Gamma_1) \cap \{\varrho > 0\}$, $0 < r_1$ *such that: There exists a continuous version of* $G_\lambda^p f$ *on* $\overline{U_{r_1}}$, U_{r_1} *defined as above, denoted by* $\widetilde{G_\lambda^p f}$. *Furthermore, there exists a constant* $C_4 < \infty$ *such that*

$$\sup_{y \in \overline{U_{r_1}}} |\widetilde{G_\lambda^p f}(y)| \leq C_4 \|f\|_{L^p(\overline{\Omega}, \mu)}. \quad (4.14)$$

(iii) *Let* $E_1 := (\Omega \cup \Gamma_1) \cap \{\varrho > 0\}$. *Then for* $p > \frac{d}{2}$ *with* $2 \leq p < \infty$ *it holds* $D(L_p) \hookrightarrow C^0(E_1)$ *and the embedding is locally continuous.*

(iv) *Assume* $p > \frac{d}{2}$ *and* $2 \leq p < \infty$. *Let* $(u_n)_{n \in \mathbb{N}}$ *be a sequence in* $D(L_p)$ *such that the sequence* $((1-L)u_n)_{n \in \mathbb{N}}$ *is uniformly bounded in* $L^\infty(\overline{\Omega}, \mu)$-*norm. Then the sequence of continuous functions* $(\widetilde{u}_n)_{n \in \mathbb{N}}$ *is equicontinuous on* E_1.

Proof. Choose $0 < r < \infty$ such that $B_r(x) \cap \Omega \subset\subset \Omega \cap \{\varrho > 0\}$. If $x \in \Omega$ choose $0 < r_0 < r$. If $x \in \Gamma_1$ choose $0 < r_0 < r$ such that $B_{r_0}(x) \cap \partial\Omega \subset\subset \Gamma_1$. Set $U_{r_0} := B_{r_0}(x) \cap \Omega$. For $f \in L^p(\overline{\Omega}, \mu)$ we have by Proposition 4.2.4 that $G_\lambda^p f \in H^{1,2}(B_{r_0}(x) \cap \Omega)$ and it holds

$$\lambda \int_{B_{r_0}(x)\cap\Omega} \varrho G_\lambda^p f \, v \, dx + \int_{B_{r_0}(x)\cap\Omega} (\varrho A \nabla G_\lambda^p f, \nabla v) \, dx$$
$$= \int_{B_{r_0}(x)\cap\Omega} \varrho f \, v \, dx \quad \text{for all } v \in C_c^1(B_{r_0}(x) \cap \overline{\Omega})$$

and the norm estimates (4.8) and (4.9) hold with $R = r$ and $R_0 = r_0$.

(i): Choose $\hat{A} := \varrho A$, $c := \lambda \varrho$. Then $\hat{A}$ is uniformly elliptic on $\overline{U_{r_0}}$ since ϱ is uniformly bounded from below on $\overline{U_{r_0}}$ and A is uniformly elliptic on $\overline{U_{r_0}}$. Set $\hat{f} = \varrho f$, $u = G_\lambda f$. The assumptions of Theorem 3.1.1 are fulfilled with $e = 0$. Hence there exists $r' > 0$, $r' < r_0$ such that $G_\lambda^p f \in H^{1,q}(B_{r'}(x) \cap \Omega)$. Note that the $L^p(B_{r_0}(x) \cap \Omega, dx)$-norm of $\hat{f}$ can be estimated by the $L^p(B_r(x) \cap \Omega, dx)$-norm of f. Moreover, the $H^{1,2}$-norm of $G_\lambda f$ on $B_{r_0}(x) \cap \Omega$ can be estimated using (4.8) by the $L^p(B_r(x) \cap \Omega, dx)$-norm of f and $G_\lambda f$. Combining this with estimate (3.2) we get estimate (4.13). Since ϱ is uniformly bounded away from zero in $\overline{B_{r_0}(x)} \cap \overline{\Omega} \cap \{\varrho > 0\}$, the $L^p(U_r, dx)$-norms in (4.13) can be replaced by the $L^p(U_r, \mu)$-norms and hence by the $L^p(\overline{\Omega}, \mu)$-norms after adapting the constant C_3 suitable.

(ii): If $p < d$, set $q := \frac{dp}{d-p}$. Then from the assumption $p > \frac{d}{2}$ it follows $q > d$. If $p \geq d$, then choose $d < q < \infty$ arbitrary. Let r' as in (i). By (i) we get $G_\lambda^p f \in H^{1,q}(B_{r'}(0) \cap \Omega)$. Let $\tilde{\Omega} := B_{r'}(x) \cap \Omega$, $\tilde{\Gamma} = B_{r'}(x) \cap \partial\Omega \subset \Gamma_1$. Then $U_{r'} = \tilde{\Omega} \cup \tilde{\Gamma}$. Set $r_1 := \frac{r'}{2}$, $\Omega_0 := B_{r_1}(x) \cap \Omega$, $\Gamma_0 := B_{r_1}(x) \cap \partial\Omega \subset\subset \tilde{\Gamma}$. Then $\Omega_0 \cup \Gamma_0 \subset\subset \tilde{\Omega} \cup \tilde{\Gamma}$. So Sobolev embedding theorem, see Theorem 7.5.12, applies. So we get that $G_\lambda^p f$ has a dx-version which is continuous on $\overline{U_{r_1}}$. Estimate (4.14) follows by (4.13) and the $L^p(\overline{\Omega}, \mu)$-contraction property of G_λ.

(iii): Denote by U'_x, $x \in E_1$, the neighborhood $U_{r'}$ provided by (i). Recall that $D(L_p) = G_1 L^p(\overline{\Omega}, \mu)$, hence $D(L_p)|_{U_{r'}} \subset H^{1,q}(B_{r'}(0) \cap \Omega)$ for $r' > 0$ and q as in (i). Then every $u \in D(L_p)$ has a continuous version $\tilde{u}$ on U'_x. The value $\tilde{u}(x)$, $x \in E_1$, is independent of the choice of the neighborhood since the intersection of two neighborhoods of x is open and non-empty and hence has strictly positive μ-measure. We have that E_1 is open in the trace topology on $\overline{\Omega}$. Define $E_1^n := \{x \in E_1 \mid \text{dist}(x, \overline{\Omega} \setminus E_1) \geq \frac{1}{n}\} \cap \overline{B_n(0)}$, $n \in \mathbb{N}$. Then all E_1^n are compact and $E_1 = \bigcup_{n \in \mathbb{N}} E_1^n$. Every E_1^n can be covered by finitely many U'_{x_i}, $x_i \in E_1^n$, $1 \leq i \leq D_n \in \mathbb{N}$. Then we get for $u \in D(L^p)$ a μ-version that is continuous on E_1^n. This yields a continuous μ-version on E_1. Thus the embedding $D(L_p) \hookrightarrow C^0(E_1)$ exists. The local continuity follows from (4.14). Indeed, set $f := (1 - L_p)u$. It holds $u = G_1^p(1 - L_p)u = G_1^p f$. Then (4.14) implies

$$\sup_{y \in \overline{U_{r'}}} |\tilde{u}(y)| \leq C_4 \|(1 - L_p)u\|_{L^p(\overline{\Omega}, \mu)} \leq C_4 \|u\|_{D(L^p)}.$$

(iv): Choose q as in (ii). Let $(u_n)_{n \in \mathbb{N}}$ be a sequence with the stated properties. Set $f_n := (1 - L_p)u_n$. Then $u_n = G_1(1 - L_p)u_n = G_1 f_n$. Since G_1 is sub-Markovian, we have that $(u_n)_{n \in \mathbb{N}}$ is also bounded in $L^\infty(\overline{\Omega}, \mu)$-norm. For

$x \in (\Omega \cup \Gamma_1) \cap \{\varrho > 0\}$, let $0 < r' < r$ as in (i) and $r_1 < r'$ as in (ii). So every u_n has a continuous version $\tilde{u}_n$ on $\overline{U_{r_1}}$.

Since ϱ is uniformly bounded from below on $\overline{U_r}$, the $L^p(U_r, dx)$-norms in (4.13) can be replaced by the $L^p(U_r, \mu)$-norm after suitable modification of the constant C_3. These norms can be replaced by the corresponding $L^\infty(U_r, \mu)$-norms which can again be replaced by the $L^\infty(\overline{\Omega}, \mu)$-norms. Since both $(u_n)_{n\in\mathbb{N}}$ and $(f_n)_{n\in\mathbb{N}}$ are bounded in $L^\infty(\overline{\Omega}, \mu)$-norm, we get that $(u_n)_{n\in\mathbb{N}}$ is a bounded sequence in $H^{1,q}(B_{r'}(0) \cap \Omega)$. Sobolev embedding yields that $(\tilde{u}_n)_{n\in\mathbb{N}}$ is bounded in $C^{0,\beta}(\overline{U_{r_1}})$ for $0 < \beta \leq 1 - \frac{d}{q}$. Thus the sequence is uniformly equicontinuous on $\overline{U_{r_1}}$. So $(\tilde{u}_n)_{n\in\mathbb{N}}$ is equicontinuous. $\qquad\square$

5 Applications

In this chapter we apply the results of Chapter 4 to concrete models in Physics. In Section 5.1 we consider finite particle systems with singular interaction in continuum. In Section 5.2 we consider the Ginzburg-Landau interface model with reflection (also called: entropic repulsion) at a hard wall. In both cases we obtain stochastic dynamics described by an $\mathcal{L}^p$-strong Feller process that gives solutions in the sense of the corresponding martingale problem for the L^p-generator.

5.1 Stochastic Dynamics for Interacting Particle Systems

In this section we apply the results from before to construct stochastic dynamics for finite particle systems. As in the works of Albeverio, Kondratiev and Röckner ([AKR03]) and Fattler and Grothaus ([FG07], [FG08], [Fat08]) we can handle potentials with very singular repulsion, i.e., the case of superstable potentials is included. We refer the reader to [AKR03] and the references therein for further applications in Mathematical Physics. Generalizing the results of [FG07] we allow as state space for the particles a general domain rather than a cuboid. This makes the geometry of the state space of the particle process a little bit more complicated. Additionally, we allow non-constant diffusion matrices.

Before we start, let us mention that Andres and von Renesse ([AR12]) use Dirichlet form methods to describe a finite system of interacting particles on the unit interval. This model differs from ours. There the particles are reflected whenever they hit each other. They construct the process as a strong Feller process and derive a criterion for the existence of a semimartingale decomposition.

Let $N \in \mathbb{N}$, $N \geq 2$, $\Omega_0 \subset \mathbb{R}^d$, $d \in \mathbb{N}$, be an open set with Lipschitz smooth boundary, $\Lambda := \overline{\Omega_0}$. Let $A : \Lambda \to \mathbb{R}^{d \times d}$ be a continuously differentiable matrix-valued mapping of symmetric strictly elliptic matrices.

Assume that $\partial\Omega_0$ is C^2-smooth except for a set of capacity zero with respect to the gradient Dirichlet form with coefficient matrix A and Lebesgue measure on Ω_0. Let $\Gamma_2 \subset \partial\Omega_0$ be C^2-smooth and open in $\partial\Omega_0$ such that it is complemented in $\partial\Omega_0$ by a set of capacity zero. So with the notation of (7.19) in Section 7.6 we have with $\Upsilon = \Omega_0$, $\mathrm{cap}_{\mathcal{E}^A,1}(\partial\Omega_0 \setminus \Gamma_2) = 0$.

Remark 5.1.1. For interesting applications we have to allow that the boundary is non-smooth at certain parts. The description in terms of the capacity of the gradient Dirichlet form with matrix A is natural since this Dirichlet form corresponds to the one-particle dynamic without interaction. Furthermore, the capacity can be characterized in terms of the Hausdorff measure, see Corollary 7.6.16. If $d \geq 2$ and Ω_0 is a cuboid, i.e., $\Omega_0 = \times_{i=1}^{d}(a_i, b_i)$ for $a_i < b_i \in \mathbb{R} \cup \{-\infty, \infty\}$, $1 \leq i \leq d$, the assumptions are fulfilled.

Let $\Psi : \mathbb{R}^d \to \mathbb{R} \cup \{\infty\}$ be a symmetric pair potential, i.e., $\Psi(-x) = \Psi(x)$ which fulfills the following conditions.

Condition 5.1.2.   (i) For $x \to 0$ it holds $|\Psi(x)| \to \infty$.

(ii) The mapping $\mathbb{R}^d \to \mathbb{R}_0^+$, $x \mapsto \exp(-\Psi(x)) =: \varrho_0(x)$ is continuous.

(iii) The function ϱ_0 is weakly differentiable on $\mathbb{R}^d$, $\exp(-\frac{\Psi}{2}) \in H^{1,2}_{\mathrm{loc}}(\mathbb{R}^d)$. Ψ is weakly differentiable on $\mathbb{R}^d \setminus \{0\}$ and there exists $p > \frac{Nd}{2}$ such that

$$\nabla\Psi \in L^p_{\mathrm{loc}}(\{\varrho_0 > 0\}, \exp(-\Psi)dx). \tag{5.1}$$

Remark 5.1.3. Note that the continuity assumption on ϱ_0 excludes potentials with $\Psi(x) = -\infty$ for some $x \in \mathbb{R}^d$.

Observe that the assumed L^p-regularity in (iii) depends on the number of particles N. The more particles the higher the assumed regularity. Nevertheless, under Condition (i) and (ii) potentials with $\Psi \in C^1(\mathbb{R}^d \setminus \{|\Psi| = \infty\})$ satisfy the integrability condition for every $1 \leq p \leq \infty$. Indeed, $\{|\Psi| = \infty\} = \{\Psi = \infty\} = \{\varrho_0 = 0\}$. Since ϱ_0 is continuous, we can find for every $x \in \{|\Psi| < \infty\}$ a compact set which is contained in $\{|\Psi| < \infty\}$. On this set Ψ and its first derivatives are bounded, hence locally integrable up to every order.

One could replace (5.1) with the stronger integrability condition $\nabla\Psi \in L^p_{\mathrm{loc}}(\mathbb{R}^d, \exp(-\Psi)dx)$, $p > Nd$, i.e., also integrability at singularities is assumed. In this case the continuity condition on ϱ_0 follows from (iii), see Remark 4.1.7.

Remark 5.1.4. A concrete example for an admissible pair potential is the so-called *Lennard-Jones(12,6)* potential (see e.g. [Sch06, Sec. 5.3.2.1]) defined by

$$V(r) := 4\varepsilon\left(\left(\frac{\sigma}{r}\right)^{12} - \left(\frac{\sigma}{r}\right)^{6}\right)$$

where $r > 0$ denotes the distance of the two interacting particles. The parameter $\varepsilon > 0$ determines the minimum of the potential and the parameter $\sigma > 0$ determines the radius at which the interaction changes from a repulsive to an attractive one.

Define $\varrho : \mathbb{R}^{Nd} \to \mathbb{R}_0^+$, by

$$x := (x_1, ..., x_N) \mapsto \frac{1}{Z} \exp\left(-\sum_{1 \leq i < j \leq N} \Psi(x_i - x_j)\right)$$

$$= \frac{1}{Z} \prod_{1 \leq i < j \leq N} \varrho_0(x_i - x_j), \quad (5.2)$$

with $0 < Z < \infty$ some constant e.g. a normalization constant. Then by continuity of ϱ_0 and Condition 5.1.2(iii) it follows that $\sqrt{\varrho} \in H_{\text{loc}}^{1,2}(\mathbb{R}^{Nd})$ and $\frac{\nabla \varrho}{\varrho} \in L_{\text{loc}}^p(\{\varrho > 0\}, \varrho dx)$ for p as in Condition 5.1.2. We construct a stochastic process describing the dynamics for an N-particle system with interaction potential $\nabla A + A\nabla\Psi$ and diffusion matrix A.

Define $\hat{A} : \Lambda^N \to \mathbb{R}^{Nd \times Nd}$ by $(x_1, ..., x_N) \mapsto \text{diag}(A(x_1), ..., A(x_N))$, i.e.,

$$\hat{A}(x_1, ..., x_N) = \begin{pmatrix} A(x_1) & & 0 \\ & \ddots & \\ 0 & & A(x_n) \end{pmatrix}.$$

Furthermore, set $\mu := \varrho\, dx$, the measure on Λ^N with density ϱ with respect to the Lebesgue measure. Define

$$\mathcal{E}(u, v) := \int_{\Lambda^N} (\hat{A}\nabla u, \nabla v)\, d\mu, \quad (5.3)$$

$$\mathcal{D} := \left\{ u \in C_c(\Lambda^N) \,\middle|\, u \in H_{\text{loc}}^{1,1}(\Omega_0^N), \mathcal{E}(u, u) < \infty \right\}.$$

Denote by $(\mathcal{E}, D(\mathcal{E}))$ the corresponding closure in $L^2(\Lambda^N, \mu)$ which exists due to Theorem 4.1.4 using that $\hat{A}$ and ϱ are continuous. By $\text{cap}_{\mathcal{E}}$ we denote the associated capacity.

Set $\Lambda_s = \Omega_0 \cup \Gamma_2$, we define the set of all *admissible configurations* Λ_{ad}^N by

$$\Lambda_{ad}^N := \{\varrho > 0\} \cap \big\{(x_1, ..., x_N) \in \Lambda_s^N \mid x_k \neq x_l \text{ for } k \neq l, 1 \leq k, l \leq N,$$
$$\text{there exists at most one } 1 \leq i \leq N \text{ such that } x_i \in \Gamma_2\big\}$$

So in an admissible configuration all particles are in Ω_0 or at the smooth boundary part Γ_2. Moreover, there are never two or more particles at the same place. Additionally, we exclude the case that two or more particles are at the boundary. This exclusion has to be done for technical reason since the boundary of the configuration space is in general not smooth if two particles are located at the boundary.

We prove that the complement of this set in Λ^N has capacity zero and that the boundary part $\Lambda_{ad}^N \cap \partial(\Lambda^N)$ is C^2-smooth and open in $\partial(\Lambda^N)$. We prove two preparatory lemmata first.

Lemma 5.1.5. *The boundary part $\Lambda_{ad}^N \cap \partial(\Lambda^N)$ is C^2-smooth and open in $\partial(\Lambda^N)$. Let Λ_s as above, $\Lambda_1^N := \{(x_1, ..., x_N) \in \Lambda_s^N \mid \text{there exists at most one } 1 \leq i \leq N \text{ such that } x_i \in \Gamma_2\}$. Then $\Lambda^N \setminus \Lambda_1^N$ has capacity zero.*

Proof. First note that $\partial(\Lambda^N)$ consists of all points $x = (x_1, ..., x_N)$ where at least one component x_i is located at $\partial\Omega_0$. Since $\Lambda_{ad}^N \subset \Lambda_s^N \cap \Lambda_1^N$, we have that each boundary point $x = (x_1, ..., x_N)$ has components located only in the interior or at a smooth boundary part. Furthermore, there exists at most one component, say x_1, which is located at the boundary. Taking the corresponding C^2-diffeomorphism ψ which flattens the boundary of $\partial\Omega_0$ at x_1 one can construct a C^2-diffeomorphism flattening the boundary of Ω_0^N at the point x. Then we get also an open neighborhood U of x_1 in $\partial\Omega_0$ which is contained in Γ_2 since Γ_2 is open in $\partial\Omega_0$. The other coordinates x_i, $2 \leq i \leq N$, are contained in the interior, so we can construct from this an open neighborhood $\widetilde{U}$ of x in $\Lambda_{ad}^N \cap \partial(\Lambda^N)$. So $\Lambda_1^N \cap \partial\Lambda^N$ is C^2-smooth.

We have

$$\Lambda^N \setminus \Lambda_1^N = \big\{(x_1, ..., x_N) \in \Lambda^N \mid \text{either there exists } 1 \leq i \leq N$$
$$\text{with } x_i \in \partial\Omega_0 \setminus \Gamma_2 \text{ or there exist at least two } 1 \leq i, j \leq N$$
$$\text{with } x_i, x_j \in \Gamma_2 \text{ and } x_k \in \Omega_0 \cup \Gamma_2 \text{ for } k \neq i, j\big\}.$$

Since $\partial\Omega_0 \setminus \Gamma_2$ has zero capacity with respect to the gradient Dirichlet form with coefficient matrix A on Ω_0, we have that $\text{cap}_{\mathcal{E}}(\overline{\Omega_0}^k \times (\partial\Omega_0 \setminus \Gamma_2) \times \overline{\Omega_0}^{N-k-1}) = 0$ for $0 \leq k \leq N - 1$, see Theorem 7.6.10 in the appendix.

It is left to handle the case that two or more particles are at smooth boundary parts. Without loss of generality we may assume that for $k \geq 2$ the first k particles are at a smooth boundary part, while the last $N - k$ particles are in the interior. So let $x = (x_1, ..., x_N)$ with the stated properties. Then for $1 \leq i \leq k$ there exist open relatively compact neighborhoods U_i of x_i, V_i of 0 and C^2-diffeomorphism $\psi_i : U_i \to V_i$ with the property $\psi_i(U_i \cap \partial\Omega_0) \subset (\mathbb{R}^{d-1} \times \{0\})$.

For $i > k$, choose $U_i \subset \Omega_0$ open and relatively compact with $x_i \in U_i$. Set $V_i := U_i - x_i$, define $\psi_i : U_i \to V_i$ by $y \mapsto y - x_i$. Define $\psi : U := \times_{i=1}^{N} U_i \to \times_{i=1}^{N} V_i =: V$ by $(y_1, ..., y_N) \mapsto (\psi_1(y_1), ..., \psi_N(y_N))$. Then

$$\psi\Big(\underset{i=1}{\overset{k}{\times}} (U_i \cap \partial\Omega_0) \times \underset{i=k+1}{\overset{N}{\times}} U_i \Big) \subset \big((\mathbb{R}^{d-1} \times \{0\})^k \times (\mathbb{R}^d)^{N-k} \big) \cap V$$

has finite $(Nd - k)$-dimensional Hausdorff measure, see Lemma 7.6.14. Since ψ is a smooth diffeomorphism, we have that also

$$\underset{i=1}{\overset{k}{\times}} (U_i \cap \partial\Omega_0) \times \underset{i=k+1}{\overset{N}{\times}} U_i \subset \psi^{-1}\big(\big((\mathbb{R}^{d-1} \times \{0\})^k \times (\mathbb{R}^d)^{N-k} \big) \cap V \big)$$

has finite $(Nd - k)$-dimensional Hausdorff measure.

Since $k \geq 2$, we may apply Corollary 7.6.16 with $\Upsilon = \Lambda^N$, $A = \hat{A}$ and ϱ to conclude $\mathrm{cap}_{\mathcal{E}}(\times_{i=1}^{k}(U_i \cap \partial\Omega_0) \times \times_{i=k+1}^{N} U_i) = 0$.

Since $\Gamma_2^k \times \Omega_0^{N-k}$ can be countably covered by such neighborhoods, we get $\mathrm{cap}_{\mathcal{E}}(\Gamma_2^k \times \Omega_0^{N-k}) = 0$. □

Lemma 5.1.6. *The set $\Lambda^N \cap \{\varrho = 0\}$ has capacity zero.*

Proof. Condition 5.1.2 together with the assumptions on A imply that $\hat{A}$ and ϱ fulfill the assumptions of Lemma 4.1.8. Hence the claim follows from this lemma. □

Theorem 5.1.7. *The complement of Λ_{ad}^N in Λ^N has capacity zero.*

Proof. If for $x \in \Lambda^N$ there exist k, l with $k \neq l$ and $x_k = x_l$ then $\varrho(x) = 0$ by Condition 5.1.2(i). Thus

$$\Lambda^N \setminus (\Lambda_{ad}^N) = (\Lambda^N \setminus \Lambda_1^N) \cup \{\varrho = 0\}.$$

So the claim follows from Lemma 5.1.5 and Lemma 5.1.6. □

Remark 5.1.8. The sets of the form $\{x \in \Lambda^N \mid x_k = x_l,\ 1 \le k, l \le N, k \ne l\}$ are contained in linear subspaces with codimension d. These sets have locally finite $(Nd - 2)$-dimensional Hausdorff measure and hence capacity zero for $d \ge 2$. So in the case $d \ge 2$ one could drop the repulsion Condition 5.1.2(i).

Altogether, we get that Λ_{ad}^N is complemented by a set of capacity zero and $\Lambda_{ad}^N \cap \partial(\Lambda^N)$ is C^2-smooth. Using Condition 5.1.2(iii) and the differentiability assumptions on A we get

$$\mathcal{D}_{\text{Neu}} := \left\{ u \in C_c^2(\Lambda_{ad}^N) \mid (\eta, \hat{A}\nabla u) = 0 \text{ on } \partial(\Lambda^N) \cap \Lambda_{ad}^N \right\} \subset D(L_{p_N})$$

with

$$L_{p_N} u\,(x) = \sum_{k=1}^N \Delta_{k, A(x^k)} u(x)$$

$$+ \left(\nabla_k A(x^k), \nabla_k u(x) \right) - \sum_{k=1}^N \sum_{\substack{l=1 \\ l \ne k}}^N \left(A(x^k) \nabla_k u(x), \nabla_k \Psi(x^k - x^l) \right), x \in \Lambda^N,$$

for $u \in \mathcal{D}_{\text{Neu}}$. For p_N as in Condition 5.1.2(iii), $D(L_{p_N})$ fulfills the assumptions of Theorem 2.1.3. We write $x = (x^1, ..., x^N)$ and $x^k = (x_1^k, ..., x_d^k)$, $1 \le k \le N$. We define $\Delta_{k, A(x^k)} u(x) = \sum_{i,j=1}^d a_{ij}(x^k) \partial_{x_i^k} \partial_{x_j^k} u\,(x)$, $\nabla_k u(x) = (\partial_{x_1^k} u(x), ..., \partial_{x_d^k} u(x))$ and $\nabla_k A(x^k) = ((\nabla_k A(x^k))_1, ..., (\nabla_k A(x^k))_d)$ with $(\nabla_k A(x^k))_i = \sum_{j=1}^d \partial_{x_j^k} a_{ij}(x^k)$, $1 \le i \le d$. So all assumptions of Theorem 4.1.14 are fulfilled with $\Omega = \Omega_0^N$, $E_1 = \Lambda_{ad}^N$, $\Gamma_2 = \Lambda_{ad}^N \cap \partial(\Lambda^N)$, $A = \hat{A}$ and $p = p_N$.

Write the process $\mathbf{X}_t$ from Theorem 4.1.14 as $\mathbf{X}_t = (\mathbf{X}_t^1, ..., \mathbf{X}_t^N)$, $t \ge 0$, with $\mathbf{X}_t^k \in \Omega \cup \Gamma_2$ for $1 \le k \le N$. Then $\mathbf{X}_t$ solves the martingale problem for $(L_{p_N}, D_{L_{p_N}})$. In particular, for $u \in \mathcal{D}_{\text{Neu}}$

$$u(\mathbf{X}_t) - u(\mathbf{X}_0) -$$

$$\int_0^t \sum_{k=1}^N \left(\left(\nabla A(\mathbf{X}_s^k), \nabla_k u(\mathbf{X}_s) \right) - \sum_{\substack{l=1 \\ l \ne k}}^N \left(A(\mathbf{X}_s^k) \nabla_k u(\mathbf{X}_s), \nabla \Psi(\mathbf{X}_s^k - \mathbf{X}_s^l) \right) \right) ds$$

$$- \int_0^t \sum_{k=1}^N \Delta_{k, A(\mathbf{X}_s^k)} u(\mathbf{X}_s)\, ds$$

is a martingale for $t \geq 0$ and every starting point $\mathbf{X}_0 \in \Lambda_{ad}^N$. Moreover, the process stays in Λ_{ad}^N, except perhaps leaving to Δ, i.e., particles will not meet each other nor hit the non-smooth boundary part.

If the state space Ω_0 is relatively compact, the constructed process is conservative. In the non relatively compact case, under certain growth conditions on the matrix A and density ϱ_0 the Dirichlet form $(\mathcal{E}, D(\mathcal{E}))$ is conservative and hence by Remark 4.1.16 also the process $(\mathbf{X}_t)_{t\geq 0}$ is conservative.

The strong local property of $(\mathcal{E}, D(\mathcal{E}))$ implies that the process is not absorbed at boundary points since these points belong to the state space. Furthermore, the time spent at the boundary has Lebesgue measure zero since $\partial \Lambda^N$ has μ-measure zero by the smoothness assumptions. In Section 6.4 we further analyze the boundary behavior of the process.

As in [FG07, Sec. 7] and [FG07, Theo. 7.6] one can construct from this a diffusion process on an N-particle configuration space.

5.2 Stochastic Dynamics for Ginzburg-Landau Interface Models

Let $d \in \mathbb{N}$, $N \in \mathbb{N}$. We construct stochastic dynamics for the Ginzburg-Landau interface model on the lattice $[-N, N]^d \cap \mathbb{Z}^d$ with reflection at a hard wall. This model was studied by Funaki and Olla (see [Fun03] and [FO01]) and Deuschel and Nishikawa, see [DN07] and the references therein. The interface model describes the behavior of an discretized interface e.g. the surface of a fluid which is conditioned to stay above a hard wall. The dynamics of the interface are driven by the competing effects of the stochastic fluctuation, the forces between adjacent parts of the interface and the repelling at the wall. At the boundary points of the lattice $[-N, N]^d \cap \mathbb{Z}^d$ the interface is statically pinned at zero, i.e., at the wall.

For an overview over different interface models and related questions we refer to [Fun05]. We follow the notation of [DN07]. For a comparison with the results therein and ours see below. Let $\Lambda_M := [-M, M]^d \cap \mathbb{Z}^d$, $M \in \mathbb{N}$. The parameter d describes the dimension of the underlying wall. Now fix an $N \in \mathbb{N}$. The height of the interface is described by functions $\phi : \Lambda_N \to \mathbb{R}_0^+$, i.e., at every lattice point $x = (x_1, ..., x_d) \in \Lambda_N$, $\phi(x)$ describes the height of the interface at x.

We define the following state spaces

$$\Phi_N := \{\phi : \Lambda_N \to \mathbb{R}\},$$

$$\Phi_N^+ := \{\phi : \Lambda_N \to \mathbb{R}_0^+\},$$

$$\Phi_{N+1,0} := \{\phi : \Lambda_{N+1} \to \mathbb{R} \mid \phi(x) = 0 \text{ if } |x|_{\max} = N + 1 \text{ for } x \in \Lambda_{N+1}\},$$

$$\Phi_{N+1,0}^+ := \{\phi : \Lambda_{N+1} \to \mathbb{R}_0^+ \mid \phi(x) = 0 \text{ if } |x|_{\max} = N + 1 \text{ for } x \in \Lambda_{N+1}\}.$$

Here we denote by $|\cdot|_{\max}$ the maximum norm of a vector in $\mathbb{R}^d$. Observe that $\dim \Phi_N = (2N + 1)^d$ and Φ_N^+ is a subset of Φ_N. Define $J : \Phi_N \to \Phi_{N+1,0}$ by

$$(J\phi)(x) = \begin{cases} \phi(x) & \text{if } x \in \Lambda_N, \\ 0 & \text{else.} \end{cases}$$

Then J is an isomorphism and $J(\Phi_N^+) = \Phi_{N+1,0}^+$. As state space we take $\Phi_{N+1,0}^+$. This corresponds to an interface which moves inside the box $[-N, N]^d$ and is pinned at the boundary points in $[-(N + 1), N + 1]^d \setminus [-N, N]^d$.

In order to incorporate the static boundary conditions in our setting of Chapter 4, we construct a process on Φ_N^+ and then map it under J onto $\Phi_{N+1,0}^+$. In view of Chapter 4 we have $\overline{\Omega} = \Phi_N^+ \subset \Phi_N \simeq \mathbb{R}^{(2N+1)^d}$ with $\Omega = \{\phi : \Lambda_N \to \mathbb{R}^+\}$. For the boundary $\partial\overline{\Omega}$ we have

$$\partial\overline{\Omega} = \{\phi \in \Phi_N^+ \mid \phi(x) = 0 \text{ for some } x \in \Lambda_N\}.$$

So the boundary points of the state space correspond exactly to states of the interface with zero height at some positions in $[-N, N]^d$. Hence the boundary of Φ_N is C^∞-smooth if the interface has height zero at *exactly* one position.

We fix a symmetric potential $V : \mathbb{R} \to \mathbb{R}$ fulfilling the following conditions.

(i) The mapping $\mathbb{R} \to \mathbb{R}^+$, $z \mapsto \exp(-V(z)) =: \varrho_0(z)$ is continuous.

(ii) The function ϱ_0 is weakly differentiable on $\mathbb{R}$, $\exp(-\frac{V}{2}) \in H_{\text{loc}}^{1,2}(\mathbb{R})$. V is weakly differentiable on $\mathbb{R}$ and there exists $p > \frac{(2N+1)^d}{2}$ such that

$$\nabla V \in L_{\text{loc}}^p(\{\varrho_0 > 0\}, \exp(-V)dx).$$

To each state in $\phi \in \Phi_{N+1}$ we associate the *potential energy*

$$\mathcal{H}_{N+1}(\phi) := \frac{1}{2} \sum_{\substack{x,y \in \Lambda_{N+1}, \\ |x-y|_{\max}=1}} V(\phi(x) - \phi(y)).$$

For $\phi \in \Phi_N$ define

$$\mathcal{H}_{N,0}(\phi) := \frac{1}{2} \sum_{\substack{x,y \in \Lambda_N, \\ |x-y|_{\max}=1}} V(\phi(x) - \phi(y))$$

$$+ \sum_{\substack{x \in \Lambda_N, \\ y \in \Lambda_{N+1} \backslash \Lambda_N, \\ |x-y|_{\max}=1}} V(\phi(x)) + \frac{1}{2} \sum_{\substack{x,y \in \Lambda_{N+1} \backslash \Lambda_N, \\ |x-y|_{\max}=1}} V(0).$$

Then for $\phi \in \Phi_N$ it holds

$$\mathcal{H}_{N+1}(J\phi) = \mathcal{H}_{N,0}(\phi). \tag{5.4}$$

On $(\Phi_{N+1}^+, \mathcal{B}(\Phi_{N+1}^+))$ define

$$\mu_{N+1,\text{pinned}} := \frac{1}{Z_N^+} \exp(-\mathcal{H}_{N+1}) \prod_{x \in \Lambda_N} d\phi(x) \prod_{x \in \Lambda_{N+1} \backslash \Lambda_N} \delta_0(\phi(x))$$

and on $(\Phi_N^+, \mathcal{B}(\Phi_N^+))$ define

$$\mu_{N,0} := \frac{1}{Z_N^+} \exp(-\mathcal{H}_{N,0}) \prod_{x \in \Lambda_N} d\phi(x),$$

with $0 < Z_N^+ < \infty$ e.g. a normalization constant, $d\phi(x)$ the Lebesgue measure and δ_0 the point measure in 0. Using (5.4) we get

$$\mu_{N+1,\text{pinned}} = \mu_{N,0} \circ J^{-1}.$$

Take as Hilbert space $L^2(\Phi_N^+, \mu_{N,0})$ and define

$$\mathcal{E}(u,v) := \int_{\Phi_N^+} \sum_{x \in \Lambda_N} \frac{\partial u(\phi)}{\partial \phi(x)} \frac{\partial u(\phi)}{\partial \phi(x)} \, d\mu_{N,0}(\phi),$$

$$\mathcal{D} := \left\{ u \in C_c(\Phi_N^+) \,|\, u \in H_{\text{loc}}^{1,1}(\Phi_N^+), \mathcal{E}(u,u) < \infty \right\}.$$

In view of Chapter 4 we have $A = 1$, $\varrho = \frac{1}{Z_N^+} \exp(-\mathcal{H}_{N,0})$. Note that the dimension of the state space is $\widetilde{D} := (2N+1)^d$ now. The assumptions on V imply $\frac{\nabla \varrho}{\varrho} \in L_{\text{loc}}^p(\Phi_N, \mu)$. Define the state space

$$\widetilde{\Phi_N^+} = \{\varrho > 0\} \cap \left\{ \phi \in \Phi_N^+ \,|\, \text{There exists at most one } x \in \Lambda^N \text{ with } \phi(x) = 0 \right\}.$$

Then we get as in Lemma 5.1.5 that the boundary part $\widetilde{\Phi_N^+} \cap \partial \Phi_N^+$ is C^2-smooth and the complement $\Phi_N^+ \setminus \widetilde{\Phi_N^+}$ has capacity zero. Hence the results of Chapter 4 can be applied. So by Theorem 4.1.14 we obtain a diffusion process $(\mathbf{X}_t)_{t\geq 0}$ with state space $\widetilde{\Phi_N^+}$. Then the image process $(J \circ \mathbf{X}_t)_{t\geq 0}$ has as state space $J\widetilde{\Phi_N^+} \subset \Phi_{N+1,0}^+$. Note that the time the process spends at states where one height variable in $[-N, N]^d$ is zero has measure zero. Hence there is no dynamical pinning effect. Define

$$\mathbb{V}(\phi) := \sum_{\substack{y \in \Lambda_N, \\ |x-y|_{\max}=1}} -V'(\phi(x) - \phi(y))$$

$$- \sum_{\substack{y \in \Lambda_{N+1}\setminus\Lambda_N \\ |x-y|_{\max}=1}} V'(\phi(x)) \quad \text{for } x \in \Lambda_N, \ \phi \in \Phi_N^+$$

and

$$\widetilde{L}u = \sum_{x \in \Lambda_N} \mathbb{V}(\phi)\frac{\partial}{\partial \phi(x)} u(\phi) + \left(\frac{\partial}{\partial \phi(x)}\right)^2 u(\phi)$$

for

$$u \in \mathcal{D}_{\text{Neu}} := \left\{ v \in C_c^2(\widetilde{\Phi_N^+}) \,\Big|\, \frac{\partial}{\partial \phi(z)} v = 0 \text{ if } \phi(z) = 0 \text{ for some } z \in \Lambda_N \right\}.$$

Then $(\mathbf{X}_t)_{t\geq 0}$ solves the martingale problem for $(\widetilde{L}, \mathcal{D}_{\text{Neu}})$. The stochastic differential equation, which is formally associated with the martingale problem, is given by

$$d\phi_t(x) = \mathbb{V}(\phi_t)dt + d\ell_t(x) + \sqrt{2}dW_t(x), \ x \in \Lambda^N, t \geq 0, \qquad (5.5)$$

with prescribed initial condition $\phi_0 \in \Phi_N^+$ or in our case $\widetilde{\Phi_N^+}$ only. Here $(W_t(x))_{t\geq 0}$, $x \in \Lambda_N$, denotes a family of independent Brownian motions, ℓ_t denotes the local time of the process $(\mathbf{X}_t)_{t\geq 0}$ at boundary points of $\widetilde{\Phi_N^+}$. In Section 6.5 we consider a similar wetting model for which we construct the boundary local time. Then we show that our construction yields also a weak solution.

In [DN07] stronger conditions on the interaction potential are posed, i.e., convexity and higher smoothness is assumed. Under these conditions existence and uniqueness of strong solutions to (5.5) can be concluded from

[Tan79]. In [DN07] these solutions are then used to construct dynamics on $\mathbb{Z}^d$. Using Dirichlet form methods, the authors derive lower and upper estimates for the interface on $\mathbb{Z}^d$. Using our results we get existence of solutions to (5.5) in a weaker sense, but under more general conditions. Since our solutions are by construction associated with a gradient Dirichlet form, one can use analytic methods for Dirichlet forms to further analyze the process. Whether similar results as in [DN07] hold under more general conditions on the potential is an open question.

6 Construction of the Local Time and Skorokhod Decomposition

In this chapter we further analyze the boundary behavior of the diffusion process $(\mathbf{X}_t)_{t \geq 0}$ constructed in Chapter 4. We keep the notion of Chapter 4, in particular Γ_2 denotes an open C^2-smooth boundary part of $\partial\Omega$. We construct the local time at the boundary part $\Gamma_2 \cap \{\varrho > 0\}$ as an additive functional of the process $(\mathbf{X}_t)_{t \geq 0}$. For this we first need to refine a construction result for additive functionals of Fukushima, Oshima and Takeda ([FOT11]) to our setting, see Theorem 6.1.11. The construction of the local time is based on boundedness properties of the corresponding potential that are proven using our elliptic regularity result of Chapter 3.1, see Theorem 6.2.1. Using the local time we provide a semimartingale decomposition of $(u(\mathbf{X}_t))_{t \geq 0}$ for $u \in C_c^2(E_1)$. This decomposition is valid under the path measure $\mathbb{P}_x$ for all x in the set of admissible starting points, see Theorem 6.2.9 and (6.18). By localization techniques we obtain a semimartingale decomposition (or Skorokhod representation) for the process itself and can identify it as a weak solution to an SDE with reflection at the boundary, see Theorem 6.3.2 and Theorem 6.3.5. We apply the results to stochastic dynamics for interacting particles (see Section 6.4) and for a Ginzburg-Landau interface model with area conservation (see Section 6.5).

Let us compare our results with prior results on reflected diffusions. For a good overview see the introduction of [LS84] and [PW94]. The one-dimensional case goes back to Skorokhod, see [Sko61], and McKean, see [McK63].

Lions and Sznitman construct unique strong solutions for Lipschitz continuous coefficients on domains, see [LS84]. They pose certain conditions on the inward normal vector field at the boundary which hold in particular for smooth domains. Saisho has generalized these results to the case of weaker assumptions on the domain, see [Sai87]. Tanaka proves existence and uniqueness of solutions for Lipschitz continuous coefficients in convex domains, see [Tan79].

Stroock and Varadhan formulate for an SDE with reflection a sub-martingale problem. They obtain solutions for smooth boundary and strictly elliptic bounded matrix coefficient and bounded drift. Furthermore, they construct also a boundary local time and prove (under some assumptions) uniqueness in law. See [SV71].

Let us now describe results that are obtained using Dirichlet form methods. Chen considers a gradient Dirichlet form with matrix and density with mild differentiability conditions. Under mild assumptions on the boundary he provides a semimartingale decomposition holding for quasi-every starting point, see [Che93]. Bass and Hsu give a pointwise semimartingale decomposition for reflected Brownian motion in Lipschitz domains, see [BH91] and [BH90]. They obtain also some results in Hölder domains.

For the elliptic diffusions, constructed by Fukushima and Tomisaki in [FT95] and [FT96], a pointwise semimartingale decomposition is obtained therein. The decomposition is based on the theory of [FOT94, Ch. 5] (also contained in [FOT11]). We apply this theory also for our results. Recall our remark from Chapter 4 that the assumptions in [FT95] and [FT96] exclude the case of singular drift. Hence our results are not covered by these works.

Pardoux and Williams (see [PW94]) as-well as Williams and Zheng (see [WZ90]) provide approximations of reflected diffusions by diffusions on $\mathbb{R}^d$ or the interior of the state space. The convergence results are obtained by Dirichlet form methods.

Pathwise uniqueness for Brownian motion (without drift) on bounded $C^{1,\alpha}$-domains, $0 < \alpha < 1$, is obtained by Bass and Hsu in [BH00].

Trutnau ([Tru03]) constructs a generalized (non-symmetric) Dirichlet form with singular non-symmetric drift term. The corresponding diffusion process is constructed and a Skorokhod decomposition is obtained for quasi-every starting point. Note that there additional techniques are needed both for the construction of the process and the Skorokhod decomposition since the classical theory of [FOT11] and [MR92] covers just the case of symmetric and coercive Dirichlet forms, respectively, and not generalized Dirichlet forms.

As already mentioned before, Fattler and Grothaus construct distorted Brownian motion with reflection at the boundary in [FG07] as $\mathcal{L}^p$-strong Feller process. However, a pointwise Skorokhod decomposition is not given there. So also in this case we obtain new results. For the distorted Brownian motion on $\mathbb{R}^d$, constructed by Albeverio, Kondratiev and Röckner in [AKR03], we can apply our results (with $\Omega = \mathbb{R}^d$ and $\partial\Omega = \emptyset$) to get in this case a pointwise semimartingale decomposition and existence of weak solutions. In particular, this gives a detailed proof of [AKR03, Rem. 5.6]. The results stated in this chapter were in the meanwhile submitted for publication, see [BG13b].

6.1 Construction of Strict Additive Functionals

In this section we construct strict (positive) continuous additive functionals. We follow [FOT11, Ch. 5]. We fix a locally compact separable metric space $(E, \mathbf{d})$, a locally finite Borel measure μ with full topological support and a regular symmetric Dirichlet form $(\mathcal{E}, D(\mathcal{E}))$ on $L^2(E, \mu)$. So we assume Condition 2.1.1 but without any locality assumptions on $(\mathcal{E}, D(\mathcal{E}))$. Denote by E^Δ the Alexandrov compactification of E. Let $E_1 \subset E$ such that $E^\Delta \setminus E_1$ is properly exceptional. Let $\mathbf{M}$ be a Hunt process $\mathbf{M}$ with state space E and cemetery Δ. Assume that $\mathbf{M}$ is associated with the symmetric regular Dirichlet form $(\mathcal{E}, D(\mathcal{E}))$ on $L^2(E, \mu)$.

In [FOT11, Ch. 5] additive functionals are constructed for restrictions of the process $\mathbf{M}$ to subsets $E \setminus N \subset E^\Delta$ with N being a properly exceptional set. In general this set N is non-empty and depends on the constructed functional. Under the additional assumption that the transition semigroup $(P_t)_{t \geq 0}$ is absolutely continuous on E, these results are refined to yield additive functionals for the original process $\mathbf{M}$, see [FOT11, Theo. 5.1.6].

We can generalize this result to the case that the semigroup $(P_t)_{t \geq 0}$ is absolutely continuous on E_1 only. So in this case the exceptional set is given in advance and fixed.

Let us briefly recall the definition of several classes of measures, see [FOT11, Ch. 2, Sec. 2]. For the notion of nests and generalized nests, see Definition 7.6.2. We say that a nest $(F_n)_{n \in \mathbb{N}}$ is *associated* with a measure ν if $\nu(F_n) < \infty$ for all $n \in \mathbb{N}$.

Definition 6.1.1. A positive locally finite Borel measure ν is called *smooth* if it charges no set of capacity zero and has an associated generalized nest, see [FOT11, p. 83]. The class of all smooth measures is denoted by S. A positive locally finite Borel measure ν is called a *finite energy integral* if the mapping

$$D(\mathcal{E}) \cap C_c(E) \ni f \mapsto \int_E f \, d\nu \in \mathbb{R}$$

is continuous in the $\mathcal{E}_1$-norm. The class of all finite energy integrals is denoted by S_0. If $\nu \in S_0$ then for $0 < \alpha < \infty$ there exists a corresponding unique α-potential $U_\alpha \nu \in D(\mathcal{E})$ such that

$$\mathcal{E}_\alpha(U_\alpha \nu, f) = \int_E f \, d\nu \quad \text{for all } f \in D(\mathcal{E}),$$

see [FOT11, Ch. 2, Sec. 2].

By S_{00} we denote the class of all finite energy integrals such that the α-potential for $0 < \alpha < \infty$ is essentially bounded. By S_1 we denote the class of all smooth measures such that for every $\nu \in S_1$ there exists an associated generalized nest $(F_n)_{n\in\mathbb{N}}$ such that $1_{F_n}\nu \in S_{00}$. Such a measure is called *smooth in the strict sense*, see [FOT11, Ch. 5, p. 238].

The theory of [FOT11, Ch. 5] connects measures, potentials of measures and additive functionals. We give the precise definition of an additive functional in the sense of [FOT11]. For this we need the notion of the restriction of a process to subsets of E^Δ.

Definition 6.1.2. Let $N \subset E$ be a properly exceptional set (see Definition 7.6.5). Let $\mathbf{M}^{E^\Delta\backslash N}$ be the restriction of $\mathbf{M}$ to $E^\Delta \backslash N$, i.e.,

$$\mathbf{M}^{E^\Delta\backslash N} = (\Omega^{E^\Delta\backslash N}, \mathcal{F}^{\Omega^{E^\Delta\backslash N}}, (\mathcal{F}_t^{\Omega^{E^\Delta\backslash N}})_{t\geq 0}, (\mathbf{X}_t^{E^\Delta\backslash N})_{t\geq 0}, (\mathbb{P}_x^{E^\Delta\backslash N})_{x\in E^\Delta\backslash N})$$

in the notion of Definition 7.3.18. A mapping $A = (A_t)_{t\geq 0} : \Omega^{E^\Delta\backslash N} \to \mathbb{R}_0^+$ is called a *positive continuous additive functional* (PCAF) with exceptional set N if $A_0 = 0$, A_t is $\mathcal{F}_t^{\Omega^{E^\Delta\backslash N}}$-adapted and the following properties hold:

(i) $A_t(\omega) < \infty$ for $t < \mathcal{X}$, $\omega \in \Omega^{E^\Delta\backslash N}$.

(ii) For every $\omega \in \Omega^{E^\Delta\backslash N}$, $[0,\infty) \ni t \mapsto A_t(\omega)$ is continuous.

(iii) $A_t(\omega) = A_\mathcal{X}(\omega)$ for $t \geq \mathcal{X}$ and $\omega \in \Omega^{E^\Delta\backslash N}$.

(iv) There exists $\Lambda \in \mathcal{F}^{\Omega^{E^\Delta\backslash N}}$ with $\theta_t\Lambda \subset \Lambda$ for $t \geq 0$ such that $A_{t+s}(\omega) = A_t(\omega) + A_s(\theta_t\omega)$ for $0 \leq s, t < \infty$ and $\omega \in \Lambda$.

Note that in order to give sense to the additivity property (iv) one has to assume shift-invariance of Λ. We call Λ the *additivity set* of $(A_t)_{t\geq 0}$.

A functional A is called *finite* if (i) holds for $t < \infty$ instead of just $t < \mathcal{X}$.

A functional A is called a *finite continuous additive functional* (finite CAF) if A is not necessarily positive, fulfills (ii), (iii), (iv) and $|A_t(\omega)| < \infty$ for $t < \infty$, $\omega \in \Omega^{E^\Delta\backslash N}$.

A mapping $A = (A_t)_{t\geq 0} : \Omega^{E^\Delta\backslash N} \to \mathbb{R} \cup \{\infty\}$ is called a *local continuous additive functional* with exceptional set N if $A_0 = 0$, A_t is $\mathcal{F}_t^{\Omega^{E^\Delta\backslash N}}$-adapted and:

(i) $|A_t(\omega)| < \infty$ for $t < \mathcal{X}$, $\omega \in \Omega^{E^\Delta\backslash N}$.

(ii) For every $\omega \in \Omega^{E^\Delta\backslash N}$, $[0, \mathcal{X}) \ni t \mapsto A_t(\omega)$ is continuous.

(iii) There exists $\Lambda \in \mathcal{F}^{\Omega^{E^\Delta \setminus N}}$ with $\theta_t \Lambda \subset \Lambda$ for $t \geq 0$ such that $A_{t+s}(\omega) = A_t(\omega) + A_s(\theta_t \omega)$ for $0 \leq s, t < \infty$ with $s + t < \mathcal{X}$ and $\omega \in \Lambda$.

We denote the class of all PCAF by A_c^+. We call a PCAF (or CAF) *strict* on E_1 if $N = E \setminus E_1$. The class of all strict PCAF on E_1 we denote by $A_{c,1}^+$. Since we only consider strict PCAF on E_1, we omit the mention of E_1. Analogously we define the class of all strict (local) CAF.

Note that measurability and adaptedness are assumed w.r.t. $\mathcal{F}^{\Omega^{E^\Delta \setminus N}}$ and $(\mathcal{F}_t^{\Omega^{E^\Delta \setminus N}})_{t \geq 0}$. These σ-algebras are in general larger than the corresponding trace σ-algebras of $\mathcal{F}$ and $(\mathcal{F}_t)_{t \geq 0}$ on $\Omega^{E^\Delta \setminus N}$.

Note that in the case of a local CAF we do not assume any continuity conditions for $t \geq \mathcal{X}$. So the study of those functionals makes sense only in the random time interval $[0, \mathcal{X})$.

Remark 6.1.3. Let A be a finite CAF. Measurability and the pathwise continuity of A imply that the mapping $\Omega^{E^\Delta \setminus N} \times [0, \infty) \ni (\omega, t) \mapsto A_t(\omega)$ is $\mathcal{F}^{\Omega^{E^\Delta \setminus N}} \otimes \mathcal{B}([0, \infty))$-measurable. This follows by a standard argument, see e.g. [KS91, Prop. 1.13].

Observe that we assume the additivity on the set Λ and Λ is chosen independently of x. Such a PCAF is called *perfect* in the sense of [BG68, Ch. IV, Def. 1.3].

Remark 6.1.4. Our definition of additive functionals differs from the definition given in [FOT11, p. 222]. There a defining set Λ for the functional is introduced and properties (i) to (iii) of Definition 6.1.2 are assumed to hold only on Λ. But such a functional can be extended in the canonical way to a functional fulfilling our definition. Indeed, let $(\widetilde{A}_t)_{t \geq 0}$ fulfill the definition of [FOT11]. Let $\Lambda \subset \Omega^{E^\Delta \setminus N}$ be the defining set. Then $A_t := 1_\Lambda \widetilde{A}_t$ defines a functional in our sense with additivity set Λ.

PCAF are constructed from measures via the so-called Revuz correspondence. For our purpose the following equivalent characterization of the Revuz correspondence is most suitable: A smooth measure ν is said to be in *Revuz correspondence* to $A \in A_c^+$ if

$$\int_{E \setminus N} h\left(U_A^\alpha f\right) d\mu = \int_{E \setminus N} (R_\alpha^{E^\Delta \setminus N} h) \, f d\nu \text{ for } h, f \in \mathcal{B}^+(E \setminus N) \text{ and } \alpha > 0$$

(6.1)

with $U_A^\alpha f(x) := \mathbb{E}_x^{(E^\Delta \setminus N)}[\int_0^\infty e^{-\alpha s} f(\mathbf{X}_s) dA_s]$, $x \in E^\Delta \setminus N$, N being the exceptional set of A. Here $(R_\alpha^{E^\Delta \setminus N})_{\alpha > 0}$ and $\mathbb{E}_\cdot^{E^\Delta \setminus N}[\cdot]$ denote the resolvent and expectation, respectively, of the restricted process $\mathbf{M}^{E^\Delta \setminus N}$. Combining Lemma 7.3.3 and Proposition 7.7.1 (and Corollary 7.7.2), we get that $U_A^\alpha f$ is $\mathcal{B}^*(E^\Delta \setminus N)$-measurable.

See [FOT11, Theo. 5.1.3] for these definitions and further equivalent descriptions of the Revuz correspondence. In the case that ν has an α-potential, i.e., $\nu \in S_0$, the Revuz correspondence is equivalent to:

$$U_A^1 1 \text{ is a } \mathcal{E}\text{-quasi-continuous version of } U_1 \nu.$$

See Definition 7.6.3 for the definition of $\mathcal{E}$-quasi-continuous.

We are interested in the construction of additive functionals which are strict on the fixed set E_1. Denote the restriction of the Markov process $\mathbf{M}$ to the set $E_1 \cup \{\Delta\}$ by
$\mathbf{M}^1 := (\Omega^1, \mathcal{F}^1, (\mathcal{F}_t^1)_{t \geq 0}, (\mathbf{X}_t^1)_{t \geq 0}, (\mathbb{P}_x^1)_{x \in E_1 \cup \{\Delta\}})$. The corresponding resolvent and semigroup we denote just by $(R_\alpha)_{\alpha > 0}$ and $(P_t)_{t \geq 0}$, respectively. In this section we do not assume Feller properties for the semigroup $(P_t)_{t \geq 0}$. We just assume absolute continuity of $(P_t)_{t > 0}$ on E_1 in the sense of the following condition.

Condition 6.1.5. There exists a $\mathcal{B}(E_1 \times E_1)$-measurable map $p_t : E_1 \times E_1 \to \mathbb{R}_0^+$, $t > 0$, such that

$$P_t f(x) = \int_{E_1} f(y) \, p_t(x, y) d\mu(y) \text{ for } x \in E_1 \text{ and } f \in \mathcal{B}^+(E_1).$$

Remark 6.1.6. In the case that $(P_t)_{t \geq 0}$ is $\mathcal{L}^p$-strong Feller on E_1 the condition is fulfilled, see Theorem 2.2.9(i).

Remark 6.1.7. The absolute continuity condition on $(P_t)_{t \geq 0}$ is in general not enough to conclude that the process $\mathbf{M}$ has right-continuous sample paths for every starting point in E_1, hence it need not to be a Hunt process. Therefore, we *assume* that the associated process $\mathbf{M}$ is a Hunt process.

According to [FOT11, Ch. 2.2] we call a function $u \in L^2(E, \mu)$ α-excessive if

$$u \geq 0 \text{ and } e^{-\alpha t} T_t u(x) \leq u(x) \quad \text{for } \mu\text{-a.e. } x.$$

The absolute continuity condition of the semigroup allows us to deduce certain pointwise statements for excessive functions.

Proposition 6.1.8. *Assume that $(P_t)_{t \geq 0}$ fulfills the absolute continuity Condition 6.1.5. Let u be α-excessive. Then*

(i) $e^{-\alpha t_2} P_{t_2} u(x) \leq e^{-\alpha t_1} P_{t_1} u(x)$ *for* $t_2 \geq t_1 > 0$ *and* $x \in E_1$.

(ii) *For every* $t > 0$ *the function* $e^{-\alpha t} P_t u$ *is also α-excessive.*

(iii) $e^{-\alpha t} P_t u(x) \uparrow u(x)$ *as* $t \downarrow 0$ *for μ-a.e. $x \in E_1$.*

(iv) $e^{-\alpha(t+h)} P_{t+h} u(x) \uparrow e^{-\alpha t} P_t u(x)$ *as* $h \downarrow 0$ *for* $t > 0$ *and* $x \in E_1$.

Proof. (i): For $h \geq 0$ it holds $e^{-\alpha h} P_h u(x) \leq u(x)$ for μ-a.e. x. Set $h := t_2 - t_1$. Then

$$e^{-\alpha t_2} P_{t_2} u(x) = e^{-\alpha t_1} \int_{E_1} e^{-\alpha h} P_h u(y) \, p_{t_1}(x,y) d\mu(y)$$

$$\leq e^{-\alpha t_1} \int_{E_1} u(y) \, p_{t_1}(x,y) d\mu(y) = e^{-\alpha t_1} P_{t_1} u(x).$$

(ii): For $t > 0$, $h \geq 0$ we have $e^{-\alpha h} P_h e^{-\alpha t} P_t u = e^{-\alpha t} P_t e^{-\alpha h} P_h u \leq e^{-\alpha t} P_t u$.
(iii): Since $(T_t)_{t \geq 0}$ is L^2-strongly continuous, we get from the sequence $(1/n)_{n \in \mathbb{N}}$ a subsequence $(n_k)_{k \in \mathbb{N}}$ with $n_k \downarrow 0$ such that $e^{-\alpha n_k} P_{n_k} u(x) \uparrow u(x)$ as $k \to \infty$ for μ-a.e. $x \in E_1$. By (i) $e^{-\alpha t} P_t u(x)$ increases if t decreases. So we get convergence in fact for every sequence converging to 0.
(iv): As in (i) we get using (iii)

$$e^{-\alpha(t+h)} P_{t+h} u(x) = e^{-\alpha t} \int_{E_1} e^{-\alpha h} P_h u(y) \, p_t(x,y) d\mu(y)$$

$$\uparrow e^{-\alpha t} \int_{E_1} u(y) \, p_t(x,y) d\mu = e^{-\alpha t} P_t u(x)$$

for every $x \in E_1$ as $h \downarrow 0$. $\qquad\square$

From this proposition we can deduce the existence of a μ-version with certain regularity properties for every excessive function.

Theorem 6.1.9. *Assume Condition 6.1.5. Let $\alpha > 0$, $u \in L^2(E,\mu)$ be α-excessive and essentially bounded on E_1. Then there exists a unique μ-version $\tilde{u}$ of u such that*

(i) $\tilde{u} \in \mathcal{B}_b(E_1)$, *i.e., boundedness holds for every point in E_1,*

(ii) $e^{-\alpha t_2} P_{t_2} \tilde{u}(x) \leq e^{-\alpha t_1} P_{t_1} \tilde{u}(x)$ *for* $t_2 \geq t_1 \geq 0$ *and* $x \in E_1$, *(iii)* $\lim_{t\to 0} P_t\widetilde{u}(x) = \widetilde{u}(x)$ *for every* $x \in E_1$.

For $\nu \in S_{00}$ *and the corresponding α-potential $U_\alpha\nu$ there exists a version* $\widetilde{U_\alpha\nu}$ *such that properties (i), (ii) and (iii) hold with $\widetilde{u} = \widetilde{U_\alpha\nu}$.*

Proof. Let $C := \|u\|_{L^\infty(E_1,\mu)}$. Then $e^{-\alpha t}P_t u(x) \le C$ for every $t \ge 0$ and $x \in E_1$. Define $\widetilde{u}(x) := \sup_{t>0} e^{-\alpha t}P_t u(x)$. Then $\widetilde{u} \in \mathcal{B}_b(E_1)$ with $\sup_{x\in E_1}\widetilde{u}(x) \le C$. From Proposition 6.1.8 we get that $e^{-\alpha t}P_t u$ is decreasing in t. So we have $\widetilde{u}(x) = \lim_{t\to 0} e^{-\alpha t}P_t u(x) \le C < \infty$ for every $x \in E_1$. Proposition 6.1.8(iii) implies $\widetilde{u}(x) = \lim_{t\to 0} e^{-\alpha t}P_t u(x) = u(x)$ for μ-a.e. $x \in E_1$. Thus $\widetilde{u}$ is a μ-version of u and property (iii) holds. For $t_1 > 0$, property (ii) follows from Proposition 6.1.8(i), replacing u by $\widetilde{u}$. For $t_1 = 0$ the property follows from definition of $\widetilde{u}$.

For $\nu \in S_{00}$ we can apply [FOT11, Theo. 2.2.1] to get that $U_\alpha\nu$ is α-excessive, $\alpha \ge 0$. Since $U_\alpha\nu$ is essentially bounded, we can apply the proven statements to $u := U_\alpha\nu$. □

We call an excessive function that fulfills properties (i),(ii) and (iii), a *strictly α-excessive function*. This definition matches the definition of α-excessive of [BG68, Ch. II, Def. 2].

The convergence property (iii) in the theorem is essential for the construction of the strict additive functional. Recall that additive functionals in the non strict sense have nice properties for quasi-every point $x \in E$ only. The absolute continuity property of the semigroup transforms statements into some that hold for every point $x \in E_1$ but for strict positive time only. The convergence property (iii) allows then to get a pointwise statement for $t = 0$.

Remark 6.1.10. If u as in Theorem 6.1.9 has a continuous version on E_1, then $\widetilde{u}$ is equal to this continuous version. To see this denote this continuous version by $\breve{u}$. We have $P_t\widetilde{u} = P_t\breve{u}$. Using the right-continuity of the paths of $\mathbf{M}^1$ and Lebesgue's dominated convergence we obtain $P_t\breve{u}(x) \xrightarrow{t\to 0} \breve{u}(x)$ for every $x \in E_1$. Together with Theorem 6.1.9(iii) we obtain $\widetilde{u}(x) = \breve{u}(x)$ for every $x \in E_1$.

We provide the construction of strict finite PCAF on E_1 now. The theorem is based on [FOT11, Theo. 5.1.6]. We give details to some convergence properties in the proof. In order to rigorously prove them we need the strictly α-excessive version of the potentials $U_\alpha\nu$, $\nu \in S_{00}$, $\alpha > 0$, provided by Theorem 6.1.9. In [FOT11] such a version is constructed from the resolvent kernels $(R_\alpha)_{\alpha>0}$ of the process, see Remark 6.1.12 below.

The proof of our theorem is a modification of the proof of [FOT11, Theo. 5.1.6] using the regularity provided by the previous theorem.

Theorem 6.1.11. *Let* $\mathbf{M}^1$ *be the restriction of* $\mathbf{M}$, *see the beginning of this section, to* $E_1 \cup \{\Delta\}$. *Denote by* $(\mathbb{P}_x^1)_{x \in E_1 \cup \{\Delta\}}$ *the path measure of* $\mathbf{M}^1$ *and by* $\mathbb{E}_x^1[\,\cdot\,]$, $x \in E_1 \cup \{\Delta\}$, *the expectation w.r.t.* $\mathbb{P}_x^1$. *We denote the semigroup and resolvent just by* $(P_t)_{t \geq 0}$ *and* $(R_\alpha)_{\alpha > 0}$, *respectively. Assume that* $(P_t)_{t > 0}$ *fulfills Condition 6.1.5.*

Let $\nu \in S_{00}$ *and let* $(U_\alpha \nu)_{\alpha > 0}$ *be the corresponding potentials. Denote by* $\widetilde{U_\alpha \nu}$ *the bounded strictly* α-*excessive version of* $U_\alpha \nu$ *provided by Theorem 6.1.9.*

Then there exists a unique finite PCAF $(\widetilde{A}_t)_{t \geq 0}$ *that is strict on* E_1 *and in Revuz correspondence to* ν. *For* $\widetilde{A}_t$ *it holds*

$$\mathbb{E}_x^1 \left[\int_0^\infty \exp(-\alpha s) d\widetilde{A}_s \right] = \widetilde{U_\alpha \nu}(x) \quad \text{for every } x \in E_1 \text{ and } \alpha > 0. \quad (6.2)$$

Remark 6.1.12. From (6.2) and absolute continuity of $(P_t)_{t \geq 0}$ one can conclude using some standard calculations that every version of $U_\alpha \nu$ that fulfills (6.2) pointwisely must be strictly α-excessive and hence is equal to our $\widetilde{U_\alpha \nu}$. In [FOT11] a μ-version of $G_\alpha \nu$ is constructed from an strictly α-excessive density of the resolvent kernels $(R_\alpha)_{\alpha > 0}$. The existence of such a density follows from the absolute continuity condition, see [FOT11, Lem. 4.2.4]. Denote this μ-version by $R_\alpha \nu$. A careful calculation shows that this version is strictly α-excessive, hence equal to the version provided by Theorem 6.1.9. So our formulation is compatible with the one of [FOT11, Theo. 5.1.6].

Proof. From [FOT11, Theo. 5.1.1] we get a finite PCAF A with exceptional set $N \subset E$ and additivity set $\Lambda \subset \Omega^{E^\Delta \backslash N}$. In view of Theorem 7.6.6 and Lemma 7.3.21 we may assume that N is in fact Borel. This functional is first defined w.r.t. the restricted process $\mathbf{M}^{E^\Delta \backslash N}$. After further restricting we get a functional w.r.t. the process $\mathbf{M}^{(E_1 \cup \{\Delta\}) \backslash N}$. Replace Λ by $\Lambda \cap \Omega^{(E_1 \cup \{\Delta\}) \backslash N} \in \mathcal{F}^{\Omega^{(E_1 \cup \{\Delta\}) \backslash N}}$. Note that $\mathbb{P}_x^{(E_1 \cup \{\Delta\}) \backslash N}(\Lambda) = 1$ for $x \in E_1 \backslash N$. Furthermore, it holds

$$\mathbb{E}_x^{(E_1 \cup \{\Delta\}) \backslash N} \left[\int_0^\infty \exp(-\alpha s) dA_s \right] = U_\alpha \nu(x) \quad \text{for } x \in E_1 \backslash N \text{ and } \alpha > 0. \quad (6.3)$$

From now on we denote the path measure of the process $\mathbf{M}^{(E_1 \cup \{\Delta\}) \backslash N}$ by $(\mathbb{P}_x^2)_{x \in (E_1 \cup \{\Delta\}) \backslash N}$, the expectation by $\mathbb{E}^2[\,\cdot\,]$ and the semigroup by $(P_t^2)_{t \geq 0}$.

After redefinition of Λ and A, we may assume $\mathbb{P}^2_\Delta(\Lambda) = 1$ and $\mathbb{P}^2_\Delta(\{\omega \in \Lambda \mid A_t(\omega) = 0 \text{ for all } t > 0\}) = 1$. Define $\varepsilon_n := \frac{1}{n}$, $n \in \mathbb{N}$. Define

$$\Lambda_0 := \mathbf{\Omega}^1 \cap \bigcap_{n \in \mathbb{N}} \theta_{\varepsilon_n}^{-1} \Lambda.$$

Then Λ_0 is shift-invariant, i.e., $\theta_t \Lambda_0 \subset \Lambda_0$ for $t \geq 0$. This follows since Λ is shift-invariant. Observe that for every $s > 0$ and $\omega \in \Lambda_0$, it holds that the mapping $t \mapsto A_t(\theta_s \omega)$ is continuous, additive and positive. Indeed, choose $0 < \varepsilon_n < s$. Then $\theta_{\varepsilon_n} \omega \in \Lambda$, hence also $\theta_s \omega \in \Lambda$ by shift-invariance of Λ.

Since $(P_t)_{t \geq 0}$ is absolutely continuous, we have $P_t(x, N) = 0$ for every $x \in E_1 \cup \{\Delta\}$. So by Lemma 7.3.22 (with $\tilde{E}_1 = E_1 \cup \{\Delta\}$ and $\tilde{E}_2 = (E_1 \cup \{\Delta\}) \setminus N$) we get $\Lambda_0 \in \mathcal{F}^1$ and $\mathbb{P}^1_x(\Lambda_0) = 1$ for all $x \in E_1 \cup \{\Delta\}$.

For $\omega \in \mathbf{\Omega}^1$ define $A^n_t(\omega) := 1_{\Lambda_0}(\omega) A_{t-\varepsilon_n}(\theta_{\varepsilon_n} \omega)$ for $t > \varepsilon_n$, $A^n_t(\omega) = 0$ for $t \leq \varepsilon_n$. Then $A^n_t(\omega)$ corresponds to the value of A on the truncated path of ω on $[\varepsilon_n, t]$. Due to Lemma 7.3.22 we have that $(A^n_t)_{t \geq 0}$ is $\mathcal{F}^1_t$-adapted for all $n \in \mathbb{N}$. For $m < n$, $t > \varepsilon_m$ we have using that A is additive on the path $\tilde{\omega} := \theta_{\varepsilon_n} \omega$,

$$A^n_t(\omega) - A^m_t(\omega) = A_{t-\varepsilon_n}(\theta_{\varepsilon_n}(\omega)) - A_{t-\varepsilon_m}(\theta_{\varepsilon_m}(\omega)) = A_{\varepsilon_m - \varepsilon_n}(\theta_{\varepsilon_n}\omega). \tag{6.4}$$

So $A^n_t(\omega)$ is increasing in n.

For $0 < t_1, t_2 < \infty$ arbitrary consider $I = [t_1, t_2]$. Choose m_0 such that $\varepsilon_{m_0} < t_1$. For $m, n > m_0$ the difference on the right hand side of (6.4) only depends on the part of ω in $(0, \varepsilon_{m_0})$. So if $(A^k_t(\omega))_{k \in \mathbb{N}}$ converges, then the convergence is uniform in $[t_1, t_2]$. So it is left to show that $(A^k_t(\omega))_{k \in \mathbb{N}}$ is bounded for $t > 0$, $\omega \in \Lambda_0$. Let $\omega \in \Lambda$ and $t > 0$. We have

$$A_t(\omega) = \int_0^t dA_s(\omega) = \int_0^t e^s e^{-s} dA_s(\omega)$$

$$\leq e^t \int_0^t e^{-s} dA_s(\omega) = e^t \int_0^\infty e^{-s} dA_s(\omega) - e^t \int_t^\infty e^{-s} dA_s(\omega).$$

By Lemma 7.7.9 we have together with (6.3)

$$e^t \mathbb{E}^2_x \left[\int_t^\infty e^{-s} dA_s \right] = P^2_t \mathbb{E}^2_\cdot \left[\int_0^\infty e^{-s} dA_s \right](x) = P^2_t U_1 \nu(x) \text{ for } x \in E_1 \setminus N. \tag{6.5}$$

Together with Lemma 7.7.8 and absolute continuity of $(P_t)_{t\geq 0}$ we get for $n \in \mathbb{N}$

$$\mathbb{E}^1_x[A^n_t] = \mathbb{E}^1_x[\mathbb{E}^2_{\mathbf{X}_{\varepsilon_n}}[A_{t-\varepsilon_n}]] = P_{\varepsilon_n}(\mathbb{E}^2_{\cdot}[A_{t-\varepsilon_n}])(x)$$
$$\leq e^{t-\varepsilon_n} P_{\varepsilon_n} \widetilde{U_1\nu}(x) - P_{\varepsilon_n} P^2_{t-\varepsilon_n} \widetilde{U_1\nu}(x)$$
$$= e^{t-\varepsilon_n} P_{\varepsilon_n} \widetilde{U_1\nu}(x) - P_t \widetilde{U_1\nu}(x), \quad x \in E_1. \quad (6.6)$$

In the last equality we used that for $x \in E_1 \setminus N$ it holds $P^2_{t-\varepsilon_n} \widetilde{U_1\nu}(x) = P_{t-\varepsilon_n} \widetilde{U_1\nu}(x)$, see also Lemma 7.3.20. By Theorem 6.1.9(iii), this expression converges for $x \in E_1$ to $e^t \widetilde{U_1\nu} - P_t \widetilde{U_1\nu}$ for $n \to \infty$. From monotone convergence we get

$$\mathbb{E}^1_x[\sup_{n\in\mathbb{N}} A^n_t] = \sup_{n\in\mathbb{N}} \mathbb{E}^1_x[A^n_t] < \infty \quad \text{for every } x \in E_1.$$

Hence $\sup_{n\in\mathbb{N}} A^n_t(\omega) < \infty$ $\mathbb{P}^1_x$-a.s. for $x \in E_1$ and thus the limit of $A^n_t(\omega)$ exists and is finite $\mathbb{P}^1_x$-a.s. Moreover, the convergence is locally uniformly in $(0, \infty)$. Define for $t > 0$ and $\omega \in \Omega^1$

$$A'_t(\omega) = \begin{cases} \lim_{n\to\infty} A^n_t(\omega) & \text{if the limit exists,} \\ 0 & \text{else.} \end{cases}$$

Then A'_t is positive. Moreover, the considerations before imply that A'_t is $\mathbb{P}^1_x$-a.s. finite and continuous in $(0, \infty)$ for $x \in E_1$.

From (6.6) and Fatou's lemma we get

$$\mathbb{E}^1_x[A'_t] \leq \liminf_{n\to\infty} \mathbb{E}^1_x[A^n_t] \leq e^t \widetilde{U_1\nu}(x) - P_t \widetilde{U_1\nu}(x) \quad \text{for } x \in E_1. \quad (6.7)$$

It is left to prove right-continuity at $t = 0$. Note that $(A'_t)_{t>0}$ is positive and increasing as the pointwise limit of the positive and increasing functionals $(A^n_t)_{t>0}$. Thus $0 \leq \mathbb{E}^1_x[A'_{t_1}] \leq \mathbb{E}^1_x[A'_{t_2}]$ for $0 < t_1 \leq t_2$. The right hand side of (6.7) converges to 0. Hence $(A'_t)_{t>0}$ converges in $L^1(\Omega^1, \mathbb{P}^1_x)$ to 0 for $x \in E_1$. Let $s_n := \frac{1}{n}$, $n \in \mathbb{N}$. Then there exists a subsequence $(s'_n)_{n\in\mathbb{N}}$ such that $(A'_{s'_n}(\omega))_{n\in\mathbb{N}}$ converges $\mathbb{P}^1_x$-a.s. to 0. Monotonicity implies that $A'_{t_n}(\omega)$ converges to 0 for any sequence $(t_n)_{n\in\mathbb{N}}$ converging to 0.

Define

$$\widetilde{\Lambda} = \{\omega \in \Lambda_0 \mid \sup_{n\in\mathbb{N}} A^n_t(\omega) < \infty \text{ for } t > 0, \, A'_{0+} = 0\}.$$

Observe that $\widetilde{\Lambda}$ is $\mathcal{F}^1$-measurable since by monotonicity of the mappings A^n_t, $n \in \mathbb{N}$, it is enough to require that the sup is finite for rational t and that the right-limit for one sequence decreasing to 0 is 0. From our considerations before we get

$$\mathbb{P}^1_x(\widetilde{\Lambda}) = 1 \quad \text{for every } x \in E_1 \cup \{\Delta\}.$$

Define for $\omega \in \mathbf{\Omega}^1$ and $t \geq 0$

$$\widetilde{A}_t := 1_{\widetilde{\Lambda}} \lim_{n \to \infty} A^n_t(\omega).$$

Then $\widetilde{A}_t(\omega) = A'_t(\omega)$ for $\omega \in \widetilde{\Lambda}$. So we have that $(\widetilde{A}_t(\omega))_{t \geq 0}$ is continuous in $[0, \infty)$ for $\omega \in \widetilde{\Lambda}$.

We show that $\widetilde{\Lambda}$ is again shift-invariant and that $(\widetilde{A}_t)_{t \geq 0}$ is additive on this set. Let $s, t > 0$ and fix $n \in \mathbb{N}$ with $\varepsilon_n < s$. Choose $\omega \in \widetilde{\Lambda}$. Set $\omega_n := \theta_{\varepsilon_n} \omega \in \Lambda$. Since A is additive on Λ we get.

$$A_{t+s-\varepsilon_n}(\omega_n) = A_{t-\varepsilon_n}(\theta_s \omega_n) + A_s(\omega_n)$$
$$= A_{t-\varepsilon_n}(\theta_s \omega_n) + A_{s-\varepsilon_n}(\omega_n) + A_{\varepsilon_n}(\theta_{s-\varepsilon_n} \omega_n)$$
$$= A_{t-\varepsilon_n}(\theta_s \omega_n) + A_{s-\varepsilon_n}(\omega_n) + A_{\varepsilon_n}(\theta_s \omega).$$

Thus

$$A^n_{t+s}(\omega) = A_{t+s-\varepsilon_n}(\omega_n) = A_{t-\varepsilon_n}(\theta_s \omega_n) + A_{s-\varepsilon_n}(\omega_n) + A_{\varepsilon_n}(\theta_s \omega)$$
$$= A^n_t(\theta_s \omega) + A^n_s(\omega) + A_{\varepsilon_n}(\theta_s \omega).$$

This implies that $\sup_{n \in \mathbb{N}} A^n_t(\theta_s \omega) < \infty$ for every $t > 0$, hence $A^n_t(\theta_s \omega)$ converges to $A'_t(\theta_s \omega)$. For $n \to \infty$ we have $\lim_{n \to \infty} A_{\varepsilon_n}(\theta_s \omega) = 0$. Thus

$$\widetilde{A_{t+s}}(\omega) = A'_t(\theta_s \omega) + \widetilde{A}_s(\omega). \tag{6.8}$$

This implies in particular $\lim_{t \to 0} A'_t(\theta_s \omega) = 0$. Thus $\theta_s \omega \in \widetilde{\Lambda}$ and $A'_t(\theta_s \omega) = \widetilde{A}_t(\theta_s \omega)$ for $t \geq 0$. From (6.8) we obtain that $(\widetilde{A}_t)_{t \geq 0}$ is additive.

It is left to prove that $\widetilde{A}_t$ is in Revuz correspondence to ν and that it is the only strict PCAF being in Revuz correspondence. Let $0 \leq t < \infty$. The product rule for the Lebesgue-Stieltjes integral, see Lemma 7.7.3(ii) below, implies

$$\int_{\varepsilon_n}^t e^{-s} dA^n_s = e^{-t} A^n_t + \int_{\varepsilon_n}^t e^{-s} A^n_s ds, n \in \mathbb{N},$$

and

$$\int_0^t e^{-s}d\widetilde{A_s} = e^{-t}\widetilde{A_t} + \int_0^t e^{-s}\widetilde{A_s}ds.$$

on $\widetilde{\Lambda}$. Using that A_t^n increases to $\widetilde{A_t}$ we get that $\int_{\varepsilon_n}^t e^{-s}dA_s^n$ converges to $\int_0^t e^{-s}d\widetilde{A_s}$. Furthermore,

$$\left| e^{-t}A_t^n + \int_{\varepsilon_n}^t e^{-s}A_s^n ds \right| = e^{-t}A_t^n + \int_{\varepsilon_n}^t e^{-s}A_s^n ds \le e^{-\varepsilon_n}A_t^n \le \widetilde{A_t} \text{ for } n \in \mathbb{N}.$$

Note that the last inequality holds only almost surely. By (6.7) we have that $\widetilde{A_t}$ is $\mathbb{P}_x$-integrable for $x \in E_1$. So Lebesgue dominated convergence yields

$$\mathbb{E}_x^1\left[\int_0^t e^{-s}d\widetilde{A_s} \right] = \lim_{n\to\infty} \mathbb{E}_x^1\left[\int_{\varepsilon_n}^t e^{-s}dA_s^n \right] \quad \text{for every } x \in E_1 \text{ and } t \ge 0.$$

For $x \in E_1$ and $T < \infty$ we have using Lemma 7.7.8 and (6.5)

$$\mathbb{E}_x^1\left[\int_{\varepsilon_n}^T e^{-s}dA_s^n \right] = \mathbb{E}_x^1\left[e^{-\varepsilon_n}\mathbb{E}_{\mathbf{X}_{\varepsilon_n}}^2\left[\int_0^{T-\varepsilon_n} e^{-s}dA_s \right] \right]$$

$$= e^{-\varepsilon_n}P_{\varepsilon_n}\left(U_1\nu - e^{-(T-\varepsilon_n)}P_{T-\varepsilon_n}^2 U_1\nu \right)(x)$$

$$= e^{-\varepsilon_n}P_{\varepsilon_n}\widetilde{U_1\nu}(x) - e^{-T}P_T\widetilde{U_1\nu}(x) \quad \text{for } n \in \mathbb{N}.$$

Observe that the right-hand side converges to $\widetilde{U_1\nu}(x) - e^{-T}P_T\widetilde{U_1\nu}(x)$ for $n \to \infty$ for $x \in E_1$. So for every $x \in E_1$,

$$\mathbb{E}_x^1\left[\int_0^T e^{-s}d\widetilde{A_s} \right] = \lim_{n\to\infty} \mathbb{E}_x^1\left[\int_{\varepsilon_n}^T e^{-s}dA_s^n \right] = \widetilde{U_1\nu}(x) - e^{-T}P_T\widetilde{U_1\nu}(x).$$

Since $\widetilde{U_1\nu}$ is bounded on E_1, $P_T\widetilde{U_1\nu}(x)$ is bounded in T for every $x \in E_1$. So letting $T \to \infty$ we obtain

$$\mathbb{E}_x^1\left[\int_0^\infty e^{-s}d\widetilde{A_s} \right] = \widetilde{U_1\nu}(x) \quad \text{for every } x \in E_1.$$

So according to [FOT11, Theo. 5.1.3(v)] the functional $(\widetilde{A_t})_{t\ge0}$ is in Revuz correspondence to ν . Uniqueness follows from Theorem 6.1.14 below.

$\square$

Remark 6.1.13. For the sets defined in the proof we have the inclusion relations $\Lambda \subset \Lambda_0$ and $\widetilde{\Lambda} \subset \Lambda_0$. So we first enlarge the set Λ to get a set Λ_0 with full $\mathbb{P}^1_x$-measure for *every* point $x \in E_1$. However, on this larger set, A might be not additive. Thus we have to make this set slightly smaller to $\widetilde{\Lambda}$ and construct a functional $\widetilde{A}$ which behaves „nice" on $\widetilde{\Lambda}$.

From now on we consider only the process $\mathbf{M}^1$. We denote the path measure just by $(\mathbb{P}_x)_{x \in E_1 \cup \{\Delta\}}$ and the expectation just by $\mathbb{E}_x[\,\cdot\,]$, $x \in E_1 \cup \{\Delta\}$.

The following uniqueness theorem and its proof are already contained in [FOT11, Theo. 5.1.6].

Theorem 6.1.14. *Let* $A^{(1)}$, $A^{(2)}$ *be two strict PCAFs that are in Revuz correspondence to a smooth measure* ν. *Then* $A^{(1)}$ *and* $A^{(2)}$ *are equivalent, i.e.,*

$$A_t^{(1)} = A_t^{(2)} \quad for\ t \geq 0 \quad \mathbb{P}_x - a.s.,\ x \in E_1.$$

Let Λ_1 *and* Λ_2 *be the additivity sets of* $A^{(1)}$ *and* $A^{(2)}$, *respectively. Then there exists a set* $\widetilde{\Lambda} \subset \Lambda_1 \cap \Lambda_2$ *with* $\mathbb{P}_x(\widetilde{\Lambda}) = 1$ *for* $x \in E_1$ *and* $A_t^{(1)}(\omega) = A_t^{(2)}(\omega)$ *for every* $t \geq 0$ *and* $\omega \in \widetilde{\Lambda}$.

Proof. Assume there exist two strict PCAFs $A^{(1)}$, $A^{(2)}$ corresponding to ν. Then [FOT11, Theo. 5.1.2] implies that for μ-a.e. $x \in E_1$ it holds $A_t^{(1)} = A_t^{(2)}$ $\mathbb{P}_x$-a.s. for $t \geq 0$. For $0 < s \leq t$ we have for every $x \in E_1$

$$\mathbb{P}_x(\{A_t^{(1)} - A_s^{(1)} = A_t^{(2)} - A_s^{(2)}\}) = \mathbb{P}_x(\{A_{t-s}^{(1)}(\theta_s \cdot) - A_{t-s}^{(2)}(\theta_s \cdot) = 0\})$$

$$= \mathbb{E}_x\left[\mathbb{E}_{\mathbf{X}_s}\left[1_{\{A_{t-s}^{(1)} = A_{t-s}^{(2)}\}}\right]\right] = P_s\left(\mathbb{E}.\left[1_{\{A_{t-s}^{(1)} = A_{t-s}^{(2)}\}}\right]\right)(x) = 1.$$

Thus $A_t^{(1)} - A_s^{(1)} = A_t^{(2)} - A_s^{(2)}$ $\mathbb{P}_x$-a.s. for $x \in E_1$. Since $A_s^{(1)}$ and $A_s^{(2)}$ converge to 0 for $s \to 0$, $\mathbb{P}_x$-a.s., we get $A_t^{(1)} = A_t^{(2)}$ for $t \geq s$ $\mathbb{P}_x$-a.s. for every $x \in E_1$. Now choose as s the sequence $s_n := \frac{1}{n}$, $n \in \mathbb{N}$. Then we get $A_t^{(1)} = A_t^{(2)}$ for $t > 0$ $\mathbb{P}_x$-a.s. for every $x \in E_1$. For $t = 0$ the claim is trivial since by construction $A_t^{(1)} = 0 = A_t^{(2)}$.

Define

$$\widetilde{\Lambda} := \Lambda_1 \cap \Lambda_2 \cap \{\omega \mid A_t^{(1)}(\omega) = A_t^{(2)}(\omega),\ t \in \mathbb{Q} \cap [0, \infty)\}$$

$$= \Lambda_1 \cap \Lambda_2 \cap \{\omega \mid A_t^{(1)}(\omega) = A_t^{(2)}(\omega),\ t \in [0, \infty)\}.$$

The first equality implies that $\widetilde{\Lambda}$ is $\mathcal{F}^1$-measurable. The second equality together with additivity implies that $\widetilde{\Lambda}$ is shift-invariant. From the proven statements it follows $\mathbb{P}_x(\widetilde{\Lambda}) = 1$ for every $x \in E_1$. $\qquad\square$

For a strict finite PCAF $(A_t)_{t>0}$ with Revuz measure ν and $f \in \mathcal{B}_b^+(E_1)$ the mapping $[0, \infty) \ni t \mapsto \int_0^t f(\mathbf{X}_s)dA_s$ defines again a strict finite PCAF. The next lemma shows that the corresponding measure is given by $f\,\nu$, i.e., multiplication with a Borel bounded function is compatible with the Revuz correspondence.

Lemma 6.1.15. *Let $(A_t)_{t \geq 0}$ be a strict finite PCAF with Revuz measure $\nu \in S_{00}$. Let $M \in \mathcal{B}(E_1)$ such that $supp[\nu] \subset M$ and $f \in \mathcal{B}_b(M)$. Then the mapping $f \cdot A := ((f \cdot A)_t)_{t \geq 0}$,*

$$(f \cdot A)_t = \int_0^t f(X_s)dA_s, \ t \geq 0,$$

defines a strict finite CAF with same additivity set as A. It holds $f \cdot A \in \mathcal{N}_c$, $\mathcal{N}_c$ as in (6.15) below and $\mathbb{E}_x[(f \cdot A)_t^2] < \infty$ for $0 \leq t < \infty$ and every $x \in E_1$. We have

$$P_s\mathbb{E}. \left[\int_0^t f(\mathbf{X}_r)\,dA_r\right](x) \xrightarrow{s \to 0} \mathbb{E}_x\left[\int_0^t f(\mathbf{X}_r)\,dA_r\right] \quad \textit{for every } x \in E_1.$$
$$(6.9)$$

If $f \in \mathcal{B}_b^+(M)$ and $f\,\nu \in S_{00}$, we have for $\alpha > 0$

$$\mathbb{E}_x\left[\int_0^\infty e^{-\alpha s}f(X_s)dA_s\right] = \widetilde{U_\alpha f\,\nu}(x) \quad \textit{for every } x \in E_1. \qquad (6.10)$$

Proof. First extend f to a function in $\mathcal{B}_b^+(E)$ by replacing f with $1_M f$. Observe that this defines indeed a measurable and bounded function on E. From Proposition 7.7.1 and Proposition 7.7.4, see below, we get that $f \cdot A$ is a strict finite PCAF with the same additivity set as A.

Next we prove $f \cdot A \in \mathcal{N}_c$. We have for $t \geq 0$ and $x \in E_1$

$$\mathbb{E}_x\left[|(f \cdot A)_t|\right] \leq \mathbb{E}_x\left[\int_0^t |f(\mathbf{X}_s)|dA_s\right] \leq \|f\|_\infty \mathbb{E}_x[A_t] \leq \|f\|_\infty e^t \widetilde{U_1\nu}(x) < \infty.$$

Note that $e((f \cdot A)) \leq \|f\|_{\sup}^2 e(A)$. Thus $e((f \cdot A)) = 0$ follows from the calculations in [FOT11, p. 245].

Similar to the calculation therein we get for $t \geq 0$ and $x \in E_1$

$$\mathbb{E}_x \left[(f \cdot A)_t^2 \right] \leq 2e^t \mathbb{E}_x \left[\int_0^t |f|(\mathbf{X}_s)\widetilde{U_1(|f|\nu)}(\mathbf{X}_s) \, dA_s \right] < \infty.$$

We prove (6.9). Using Lebesgue's dominated convergence we get

$$\lim_{s \to 0} \int_s^{t+s} f(\mathbf{X}_r(\omega)) \, dA_r(\omega) = \int_0^t f(\mathbf{X}_r(\omega)) \, dA_r(\omega) \quad \text{for } \omega \in \mathbf{\Omega}^1.$$

Note that the Markov property implies

$$P_s \mathbb{E}. \left[\int_0^t f(\mathbf{X}_r) \, dA_r \right] (x) = \mathbb{E}_x \left[\int_s^{t+s} f(\mathbf{X}_r) \, dA_r \right], x \in E_1,$$

see also Lemma 7.7.9. Define $F_s(t) := \int_s^{t+s} f(\mathbf{X}_r(\omega)) \, dA_r(\omega)$. It holds

$$|F_s(t)(\omega)| \leq \|f\|_{\sup} A_{t+s}(\omega) < A_{2t}(\omega) < \infty$$

for $0 \leq s \leq t$. Since $\mathbb{E}_x[A_{2t}] < \infty$, $x \in E_1$, we can apply again Lebesgue's dominated convergence to obtain for $x \in E_1$

$$\lim_{s \to 0} P_s \mathbb{E}. \left[\int_0^t f(\mathbf{X}_r) \, dA_r \right] (x) = \lim_{s \to 0} \mathbb{E}_x[F_s(t)] = \mathbb{E}_x[F_0(t)]$$

$$= \mathbb{E}_x \left[\int_0^t f(\mathbf{X}_r) \, dA_r \right].$$

Assume now that f is positive. From [FOT11, Lem. 5.1.3] we get for $\alpha > 0$,

$$\mathbb{E}_x \left[\int_0^\infty e^{-\alpha s} f(\mathbf{X}_s) dA_s \right] = \widetilde{U_\alpha f \, \nu}(x) \quad \text{for quasi-every } x \in E_1. \quad (6.11)$$

So $f \cdot A$ is in Revuz correspondence to $f \nu$ by [FOT11, Theo. 5.1.3(v)]. By Theorem 6.1.14 this is the only strict finite PCAF associated with $f \nu$ and hence by Theorem 6.1.11, (6.10) holds.

$\square$

6.2 Construction of the Local Time and the Martingale Problem for C_c^2-functions

In this and the next section we consider the gradient Dirichlet form (4.1) of Section 4.1 with coefficient matrix A and density ϱ. We assume Conditions

4.1.1, 4.1.2, 4.1.6 and 4.1.10. Recall that Γ_2 denotes an open (in $\partial\Omega$) subset of the C^2-smooth boundary part of $\partial\Omega$ and is complemented by a set of capacity zero. Define $E_1 = (\Omega \cup \Gamma_2) \cap \{\varrho > 0\}$. Note that as before we assume for the space dimension d that $d \geq 2$. Theorem 4.1.14 provides us with an $\mathcal{L}^p$-strong Feller process with p as in Condition 4.1.6. Let $\mathbf{M}^1 = (\mathbf{\Omega}^1, \mathcal{F}^1, (\mathcal{F}_t^1)_{t\geq 0}, (\mathbf{X}_t)_{t\geq 0}, (\mathbb{P}_x)_{x\in E_1\cup\{\Delta\}})$ be the restriction of $\mathbf{M}$ to $E_1 \cup \{\Delta\}$, see Definition 7.3.18.

Using the results of [FOT11, Ch. 5] and the refined construction theorem from the previous section, we construct a boundary local time at $\Gamma_2\cap\{\varrho > 0\}$. The local time is constructed as a strict PCAF on E_1 which grows only when the process is at $\Gamma_2 \cap \{\varrho > 0\}$, see Remark 6.2.5 below. Note that there might be several functionals that have these property. So the term local time does not refer to a specific functional. Nevertheless, we call the functional that we construct *the boundary local time*.

An important ingredient for the construction is a regularity result for certain potentials of the surface measure at compact boundary parts. We prove this using our regularity result from Chapter 3.

We use the local time as a building block for a Skorokhod decomposition of a sufficiently large class of functions. For our purpose the set $C_c^2(E_1)$ is large enough since we can locally approximate the coordinate functions $x^{(i)}$, $1 \leq i \leq d$, by functions in $C_c^2(E_1)$.

In [FOT11, Ch. 5] an extended semimartingale decomposition (in the meanwhile also called *Fukushima decomposition*) for functions in $D(\mathcal{E})$ is given. This decomposition is given in terms of additive functionals. They have properties that naturally generalize the properties of the corresponding objects in the classical semimartingale decomposition to the $\mathcal{E}$-quasi-everywhere setting in Dirichlet forms. More precisely, for the $\mathcal{E}$-quasi-continuous version $\widetilde{u}$ of $u \in D(\mathcal{E})$ it holds

$$\widetilde{u}(\mathbf{X}_t) - \widetilde{u}(\mathbf{X}_0) = N_t^{[u]} + M_t^{[u]}, \quad t \geq 0,$$

where $N^{[u]}$ and $M^{[u]}$ are finite CAFs (not necessarily strict) having certain properties for $\mathcal{E}$-quasi-every point. In particular, $M^{[u]}$ is a square-integrable martingale under $\mathbb{P}_x$ for $\mathcal{E}$-quasi-every starting point. Under additional assumptions on u these results are refined to pointwise statements there, in particular the martingale property holds for every point.

We apply [FOT11, Theo. 5.2.4] and [FOT11, Theo. 5.2.3] to identify $N^{[u]}$ and $M^{[u]}$, respectively, $u \in C_c^2(E_1)$. Using methods of [FOT11, Theo. 5.2.5] combined with an additional analysis we deduce a *pointwise* Skorokhod decomposition for $u \in C_c^2(E_1)$, formulated as a classical semimartingale decomposition.

The process $N^{[u]}$ contains an integral w.r.t. the deterministic time scale t and an integral w.r.t. the local time. The latter shows then the reflection at the boundary. Due to the singular drift terms we have to take special care of integrability issues, these are solved using the $\mathcal{L}^p$-strong Feller property of the resolvent, see e.g. Theorem 6.2.7 below.

In order to construct the local time at $\Gamma_2 \cap \{\varrho > 0\}$ we need a suitable generalized nest of compact sets. This nest will be also used later in the localization technique to prove existence of weak solutions. Define

$$U_n := \left\{\varrho > \frac{1}{n}\right\} \cap \left\{x \in \overline{\Omega} \,|\, \mathrm{dist}(x, \partial\Omega \setminus \Gamma_2) > \frac{1}{n}\right\} \cap B_n(0), n \in \mathbb{N}, \quad (6.12)$$

and $K_n := \overline{U_n}$. Then K_n is compact and $U_n \subset K_n \subset U_{n+1}$. Since Γ_2 is assumed to be open in $\partial\Omega$, we get

$$E_1 = \bigcup_{n \in \mathbb{N}} U_n = \bigcup_{n \in \mathbb{N}} K_n.$$

Since $(U_n)_{n \in \mathbb{N}}$ increases to E_1, we have

$$\inf_{n \in \mathbb{N}} \mathrm{cap}_{\mathcal{E}}(K \setminus U_n) = \mathrm{cap}_{\mathcal{E}}\left(K \setminus \bigcup_{n \in \mathbb{N}} U_n\right) = 0$$

for every compact set $K \subset \overline{\Omega}$ by Lemma 7.6.1(ii). Hence also $\lim_{n \to \infty} \mathrm{cap}_{\mathcal{E}}(K \setminus K_n) = 0$. So $(K_n)_{n \in \mathbb{N}}$ is a generalized compact nest and associated with μ since $\mu(K_n) < \infty$ for all $n \in \mathbb{N}$. By Theorem 2.3.10(iii) we get

$$\mathbb{P}_x\left(\lim_{n \to \infty} \tau_n \geq \mathcal{X}\right) = 1 \text{ for every } x \in E_1$$

with τ_n being the exit time of K_n, $n \in \mathbb{N}$.

The results of the last section give that for every measure in S_{00} we get a unique strict finite PCAF. We apply this result to construct strict finite PCAF corresponding to $\sigma_n := 1_{K_n} \sigma$, $n \in \mathbb{N}$. So we have to show that these measures are of finite energy and that the corresponding α-potential is essentially bounded on E_1. Our results yield in fact that the potential has a continuous bounded version on E_1. However, we do not need this additional regularity for our considerations.

Since we consider later also measures of the form $f 1_{K_N} \sigma$, $N \in \mathbb{N}$, with $f \in \mathcal{B}_b^+(E_1)$, we formulate a more general theorem.

Theorem 6.2.1. *Let $N \in \mathbb{N}$, $f \in \mathcal{B}_b^+(E_1)$. Then the measure $f \sigma_N (= f 1_{K_N} \sigma)$ is of finite energy. For $\alpha > 0$ the corresponding potential $U_\alpha f \sigma_N$ has a continuous bounded version $\widetilde{U_\alpha f \sigma_N}$ on E_1. Hence $f \sigma_N \in S_{00}$.*

Proof. Let $N \in \mathbb{N}$ and $f \in \mathcal{B}_b^+(E_1)$. By Lemma 4.2.2(ii) the restriction map $\iota : D(\mathcal{E}) \to H^{1,2}(U_{N+1} \cap \Omega)$, $v \mapsto v|_{U_{N+1} \cap \Omega}$ is well-defined and continuous. So for $v \in D(\mathcal{E}) \cap C_c(E)$, it holds $v \in H^{1,2}(U_{N+1} \cap \Omega)$ and $Tr(v) \in L^2(\Gamma^{(N)}, \sigma_N)$ with $\Gamma^{(N)} := \partial\Omega \cap K_N$ and $Tr : H^{1,2}(U_{N+1} \cap \Omega) \to L^2(\Gamma^{(N)}, \sigma_N)$ the trace operator, see Theorem 7.5.17. We have the estimate

$$\|Tr(v)\|_{L^2(\Gamma^{(N)}, \sigma_N)} \le K_1 \|v\|_{H^{1,2}(U_{N+1} \cap \Omega)} \le K_2 \|v\|_{\mathcal{E}_1}$$

for constants $K_1, K_2 < \infty$. Because $v \in C_c(E)$ it holds $Tr(v) = v$ σ-a.e., see also Theorem 7.5.17. Thus

$$\int_{\Gamma^{(N)}} |v| \, f d\sigma_N = \int_{\Gamma^{(N)}} |Tr(v)| \, f d\sigma_N \le \sigma(\Gamma^{(N)})^{1/2} \|f\|_{\sup} \|Tr(v)\|_{L^2(\Gamma^{(N)}, \sigma_N)}$$

$$\le K_2 \sigma(\Gamma^{(N)})^{1/2} \|f\|_{\sup} \|v\|_{\mathcal{E}_1}.$$

So $f \sigma_N$ is of finite energy with α-potential $U_\alpha f \sigma_N \in D(\mathcal{E})$ for $0 < \alpha < \infty$. Thus $U_\alpha f \sigma_N \in H^{1,2}(U_k \cap \Omega)$ for every $k \in \mathbb{N}$ by Lemma 4.2.2(ii).

Next we show that $U_\alpha f \sigma_N$ has a bounded continuous version on E_1. Fix $0 < \alpha < \infty$. Let $k \in \mathbb{N}$ with $k \ge N + 1$. Since $Tr : H^{1,q^*}(U_k \cap \Omega) \to L^{q^*}(\Gamma^{(N)}, \sigma_N)$ is continuous for $1 \le q^* < \infty$, the mapping $T : H^{1,q^*}(U_k \cap \Omega) \to \mathbb{R}$, $v \mapsto \int_{\Gamma^{(N)}} Tr(v) \, f d\sigma_N$ is a continuous linear functional. Choose $2 \le d < q < \infty$ such that $q' := \frac{q}{q-1} > 1$ and $q' < \infty$. We have

$$\alpha \int_{U_k} U_\alpha f \sigma_N \, v \, d\mu + \int_{U_k} (A\nabla U_\alpha f \sigma_N, \nabla v) \, d\mu = \mathcal{E}_\alpha(U_\alpha f \sigma_N, v)$$

$$= \int_{\Gamma^{(N)}} v f d\sigma_N = T(v) \quad \text{for all } v \in C_c^1(U_k).$$

Since $T \in (H^{1,q'}(U_k \cap \Omega))'$, there exist $g \in L^q(U_k, dx)$ and $(e_i)_{1 \le i \le d} \in (L^q)^d(U_k, dx)$ such that

$$\int_{\Gamma^{(N)}} Tr(v) f d\sigma_N = T(v)$$

$$= \int_{U_k} g \, v \, dx + \int_{U_k} \sum_{i=1}^d e_i \, \partial_i v \, dx \text{ for } v \in H^{1,q'}(U_k \cap \Omega).$$

Moreover, $\|g\|_{L^q(U_k,dx)} + \|e\|_{(L^q)^d(U_k,dx)} \leq \tilde{C}\|T\|_{(H^{1,q'}(U_k \cap \Omega))'}$ for some $\tilde{C} < \infty$, see [AD75, Theo. 3.8]. So $U_\alpha f \sigma_N$ solves

$$\alpha \int_{U_k} U_\alpha f \sigma_N \, v \, d\mu + \int_{U_k} (A\nabla U_\alpha f \sigma_N, \nabla v) \, d\mu$$

$$= \int_{U_k} g \, v \, dx + \int_{U_k} \sum_{i=1}^{d} e_i \, \partial_i v \, dx \quad \text{for all } v \in C_c^1(U_k).$$

Choose $0 < r < \infty$ in the following way: If $x \in U_k \cap \Omega$, choose r such that $B_r(x) \subset U_k \cap \Omega$. If $x \in U_k \cap \partial\Omega$, choose r such that $B_r(x) \cap \overline{\Omega} \subset U_k$ and $B_r(x) \cap \partial\Omega \subset \Gamma_2$. Note that in both cases $C_c^1(B_r(x) \cap \overline{\Omega})$ embeds into $C_c^1(U_k)$. Applying Theorem 3.1.1 with $p = q$ we get $0 < r' < r$ such that $U_\alpha f \sigma_N \in H^{1,q}(B_{r'}(x) \cap \Omega)$. Choosing $0 < r_1 < r'$ we get that $U_\alpha f \sigma_N$ has a continuous bounded version $\widetilde{U_\alpha f \sigma_N}$ on $B_{r_1}(x) \cap \overline{\Omega}$, compare the proof of Theorem 4.2.5(iii).

Since $K_n \subset U_{N+1 \vee n+1}$ and K_n is compact, $U_\alpha f \sigma_N$ has a bounded continuous version on every K_n, $n \in \mathbb{N}$. Thus there exists a continuous version $\widetilde{U_\alpha f \sigma_N}$ on E_1.

However, the function is only locally bounded. To prove the global boundedness we apply a weak maximum principle, see [FOT11, Lem. 2.2.4]. Choose $n > N$, let $M_n := \sup_{x \in K_n} |\widetilde{U_\alpha f \sigma_N}(x)| < \infty$. Since $\text{supp}[f \sigma_N] \subset \text{supp}[\sigma_N] \subset \Gamma^{(N)} \subset K_n$, it holds $|\widetilde{U_\alpha f \sigma_N}(x)| \leq M_n$ σ_N-a.e. By the weak maximum principle we get $\widetilde{U_\alpha \sigma_N} \leq M_n$ μ-a.e. hence by continuity everywhere on E_1. Thus $\widetilde{U_\alpha \sigma_N}$ is bounded on E_1 and $\|U_\alpha f \sigma_N\|_{L^\infty(E,\mu)} < \infty$. □

Applying Theorem 6.2.1 with $f = 1$ we get by Theorem 6.1.9 the following corollary.

Corollary 6.2.2. *For each $n \in \mathbb{N}$ there exists a unique strict finite PCAF corresponding to $1_{K_n} \sigma$. These we denote by $(\ell_t^n)_{t \geq 0}$, $n \in \mathbb{N}$.*

We apply the previous results to construct the local time at the boundary.

Theorem 6.2.3. *The restricted surface measure $1_{\Gamma_2} 1_{\{\varrho > 0\}} \sigma$ is smooth in the strict sense. There exists a corresponding strict PCAF denoted by $(\ell_t)_{t \geq 0}$ and called the local time (at $1_{\Gamma_2} 1_{\{\varrho > 0\}}$). For the additivity set Λ it holds $\Lambda \subset \Lambda_n$, $n \in \mathbb{N}$, Λ_n being the additivity sets of $(\ell_t^n)_{t \geq 0}$ from Corollary 6.2.2.*

Let $f \in \mathcal{B}_b^+([0,\infty) \times E_1)$ with $supp[f] \subset [0,\infty) \times K_m$ for some $m \in \mathbb{N}$. Then it holds

$$\int_0^\infty f(s, \mathbf{X}_s(\omega)) d\ell_s(\omega) = \int_0^\infty f(s, \mathbf{X}_s(\omega)) d\ell_s^m(\omega) \quad \text{for all } \omega \in \Lambda$$

and

$$\int_0^t f(s, \mathbf{X}_s(\omega)) d\ell_s(\omega) = \int_0^t f(s, \mathbf{X}_s(\omega)) d\ell_s^m(\omega) \quad \text{for all } \omega \in \Lambda.$$

Proof. Set $\nu := 1_{\Gamma_2} 1_{\{\varrho > 0\}} \sigma$. Since $\nu(K_n) = \sigma_n(K_n) < \infty$, we have that $(K_n)_{n \in \mathbb{N}}$ is a generalized nest associated with ν. Furthermore, $1_{K_n} \nu = \sigma_n \in S_{00}$ for all $n \in \mathbb{N}$ by Corollary 6.2.2. It is left to check that ν is smooth. Let $A \subset E$ with $cap_{\mathcal{E}}(A) = 0$. We have

$$\nu(A) = \nu(A \cap \{\varrho > 0\} \cap \Gamma_2) = \sup_{n \in \mathbb{N}} \nu(K_n \cap A) = \sup_{n \in \mathbb{N}} \sigma_n(A) = 0$$

since σ_n is smooth for all $n \in \mathbb{N}$.

So we can apply [FOT11, Theo. 5.1.7(i)] to construct a corresponding strict PCAF $(\ell_t)_{t \geq 0}$ with additivity sets $\Lambda \in \mathcal{F}^1$. The functional is constructed using the local times $(\ell^n)_{n \in \mathbb{N}}$ at compact boundary parts from Corollary 6.2.2, see the proof of [FOT11, Theo. 5.1.7(i)].

From the construction it follows $\Lambda \subset \Lambda_n$, $n \in \mathbb{N}$. Moreover, Λ is chosen such that for $\omega \in \Lambda$ and $t \geq 0$

$$\ell_t^m(\omega) = (1_{K_m} \cdot \ell^n)_t(\omega) \quad \text{for } n > m.$$

The definition of $(\ell_t)_{t \geq 0}$ therein implies

$$\ell_t(\omega) = \begin{cases} \ell_t^n(\omega), & \tau_{n-1}(\omega) < t \leq \tau_n(\omega), \quad n \in \mathbb{N} \\ \ell_{\mathcal{X}^-}(\omega), & t \geq \mathcal{X}(\omega), \end{cases}$$

with τ_n the exit times of K_n, $n \in \mathbb{N}$, and $\tau_0 := 0$. So let $f \in \mathcal{B}_b^+([0,\infty) \times E_1)$ with $supp[f] \subset [0,\infty) \times K_m$ for some $m \in \mathbb{N}$. For $n \geq m$ and $\omega \in \Lambda$, we have

$$\int_0^{\tau_n} f(s, \mathbf{X}_s(\omega)) \, d\ell_s(\omega) = \int_0^{\tau_n} f(s, \mathbf{X}_s(\omega)) \, d\ell_s^n(\omega)$$

$$= \int_0^{\tau_n} 1_{K_m}(\mathbf{X}_s(\omega)) f(s, \mathbf{X}_s(\omega)) \, d\ell_s^n(\omega) = \int_0^{\tau_n} f(s, \mathbf{X}_s(\omega)) \, d\ell_s^m(\omega).$$

Here we used the associative law for the Lebesgue-Stieltjes integral, see Lemma 7.7.3(i).

Letting n tend to ∞ we get the claim with upper bound $\mathcal{X}$. Since both ℓ and ℓ^m are constant for $t \geq \mathcal{X}$ the upper bound can be replaced by ∞. The claim with upper bound t follows just by multiplying f with $1_{[0,t]}$. $\square$

Remark 6.2.4. Note that the only possibility for $l_t(\omega)$ being infinite is that $t \geq \mathcal{X}(\omega)$.

Remark 6.2.5. We can conclude from the construction of $(\ell_t)_{t \geq 0}$ that the functional grows only when $\mathbf{X}_t$, $t \geq 0$, is at the boundary part $\Gamma_2 \cap \{\varrho > 0\}$. Indeed, let $n \in \mathbb{N}$. Then by Theorem 6.2.3 we get for $t \geq 0$

$$\int_0^t 1_{\Gamma_2 \cap \{\varrho > 0\}}(\mathbf{X}_s) 1_{K_n}(\mathbf{X}_s)\, d\ell_s = \int_0^t 1_{\Gamma_2 \cap \{\varrho > 0\}}(\mathbf{X}_s) d\ell_s^n = (1_{\Gamma_2 \cap \{\varrho > 0\}} \cdot \ell^n)_t.$$

By (6.10) in Lemma 6.1.15 we get that $(1_{\Gamma_2 \cap \{\varrho > 0\}} \cdot \ell^n)$ is in Revuz correspondence to $1_{\Gamma_2 \cap \{\varrho > 0\}} 1_{K_n} \sigma = 1_{K_n} \sigma$. Thus $(1_{\Gamma_2 \cap \{\varrho > 0\}} \cdot \ell^n) = \ell^n$.

Altogether, we get for $t \geq 0$

$$\int_0^t 1_{\Gamma_2 \cap \{\varrho > 0\}}(\mathbf{X}_s) 1_{K_n}(\mathbf{X}_s)\, d\ell_s = \ell_t^n.$$

Letting n tend to ∞ we obtain for $t \geq 0$

$$\int_0^t 1_{\Gamma_2 \cap \{\varrho > 0\}}(\mathbf{X}_s)\, d\ell_s = \ell_t.$$

Thus for $t \geq 0$

$$\int_0^t 1_{\Omega \cup (\partial\Omega \setminus \Gamma_2 \cap \{\varrho > 0\})}(\mathbf{X}_s)\, d\ell_s = 0.$$

To discuss the martingale solution property we also need to consider integration of functions on paths of the process with respect to the deterministic time. Here we can allow certain singularities. First note that μ is in Revuz correspondence to the additive functional $(\omega, t) \mapsto t$. Indeed, we have $U_t^\alpha f = R_\alpha f$ for every $f \in \mathcal{B}_b^+(E)$, $\alpha > 0$. Since R_α is symmetric, we get by (6.1) the Revuz correspondence.

We introduce the notion of bounded variation.

Definition 6.2.6. Let $g : \mathbb{R}^+ \to \mathbb{R}$ be a function. We say that g is of *bounded variation* up to time T, $0 \leq T \leq \infty$ if

$$\sup_{n \in \mathbb{N}} \left\{ \sum_{i=0}^{n-1} |g(t_{i+1}) - g(t_i)| \;\middle|\; 0 = t_0 \leq t_1 \leq \ldots \leq t_n = T \right\} < \infty.$$

Compare [Kle06, Def. 21.52].

Let $(\Omega, \mathcal{F}, \mathbf{Q})$ be a probability space. Let $\mathbf{G} := (\mathbf{G}_t)_{t \geq 0}$ be an $\mathbb{R}$-valued stochastic process defined on Ω. We say that $\mathbf{G}$ is *locally of bounded variation* if for every $0 \leq T < \infty$ it holds that the function $[0, \infty) \ni t \mapsto \mathbf{G}_t(\omega) \in \mathbb{R}$ is of bounded variation up to T for $\mathbf{Q}$-a.e. $\omega \in \Omega$.

Let $(\mathcal{F}_t)_{t \geq 0}$ be an filtration and τ be an $\mathcal{F}_t$-stopping time. We say that $\mathbf{G}$ is *locally of bounded variation up to* τ if there exists a sequence of $\mathcal{F}_t$-stopping times $(\tau_n)_{n \in \mathbb{N}}$ with $\tau_n \uparrow \tau$ such that $\mathbf{G}_{\cdot \wedge \tau_n}$ is locally of bounded variation for $n \in \mathbb{N}$.

Theorem 6.2.7. *(i) Let $f \in L^p(E, \mu)$. Define $A := (f \cdot t) := (f \cdot t)_{t \geq 0}$ by*

$$A_t := (f \cdot t)_t := \int_0^t f(\mathbf{X}_s) ds, \tag{6.13}$$

in the sense of Definition 7.7.5. Then $f \cdot t$ is a strict finite CAF on E_1. Furthermore, $\mathbb{E}_x[|A_t|] < \infty$ for every $0 \leq t < \infty$, $x \in E_1$, and $f \cdot t$ is locally of bounded variation.

If f is positive, $(f \cdot t)$ is in Revuz correspondence to $f \mu$.

(ii) Let $f \in L^p(E, \mu)$, $supp[f] \subset\subset U_n$ for one $n \in \mathbb{N}$, U_n as in (6.12). Then

$$\mathbb{E}_x\left[(f \cdot t)_t^2\right] < \infty \quad \text{for } x \in E_1.$$

(iii) If $f \in L_{loc}^p(E_1, \mu)$, then there exists a local strict CAF $A := (A_t)_{t \geq 0}$ that is $\mathbb{P}_x$-a.s. equal to the integral in (6.13) for $t < \mathcal{X}$. Moreover, A is locally of bounded variation up to $\mathcal{X}$.

Proof. (i): We have for $x \in E_1$

$$\mathbb{E}_x\left[\int_0^T |f|(\mathbf{X}_r) dr\right] \leq e^T \mathbb{E}_x\left[\int_0^\infty e^{-r} |f|(\mathbf{X}_r) dr\right] = e^T R_1 |f|\,(x) < \infty.$$

So $\mathbb{E}_x[\int_0^T |f|(\mathbf{X}_r) dr] < \infty$ and $\int_0^T |f|(\mathbf{X}_r) dr < \infty$ $\mathbb{P}_x$-a.s. for $x \in E_1$ and $0 \leq T < \infty$. This holds for every $0 \leq T < \infty$. So f is locally t-integrable in the sense of Definition 7.7.5. Hence by Proposition 7.7.6 $(f \cdot t)$ defines a strict finite CAF on E_1 with additivity set

$$\Lambda := \left\{ \omega \in \Omega^1 \mid \int_0^N |f|(\mathbf{X}_r(\omega))dr < \infty \text{ for all } N \in \mathbb{N} \right\}.$$

For every partition $0 = t_0 \leq t_1 \leq ... \leq t_n = t < \infty$, $n \in \mathbb{N}$ we have

$$\sum_{i=0}^{n-1} \left| \int_{t_i}^{t_{i+1}} f(\mathbf{X}_s)\, ds \right| \leq \int_0^t |f(X_s)|\, ds < \infty \quad \mathbb{P}_x - \text{a.s. for } x \in E_1.$$

So $f \cdot t$ is locally of bounded variation.

Assume that f is positive. Let $g \in \mathcal{B}^+(E_1)$. We have

$$U_{f \cdot t}^\alpha g\,(x) = \mathbb{E}_x \left[\int_0^\infty \exp(-\alpha s)\, g(\mathbf{X}_s)\, d(f \cdot t)_s \right]$$

$$= \mathbb{E}_x \left[\int_0^\infty \exp(-\alpha s)\, (fg)(\mathbf{X}_s)\, ds \right] = R_\alpha f g\,(x) \quad \text{for } \alpha > 0 \text{ and } x \in E_1.$$

Since $(R_\alpha)_{\alpha>0}$ is symmetric, we have for all $h \in \mathcal{B}_b^+(E_1)$ and $\alpha > 0$:

$$\int_{E_1} h\,(U_{f \cdot t}^\alpha g)\, d\mu = \int_{E_1} h\, R_\alpha f g\, d\mu = \int_{E_1} R_\alpha h\, g f\, d\mu.$$

So by (6.1) we get that $(f \cdot t)$ is in Revuz correspondence to $f\mu$.

(ii): With a similar calculation as in [FOT11, p. 245] we get for $0 \leq t < \infty$ and $x \in E_1$

$$\mathbb{E}_x \left[(f \cdot t)_t^2 \right] \leq 2e^t \mathbb{E}_x \left[\int_0^t |f|(\mathbf{X}_s)\, R_1 |f|(\mathbf{X}_s)\, ds \right].$$

Set $h(x) := |f| R_1 |f|\,(x)$. Since $f \in L^p(E, \mu)$ and $R_1|f|$ is bounded on the support of f, we have that $h \in L^p(E, \mu)$. So as above $\mathbb{E}_x[\int_0^t h(\mathbf{X}_s)ds] < \infty$ for $0 \leq t < \infty$ and $x \in E_1$.

(iii): Now assume that $f \in L_{\text{loc}}^p(E_1, \mu)$. Set $f_n = 1_{K_n} f$, $n \in \mathbb{N}$, K_n as in (6.12). Define A^{f_n} to be the corresponding additive functional from (6.13). Since f_n, $n \in \mathbb{N}$, has compact support in K_n, we get that A^{f_n} is a continuous additive functional with additivity set Λ_{f_n}. Let $\Lambda := \bigcap_{n \in \mathbb{N}} \Lambda_{f_n} \cap \{\omega \in \Omega^1 \mid \lim_{n \to \infty} \tau_n \geq \mathcal{X}\}$, τ_n the exit time of K_n, $n \in \mathbb{N}$. Define

$$A_t := 1_\Lambda 1_{\{t < \mathcal{X}\}} \lim_{n \to \infty} A_t^{f_n}, t \geq 0.$$

Let $t > 0$ and $\omega \in \Lambda \cap \{t < \mathcal{X}\}$. There exists $n_0 \in \mathbb{N}$ such that $t < \tau_{n_0} < \mathcal{X}$. Then

$$\lim_{n \to \infty} A_t^{f_n}(\omega) = A_t^{f_{n_0}}(\omega).$$

So $(A_t)_{t \geq 0}$ is well-defined and $\mathcal{F}_t^1$-adapted.

Since $A^f(\omega)$ equals $A^{f_{n_0}}(\omega)$ on $[0, t]$ for $t < \tau_{n_0}$, it is therefore continuous and additive on $[0, t]$. So A^f is a local strict CAF. Furthermore, $(f \cdot t)_{.\wedge \tau_n} = (f_n \cdot t)_{.\wedge \tau_n}$, $n \in \mathbb{N}$. So $(f \cdot t)_t$ is locally of bounded variation up to $\mathcal{X}$. $\square$

Lemma 6.2.8. *Let g be $\mathcal{B}(\Gamma_2)$-measurable and bounded, $\mathrm{supp}[g] \subset\subset U_n$, for one $n \in \mathbb{N}$. Then $g \cdot \ell$, defined by*

$$(g \cdot \ell)_t := \int_0^t g(\mathbf{X}_s)d\ell_s, \quad t \geq 0,$$

is a strict finite CAF with additivity set of ℓ and it holds

$$\mathbb{E}_x\left[|(g \cdot \ell)_t|\right] < \infty \text{ and } \mathbb{E}_x\left[(g \cdot \ell)_t^2\right] < \infty \quad \text{for } 0 \leq t < \infty \text{ and } x \in E_1.$$
(6.14)

Furthermore, $g \cdot \ell$ is locally of bounded variation. Assume that $g \in \mathcal{B}(\Gamma_2)$ is only locally bounded. Then $g \cdot \ell$ is a local strict CAF and locally of bounded variation up to $\mathcal{X}$.

Proof. First extend g to a function in $\mathcal{B}(E)$ in the trivial way, i.e., replace g by $1_{\Gamma_2}g$. From Theorem 6.2.3 we get $g \cdot \ell = g \cdot \ell^n$. So by Lemma 6.1.15 we get that $g \cdot \ell$ is a strict finite PCAF on E_1 and (6.14) holds. That $g \cdot \ell$ is locally of bounded variation, follows similarly as in the proof of Lemma 6.2.7. The statements for g being only locally bounded follow now with the same localizing procedure as in the proof of Theorem 6.2.7(iii). $\square$

Let us introduce two classes of functionals, according to [FOT11] but refined to pointwise properties. The Skorokhod decomposition is formulated in terms of these classes. Define

$$\mathcal{M}_c := \left\{ M : \Omega^1 \to \mathbb{R} \,\middle|\, M \text{ is a strict finite CAF}, \mathbb{E}_x[M_t^2] < \infty, \right.$$

$$\left. \mathbb{E}_x[M_t] = 0 \text{ for every } t \geq 0 \text{ and } x \in E_1 \right\}$$

and

$$\mathcal{N}_c := \left\{ N : \Omega^1 \to \mathbb{R} \; \middle| \; N \text{ is a strict finite CAF, } e(N) = 0, \right.$$

$$\left. \mathbb{E}_x[|N_t|] < \infty \text{ for every } t \geq 0 \text{ and } x \in E_1 \right\} \quad (6.15)$$

with

$$e(N) := \lim_{t \downarrow 0} \frac{1}{2t} \mathbb{E}_\mu[N_t^2].$$

The term $e(N)$ is called the *energy* of N. Note that the properties required in $\mathcal{M}_c$ and $\mathcal{N}_c$ are pointwise properties except for the zero energy requirement. If $M \in \mathcal{M}_c$, then additivity together with $\mathbb{E}_x[M_t] = 0$ imply that M is a martingale under $\mathbb{P}_x$ for every $x \in E_1$. Recall the definition of the operator $\hat{L}$ on $\mathcal{D}_{\mathrm{Neu}}$. We may extend this definition to all functions $u \in C_c^2(E_1)$ and define

$$\hat{L}u = \sum_{i,j=1}^d a_{ij} \partial_i \partial_j u + \sum_{j=1}^d \left(\sum_{i=1}^d \partial_i a_{ij} + \sum_{i=1}^d \frac{1}{\varrho} a_{ji} \partial_i \varrho \right) \partial_j u. \quad (6.16)$$

We obtain the following theorem using [FOT11, Theo. 5.2.4] and [FOT11, Theo. 5.2.5].

Theorem 6.2.9. *Let* $u \in C_c^2(E_1)$. *Let* $N^{[u]} := (N_t^{[u]})_{t \geq 0}$ *with*

$$N_t^{[u]} := \int_0^t \hat{L}u\,(\mathbf{X}_s)\,ds - \int_0^t (A\nabla u, \eta) \varrho\,(\mathbf{X}_s) d\ell_s, \quad t \geq 0.$$

Then $N^{[u]} \in \mathcal{N}_c$ *and is locally of bounded variation. Define* $M^{[u]} := (M_t^{[u]})_{t \geq 0}$ *with*

$$M_t^{[u]} := u(\mathbf{X}_t) - u(\mathbf{X}_0) - N_t^{[u]}.$$

Then $M^{[u]} \in \mathcal{M}_c$, *in particular it is an square-integrable* $\mathcal{F}_t$-*martingale starting at zero.*

The integrals are defined in the sense of Theorem 6.2.7 and Lemma 6.2.8. As additivity set Λ for $M^{[u]}$ we take the intersection of the additivity sets of $\widetilde{L}u \cdot t$ and ℓ. Due to Theorem 6.2.7 and Lemma 6.2.8 $M^{[u]}$ is additive on this set and $\mathbb{P}_x(\Lambda) = 1$ for every $x \in E_1$.

Proof. Let $u \in C_c^2(E_1)$. From Lemma 6.2.8 and Theorem 6.2.7 together with the calculations on [FOT11, p. 244] we have $N^{[u]} \in \mathcal{N}_c$ and $N^{[u]}$ is locally of bounded variation.

Choose K_N such that $\text{supp}[u] \subset\subset U_N \subset K_N$. Let $v \in D(\mathcal{E})$. Then $v \in H^{1,2}(U_N \cap \Omega)$. Since the support of u has positive distance to the non-smooth boundary part of $\partial\Omega$, we can apply the divergence theorem, Theorem 7.5.18, to obtain

$$\mathcal{E}(u,v) = \int_\Omega (A\nabla u, \nabla v)d\mu = -\int_\Omega \hat{L}u\, v\, d\mu + \int_{\partial\Omega} Tr(v)(A\nabla u, \eta)\varrho d\sigma.$$

Set either $g := (A\nabla u, \eta)^+ \varrho$ or $g := (A\nabla u, \eta)^- \varrho$.

Note that $g \cdot \ell = g \cdot \ell^N$ is in Revuz correspondence to $g\sigma$. Let $h \in L^1(E,\mu) \cap \mathcal{B}_b^+(E)$, set $v := R_1 h \in D(\mathcal{E}) \cap C^0(E_1)$. By [FOT11, Theo. 5.1.3(vi)] we get,

$$\lim_{t\to 0} \frac{1}{t}\mathbb{E}_{v\mu}\left[(g \cdot \ell)_t\right] = \lim_{t\to 0} \frac{1}{t}\mathbb{E}_{v\mu}\left[(g \cdot \ell^N)_t\right] = \int_{\partial\Omega} vg1_{K_N} d\sigma$$

$$= \int_{\partial\Omega} vg d\sigma = \int_{\partial\Omega} Tr(v)\, g\, d\sigma.$$

Here $\mathbb{E}_{v\mu}[\,\cdot\,] := \int_E \mathbb{E}_x[\,\cdot\,]v(x)\,d\mu(x)$.

By Theorem 6.2.7 we have that $(\hat{L}u \cdot t)_t = \int_0^t \hat{L}u\,(\mathbf{X}_s)\,ds$, $t \geq 0$, is a strict finite CAF. Moreover, $(\hat{L}u)^{+/-} \cdot t$ is in Revuz correspondence to $(\hat{L}u)^{+/-}\mu$. So we get

$$\lim_{t\to 0} \frac{1}{t}\mathbb{E}_{v\mu}\left[(\hat{L}u \cdot t)_t - ((A\nabla u, \eta)\varrho \cdot \ell_t)_t\right]$$

$$= \int_\Omega \hat{L}u\, v\, d\mu - \int_{\partial\Omega} Tr(v)(A\nabla u, \eta)\varrho d\sigma = -\mathcal{E}(u,v).$$

Thus from [FOT11, Theo. 5.2.4] we obtain

$$\mathbb{E}_x\left[N_t^{[u]}\right] = P_t u(x) - u(x) \quad \text{for } \mu\text{-a.e. } x \in \overline{\Omega}. \tag{6.17}$$

Using the absolute continuity of $(P_t)_{t>0}$ on E_1 we get

$$P_s \mathbb{E}.\left[N_t^{[u]}\right](x) = P_s(P_t u - u)(x) \quad \text{for every } x \in E_1.$$

The right-hand side converges to $P_t u(x) - u(x)$, as $s \to 0$, for every $x \in E_1$. This follows since $P_t u - u$ is a continuous bounded function on E_1 and the paths of $\mathbf{M}^1$ are right-continuous at zero. Recall the splitting $\hat{L}u \cdot t = (\hat{L}u)^+ \cdot t - (\hat{L}u)^- \cdot t$, the analogous property holds for $(A\nabla u, \eta)\varrho \cdot \ell$. Applying (6.9) in Lemma 6.1.15 with $f = 1$ and $A = (\hat{L}u)^{+/-}$ or $(A\nabla u, \eta)^{+/-}\varrho \cdot \ell$ we get convergence of the left-hand side. So altogether, we get that (6.17) holds for every $x \in E_1$.

Using the Markov property of $\mathbf{M}^1$ we get from this that $M^{[u]}$ is a martingale starting at zero. Since u is bounded and $\mathbb{E}_x[(\hat{L} \cdot t)_t^2]$, $\mathbb{E}_x[(g \cdot \ell)_t^2] < \infty$ for $0 \leq t < \infty$ and $x \in E_1$, we have that $M^{[u]}$ is square-integrable for every $x \in E_1$.   □

Theorem 6.2.9 yields that $M_t^{[u]} := u(\mathbf{X}_t) - u(\mathbf{X}_0) - N_t^{[u]}$ is a continuous square-integrable martingale. Hence $M^{[u]} \in \mathcal{M}_c$. Next we further analyze this martingale by considering the quadratic variation process. We introduce the notion of local martingales, see [Kle06, Def. 21.66].

Definition 6.2.10. Let $(\Omega, \mathcal{F}, \mathbf{Q})$ be a probability space with filtration $(\mathcal{F}_t)_{t \geq 0}$. We say that a stochastic process $M = (M_t)_{t \geq 0}$ is a *continuous local martingale* up to a $\mathcal{F}_t$-stopping time τ if there exists a sequence of $\mathcal{F}_t$-stopping times $(\tau_n)_{n \in \mathbb{N}}$ with $\tau_n \uparrow \tau$, $\tau_n < \tau$ such that for every $n \in \mathbb{N}$ the stopped process $M^{\tau_n} = (M_{t \wedge \tau_n})_{t \geq 0}$ is a continuous $\mathcal{F}_t$-martingale. We say that $(\tau_n)_{n \in \mathbb{N}}$ *reduces* M.

By a proper modification of the proof of [Kle06, Satz 21.70] we obtain the following theorem.

Theorem 6.2.11. *Let $(M_t)_{t \geq 0}$ be a local martingale up to a stopping time τ starting at zero that is continuous in $[0, \tau)$. Then there exists an adapted process $\langle M \rangle = (\langle M \rangle_t)_{t \geq 0}$, unique up to time τ, the quadratic variation process, with the following properties.*

(i) $\langle M \rangle_0 = 0$ and $\langle M \rangle$ is increasing.

(ii) $\langle M \rangle$ is continuous in $[0, \tau)$.

(iii) $(M_t^2 - \langle M \rangle_t)_{t \geq 0}$ is a local martingale up to time τ.

If $(M_t)_{t \geq 0}$ is square-integrable and continuous in $[0, \infty)$, then the process $(M_t^2 - \langle M \rangle_t)_{t \geq 0}$ is a martingale.

So for $u \in C_c^2(E_1)$, we get for $M^{[u]}$ an associated quadratic variation process $\langle M^{[u]} \rangle$. Note that $\langle M^{[u]} \rangle$, obtained from Theorem 6.2.11, is constructed for each $x \in E_1$ separately since we consider the measurable space endowed with the different probability measures $\mathbb{P}_x$, $x \in E_1$. Following [FOT11,

Theo. A.3.17], however, we can construct from this a process $\langle M^{[u]} \rangle$ that is a strict additive functional with common additivity set for all $x \in E_1$.

From [FOT11, Theo. 5.2.3] we get that $\langle M^{[u]} \rangle$ has as Revuz measure the energy measure of u. From this we get an explicit representation for $\langle M^{[u]} \rangle$.

Theorem 6.2.12. *Let $u \in C_c^2(E_1)$. Then $\langle M^{[u]} \rangle = 2(A\nabla u, \nabla u) \cdot t$, i.e.,*

$$M_t^{[u]^2} - 2 \int_0^t (A\nabla u, \nabla u)(\mathbf{X}_s)ds \quad \text{is an } \mathcal{F}_t^1\text{-martingale under } \mathbb{P}_x \text{ for } x \in E_1.$$

Proof. From Theorem 6.2.9 we get that $M^{[u]}$ is a continuous square-integrable martingale and a strict finite CAF which is strict on E_1. Invoking Theorem 6.2.11 and Theorem [FOT11, Theo. A.3.17] we get an associated quadratic variation process $\langle M^{[u]} \rangle$ which is a strict PCAF as well.

Since $u \in C_c^2(E_1) \subset \mathcal{D}$ (domain of the pre-Dirichlet form, see (4.1)), we find that for the energy measure $\mu_{\langle u \rangle}$, it holds $\mu_{\langle u \rangle} = 2(A\nabla u, \nabla u)\mu$, see e.g. [FOT11, p. 254]. By Theorem 6.2.7 the strict PCAF $2(A\nabla u, \nabla u) \cdot t$ is associated with $\mu_{\langle u \rangle}$.

The calculation in [FOT11, Theo. 5.2.3] yields that the Revuz measure of $\langle M^{[u]} \rangle$ is $\mu_{\langle u \rangle}$. But the unique strict PCAF associated with this measure is $2(A\nabla u, \nabla u) \cdot t$. Thus by Theorem 6.1.14 we find a set $\widetilde{\Lambda} \subset \mathbf{\Omega}^1$ with $2((A\nabla u, \nabla u) \cdot t)(\omega) = \langle M^{[u]} \rangle(\omega)$ for all $\omega \in \widetilde{\Lambda}$ and $\mathbb{P}_x(\widetilde{\Lambda}) = 1$ for all $x \in E_1$. $\qquad\square$

So we get for $u \in \mathcal{D}_1$ the *Skorokhod decomposition*

$$u(\mathbf{X}_t) - u(\mathbf{X}_0) = N_t^{[u]} + M_t^{[u]} \text{ for } t \geq 0 \tag{6.18}$$

with $N^{[u]}$, $M^{[u]}$ as in Theorem 6.2.9 and $\langle M^{[u]} \rangle$ as in Theorem 6.2.12. In particular, $(u(\mathbf{X}_t))_{t \geq 0}$ is a semimartingale, see Definition 6.3.1 below.

In order to study the behavior of the process $(\mathbf{X}_t)_{t \geq 0}$ we need also information of the joint behavior of $(u_1(\mathbf{X}_t))_{t \geq 0}$ and $(u_2(\mathbf{X}_t))_{t \geq 0}$ for $u_1, u_2 \in C_c^2(E_1)$. We state the following linearity result which can be easily proven.

Lemma 6.2.13. *Let $u_1, u_2 \in C_c^2(E_1)$, $\alpha, \beta \in \mathbb{R}$. Then $N^{[\alpha u_1 + \beta u_2]} = \alpha N^{[u_1]} + \beta N^{[u_2]}$ and $M^{[\alpha u_1 + \beta u_2]} = \alpha M^{[u_1]} + \beta M^{[u_2]}$. More precisely,*

$$M_t^{[\alpha u_1 + \beta u_2]} = \alpha M_t^{[u_1]} + \beta M_t^{[u_2]} \quad \text{for all } t \geq 0 \; \mathbb{P}_x - a.s., \; x \in E_1,$$

the analogous equality holds for $N^{[u_1]}$ and $N^{[u_2]}$.

Using this linearity, we can discuss the quadratic *covariation process*. For two martingales $M^{(1)}$ and $M^{(2)}$ define

$$\langle M^{(1)}, M^{(2)} \rangle := \frac{1}{2} \left\{ \langle M^{(1)} + M^{(2)} \rangle - \langle M^{(1)} \rangle - \langle M^{(2)} \rangle \right\}.$$

Observe that $\langle \alpha M^{(1)} \rangle = \alpha^2 \langle M^{(1)} \rangle$ for $\alpha \in \mathbb{R}$. Thus $\langle M^{(1)}, M^{(1)} \rangle = \langle M^{(1)} \rangle$. Using the previous results we can easily identify the covariation process.

Lemma 6.2.14. *Let* $u_1, u_2 \in C_c^2(E_1)$. *Then*

$$\langle M^{[u_1]}, M^{[u_2]} \rangle = 2((A \nabla u_1, \nabla u_2) \cdot t),$$

i.e.,

$$\langle M^{[u_1]}, M^{[u_2]} \rangle_t = 2 \int_0^t (A \nabla u_1, \nabla u_2)(X_s) \, ds, \quad t \geq 0, \ \mathbb{P}_x - a.s., \ \text{for } x \in E_1.$$

Proof. From Lemma 6.2.13 we get $M^{[u_1+u_2]} = M^{[u_1]} + M^{[u_2]}$. So together with Theorem 6.2.12 we get

$$\langle M^{[u_1]}, M^{[u_2]} \rangle_t = \frac{1}{2} \left(\langle M^{[u_1+u_2]} \rangle_t - \langle M^{[u_1]} \rangle_t - \langle M^{[u_2]} \rangle_t \right)$$

$$= \int_0^t (A \nabla(u_1 + u_2), \nabla(u_1 + u_2))(\mathbf{X}_s) \, ds - \int_0^t (A \nabla u_1, \nabla u_1)(\mathbf{X}_s) \, ds$$

$$- \int_0^t (A \nabla u_2, \nabla u_2)(\mathbf{X}_s) \, ds$$

$$= 2 \int_0^t (A \nabla u_1, \nabla u_2)(\mathbf{X}_s) \, ds \quad \text{for all } t \geq 0 \ \mathbb{P}_x - a.s., \ x \in E_1.$$

$\square$

6.3 Semimartingale Structure and Weak Solutions

Recall that we consider the $\mathcal{L}^p$-strong Feller process $\mathbf{M}^1$ and assume Conditions 4.1.1, 4.1.2, 4.1.6 and 4.1.10, see the beginning of Section 6.2. In this section we study the coordinates of the process $\mathbf{X}_t = (\mathbf{X}_t^{(1)}, ..., \mathbf{X}_t^{(d)})$, rather than functions of $(\mathbf{X}_t)_{t\geq 0}$. We show that $(\mathbf{X}_t)_{t\geq 0}$ is a semimartingale up to the lifetime $\mathcal{X}$. Then we prove that the process yields a weak solution to an SDE with reflection. So we recall the definition of semimartingales first.

Definition 6.3.1. Let $(\Omega, \mathcal{F}, \mathbf{Q})$ be a probability space with filtration $(\mathcal{F}_t)_{t \geq 0}$. Let $(\mathbf{X}_t)_{t \geq 0}$ be a stochastic process. Let τ be a stopping time. We say that $(\mathbf{X}_t)_{t \geq 0}$ is a *continuous semimartingale* up to τ if there exists $\mathcal{F}_t$-adapted processes $(M_t)_{t \geq 0}$ and $(N_t)_{t \geq 0}$, continuous in $[0, \tau)$, such that $(M_t)_{t \geq 0}$ is a local $\mathcal{F}_t$-martingale up to τ and $(N_t)_{t \geq 0}$ is locally of bounded variation up to τ and

$$\mathbf{X}_{t \wedge \tau} = \mathbf{X}_0 + N_{t \wedge \tau} + M_{t \wedge \tau} \quad \text{for } t \geq 0 \; \mathbf{Q} - \text{a.s.}$$

In order to apply our previous results, we have to transfer the properties from $u(\mathbf{X}_t)$, $u \in C_c^2(E_1)$, to $\mathbf{X}_t^{(i)}$, $1 \leq i \leq d$. This is done using localization arguments.

For $1 \leq i \leq d$ define $N^{(i)}$ by

$$N_t^{(i)} = \int_0^t b_i(\mathbf{X}_s) \, ds - \int_0^t \left((e_i, A\eta)_{\varrho} \right)(\mathbf{X}_s) \, d\ell_s, \quad t \geq 0, \tag{6.19}$$

where

$$b_i(x) := \sum_{j=1}^d \partial_j a_{ij}(x) + \sum_{j=1}^d \left(\frac{1}{\varrho} a_{ij} \partial_j \varrho \right)(x), \quad x \in E_1, 1 \leq i \leq d, \tag{6.20}$$

are the first-order coefficients of $\hat{L}$ from (6.16). Define $M^{(i)}$, $1 \leq i \leq d$, by

$$M_t^{(i)} := \mathbf{X}_t^{(i)} - \mathbf{X}_0^{(i)} - N_t^{(i)}, \quad t \geq 0. \tag{6.21}$$

Recall that b_i is locally $L^p(E_1, \mu)$-integrable and $(e_i, A\eta)_{\varrho}$ is locally bounded for $1 \leq i \leq d$. By e_i, $1 \leq i \leq d$, we denote the i-th unit vector.

Denote by Λ the additivity set of the boundary local time from Theorem 6.2.3. According to Theorem 6.2.7 and Lemma 6.2.8 all $N^{(i)}$ and $M^{(i)}$, $1 \leq i \leq d$, form strict local CAF and we can find a common additivity set $\tilde{\Lambda} \subset \Lambda$.

Theorem 6.3.2. *The processes $N^{(i)}$ and $M^{(i)}$, $1 \leq i \leq d$, are local strict CAF up to $\mathcal{X}$. The processes $N^{(i)}$ are locally of bounded variation up to $\mathcal{X}$ $\mathbb{P}_x$-a.s. for every $x \in E_1$. The processes $M^{(i)}$ are continuous local $\mathcal{F}_t^1$-martingales up to $\mathcal{X}$ under $\mathbb{P}_x$, $x \in E_1$, with reducing sequence $\tilde{\tau}_n := \tau_n \wedge n \wedge \mathcal{X}$ where τ_n is the exit time of K_n, $n \in \mathbb{N}$, defined after (6.12). The quadratic variation and covariation processes (up to $\mathcal{X}$) are given by*

$$\langle M^{(i)}, M^{(j)} \rangle_{\cdot \wedge \mathcal{X}} = 2 \left(a_{ij} \cdot t \right)_{\cdot \wedge \mathcal{X}}, \quad 1 \leq i, j \leq d, \tag{6.22}$$

and have the same reducing sequence. In particular, $(\mathbf{X}_t^{(i)})_{t \geq 0}$, $1 \leq i \leq d$, are semimartingales up to $\mathcal{X}$ with

$$\mathbf{X}_{t \wedge \mathcal{X}}^{(i)} = \mathbf{X}_0^{(i)} + N_{t \wedge \mathcal{X}}^{(i)} + M_{t \wedge \mathcal{X}}^{(i)}$$

for $0 \leq t < \infty$ $\mathbb{P}_x$-a.s. for $x \in E_1$ and $1 \leq i \leq d$.

There exists a set $\hat{\Lambda}$ with $\mathbb{P}_x(\hat{\Lambda}) = 1$ for $x \in E_1$ with the following properties: The set $\hat{\Lambda}$ is contained in the additivity sets of $N^{(i)}$ and $M^{(i)}$ for $1 \leq i \leq d$. Moreover, (6.22) hold on $\hat{\Lambda}$ and $N_{t \wedge \mathcal{X}}^{(i)}$ is given by the defining integral of (6.19) on $\hat{\Lambda}$. Furthermore, the paths $[0, \mathcal{X}) \ni t \mapsto N_t^{(i)}(\omega)$, $1 \leq i \leq d$, are locally of bounded variation for $\omega \in \hat{\Lambda}$.

Proof. Recall that b_i is locally $L^p(E_1, \mu)$-integrable and $(e_i, A\eta)\varrho$ is locally bounded for $1 \leq i \leq d$. So according to Theorem 6.2.7 and Lemma 6.2.8 $N^{(i)}$, $1 \leq i \leq d$, define local strict CAF up to $\mathcal{X}$ and we can find a common additivity set Λ. From the construction of the parts of $N^{(i)}$ it follows that the paths of $N^{(i)}$ are locally of bounded variation up to $\mathcal{X}$ on Λ, compare the proof of Theorem 6.2.7. The definition of $M^{(i)}$, $1 \leq i \leq d$, yields that they are also additive on Λ.

Choose a sequence of cutoff functions ϕ_n, $n \in \mathbb{N}$, with $\phi_n = 1$ on K_{n+1} and $\mathrm{supp}[\phi_n] \subset\subset U_{n+2}$, $(K_n)_{n \in \mathbb{N}}$ and $(U_n)_{n \in \mathbb{N}}$ as in (6.12).

Define $u_i^{(n)}$, $n \in \mathbb{N}$, $1 \leq i \leq d$, with $u_i^{(n)}(x) = \phi_n(x)x_i$. Then $u_i^{(n)}(x) = x_i$ in a neighborhood of U_n since $\phi_n = 1$ on U_{n+1} for $n \in \mathbb{N}$. So for $x \in U_n$ it holds $\partial_j u_i^{(n)}(x) = \delta_{ij}$ and $\partial_j \partial_k u_i^{(n)}(x) = 0$ for $1 \leq i, j, k \leq d$. Thus $\hat{L}u_i^{(n)}(x) = b_i(x)$ for $x \in U_n$ and $n \in \mathbb{N}$. So we have for all $n \in \mathbb{N}$, $t \geq 0$ and $1 \leq i \leq d$

$$N_{t \wedge \tau_n}^{[u_i^{(n)}]} = \int_0^{t \wedge \tau_n} b_i(\mathbf{X}_s)\, ds - \int_0^{t \wedge \tau_n} (\varrho(e_i, A\eta))(\mathbf{X}_s)\, d\ell_s = N_{t \wedge \tau_n}^{(i)}. \quad (6.23)$$

Using that $\tilde{\tau}_n \leq \tau_n$ we get from construction of $u_n^{(i)}$ together with (6.23) for all $0 \leq t < \infty$ and $n \in \mathbb{N}$

$$M_{t \wedge \tilde{\tau}_n}^{(i)} = \mathbf{X}_{t \wedge \tilde{\tau}_n}^{(i)} - \mathbf{X}_0^{(i)} - \int_0^{t \wedge \tilde{\tau}_n} b_i(\mathbf{X}_s)\, ds + \int_0^{t \wedge \tilde{\tau}_n} (\varrho(e_i, A\eta))(\mathbf{X}_s)\, d\ell_s$$

$$= M_{t \wedge \tilde{\tau}_n}^{[u_i^{(n)}]}$$

for $1 \leq i \leq d$. Since $M^{[u_i^{(n)}]}$ is a continuous martingale and $\tilde{\tau}_n$ is bounded, we have by the optional sampling theorem that the stopped process $M_{\cdot \wedge \tilde{\tau}_n}^{[u_i^{(n)}]}$

is also a martingale. So $(M^{(i)}_{t \wedge \tilde{\tau}_n})_{t \geq 0}$ is a martingale for every $n \in \mathbb{N}$ and $1 \leq i \leq d$. Thus $M^{(i)}$ is a continuous local martingale with reducing sequence $(\tilde{\tau}_n)_{n \in \mathbb{N}}$ for $1 \leq i \leq d$.

Define $A^{(i)}_t := 2 \int_0^t a_{ii}(\mathbf{X}_s) ds$, $1 \leq i \leq d$, according to Lemma 6.2.7. From this lemma we get that $A^{(i)}$, $1 \leq i \leq d$, is a local strict CAF up to $\mathcal{X}$. We claim that $A^{(i)}$ is the quadratic variation of $M^{(i)}$, which exists uniquely due to Theorem 6.2.11. So it is to check that $((M^{(i)}_t)^2 - A^{(i)}_t)_{t \geq 0}$ is a local martingale for $1 \leq i \leq d$. Let $(\tilde{\tau}_n)_{n \in \mathbb{N}}$ as above. We have by definition of the corresponding objects and Theorem 6.2.12

$$(M^i_{t \wedge \tilde{\tau}_n})^2 - A^{(i)}_{t \wedge \tilde{\tau}_n} = (M^{[u^{(n)}_i]})^2_{t \wedge \tilde{\tau}_n} - \langle M^{[u^{(n)}_i]} \rangle_{t \wedge \tilde{\tau}_n} \qquad (6.24)$$

for $t \geq 0$, $1 \leq i \leq d$ and $n \in \mathbb{N}$. This equality holds for all ω for which $\langle M^{[u^{(n)}_i]} \rangle_t = 2 \int_0^t (A \nabla u^{(n)}_i, \nabla u^{(n)}_i)(\mathbf{X}_s) ds$. By optional sampling theorem, we have that the right-hand side of (6.24) is a martingale. Since this holds for every $n \in \mathbb{N}$, we have that the quadratic variation of $(M^{(i)}_t)_{t \geq 0}$ is given by $A^{(i)}$ for $1 \leq i \leq d$. Now let $1 \leq i, j \leq d$, $i \neq j$. Then $M^i + M^j$ is a continuous local martingale as well. Define $\tilde{A}^{(i,j)}_t := 2 \int_0^t a_{ii}(\mathbf{X}_s) + a_{jj}(\mathbf{X}_s) + 2a_{ij}(\mathbf{X}_s) \, ds$, $t \geq 0$. By Theorem 6.2.12

$$\langle M^{[u^{(n)}_i + u^{(n)}_j]} \rangle_{t \wedge \tilde{\tau}_n} = 2 \int_0^{t \wedge \tilde{\tau}_n} (A \nabla(u^{(n)}_i + u^{(n)}_j), \nabla(u^{(n)}_i + u^{(n)}_j))(\mathbf{X}_s) \, ds$$

$$= 2 \int_0^{t \wedge \tilde{\tau}_n} (a_{ii} + a_{jj} + 2a_{ij})(\mathbf{X}_s) ds = \tilde{A}^{(i,j)}_{t \wedge \tilde{\tau}_n} \text{ for } t \geq 0 \text{ and } n \in \mathbb{N}.$$

Thus using linearity we get for $t \geq 0$ and $n \in \mathbb{N}$

$$(M^i + M^j)^2_{t \wedge \tilde{\tau}_n} - \tilde{A}^{(i,j)}_{t \wedge \tilde{\tau}_n} = (M^{[u^{(n)}_i]} + M^{[u^{(n)}_j]})^2_{t \wedge \tilde{\tau}_n} - \tilde{A}^{(i,j)}_{t \wedge \tilde{\tau}_n}$$

$$= (M^{[u^{(n)}_i + u^{(n)}_j]})^2_{t \wedge \tilde{\tau}_n} - \langle M^{[u^{(n)}_i + u^{(n)}_i]} \rangle_{t \wedge \tilde{\tau}_n}. \qquad (6.25)$$

The right-hand side of (6.25) is a martingale for every $n \in \mathbb{N}$, so $(M^i + M^j)^2 - \tilde{A}^{(i,j)}$ is a local martingale. Choose $\hat{\Lambda} \subset \Lambda$ such that for all $\omega \in \hat{\Lambda}$ it holds $\langle M^i \rangle = A^{(i)}$ and $\langle M^i + M^j \rangle = \tilde{A}^{(i,j)}$. Altogether, we get for $0 \leq t \leq \mathcal{X}$

$$\langle M^{(i)}, M^{(j)} \rangle_t = \frac{1}{2}(\langle M^i + M^j, M^i + M^j \rangle_t - \langle M^i \rangle_t - \langle M^j \rangle_t)$$

$$= 2 \int_0^t a_{ij}(\mathbf{X}_s) \, ds.$$

$\square$

Now we prove existence of weak solutions. We use the following notation for weak solutions.

We formulate the stochastic differential equation with reflection at the boundary in the following sense.

Definition 6.3.3. Let $\Omega \subset \mathbb{R}^d$ be open, $E_1 \subset \overline{\Omega}$, $\widetilde{\Gamma} := E_1 \cap \partial\Omega$, $\sigma : E_1 \to \mathbb{R}^{d \times r}$, $r \in \mathbb{N}$, $b : E_1 \to \mathbb{R}^d$, $g : \widetilde{\Gamma} \to \mathbb{R}^d$ Borel measurable. Let $(\Omega, \mathcal{F}, \mathbb{P})$ be a probability space with filtration $(\mathcal{F}_t)_{t \geq 0}$. A triple $((\mathbf{X}_t)_{t \geq 0}, (W_t)_{t \geq 0}, (\ell_t)_{t \geq 0})$ of $\mathcal{F}_t$-adapted processes is called a *weak solution with reflection* at $\widetilde{\Gamma}$ and coefficients (σ, b, g) with initial distribution $\mu_0 \in \mathcal{B}(E_1)$ if the following properties are fulfilled:

(i) $\mathbf{X}_t \in E_1$ for all $t \geq 0$, $\mathcal{L}(\mathbf{X}_0) = \mu_0$ and $(\mathbf{X}_t)_{t \geq 0}$ has continuous paths.

(ii) $(\ell_t)_{t \geq 0}$ is continuous, increasing, $\ell_0 = 0$ and $\int_0^t 1_{\widetilde{\Gamma}}(\mathbf{X}_s) d\ell_s = \ell_t$ for $t \geq 0$.

(iii) $(W_t)_{t \geq 0}$ is an $\mathbb{R}^r$-valued Brownian motion w.r.t. the filtration $(\mathcal{F}_t)_{t \geq 0}$.

(iv) $\int_0^t |b_i(\mathbf{X}_s)| \, ds < \infty$, $\int_0^t |g_i(\mathbf{X}_s)| d\ell_s < \infty$, $\int_0^t \sigma_{ij}^2(\mathbf{X}_s) \, ds < \infty$, $t \geq 0$, $1 \leq i \leq d$, $1 \leq j \leq r$, $\mathbb{P}$-a.s. and

$$\mathbf{X}_t = \mathbf{X}_0 + \int_0^t b(\mathbf{X}_s) \, ds + \int_0^t g(\mathbf{X}_s) d\ell_s + \int_0^t \sqrt{2} \, \sigma(\mathbf{X}_s) \, dW_s \quad \text{for } t \geq 0.$$

The last integral is understood as a vector valued integral w.r.t. the Brownian motion $W = (W^1, ..., W^r)$. More precisely, it holds

$$\left(\int_0^t \sigma(\mathbf{X}_s) \, dW_s \right)_i = \sum_{k=1}^r \int_0^t \sigma_{ik}(\mathbf{X}_s) \, dW_s^k, t \geq 0, 1 \leq i \leq d.$$

Here $\int_0^t \cdot dW_s^k$ denotes the Ito integral w.r.t. the Brownian motion W^k, $1 \leq k \leq r$.

Remark 6.3.4. The integrability condition in (iv) ensures that $\int_0^t b(\mathbf{X}_s) \, ds + \int_0^t g(\mathbf{X}_s) d\ell_s$, $t \geq 0$, is locally of bounded variation. Since $(\int_0^t \sigma(\mathbf{X}_s) \, dW_s)_{t \geq 0}$ is a martingale, we get that $(\mathbf{X}_t)_{t \geq 0}$ is a semimartingale. Property (ii) of the previous definition means that the path functional ℓ_t grows only at the boundary. From this one can deduce that for the integral in (iv) w.r.t. to ℓ_t only those values of the integrand are relevant when $\mathbf{X}_t$ is at the boundary part $\widetilde{\Gamma}$. So this term covers the boundary interaction. Our definition of a weak solution is adapted from [IW81, Def. 7.3].

As before we consider the process $\mathbf{M}^1$ obtained as the restriction of the $\mathcal{L}^p$-strong Feller process $\mathbf{M}$ from Theorem 4.1.14 to $E_1 \cup \{\Delta\}$. Note that if $\mathbf{M}$ is conservative then also $\mathbf{M}^1$ is conservative. Conservativity of $\mathbf{M}$ holds e.g. if the coefficients fulfill certain growth conditions, see Remark 4.1.16. So under the assumption that $\mathbf{M}^1$ is conservative, we get the following existence result.

Theorem 6.3.5. *Let* $\Omega \subset \mathbb{R}^d$ *open,* $\Gamma_2 \subset \partial\Omega$, $\sigma : \overline{\Omega} \to \mathbb{R}^{d \times r}$, $r \in \mathbb{N}$, $\varrho : \overline{\Omega} \to \mathbb{R}_0^+$. *Define* $A := \sigma\sigma^\top$. *Assume that* A *and* ϱ *satisfy Condition 4.1.1, Condition 4.1.2 and Condition 4.1.6. Assume additionally that the corresponding Dirichlet form (closure of (4.1)) is conservative and that* Γ_2 *and* A *satisfy Condition 4.1.10. Define* $E_1 := (\Omega \cup \Gamma_2) \cap \{\varrho > 0\}$ *and* $\widetilde{\Gamma} := \Gamma_2 \cap \{\varrho > 0\}$. *Let* $\mu_0 \in \mathcal{P}(E_1)$. *Let* $b := (b_1, ..., b_d)$ *with* $b_i = \sum_{j=1}^d \partial_j a_{ij} + \sum_{j=1}^d \frac{1}{\varrho} a_{ij} \partial_j \varrho$ *and* $g = -\varrho A \eta$. *Then there exists a triple* $((\mathbf{X}_t)_{t\geq 0}, (W_t)_{t\geq 0}, (\ell_t)_{t\geq 0})$ *that is a weak solution in the sense of Definition 6.3.3 with coefficients* (σ, b, g), *i.e.,*

$$\mathbf{X}_t = \mathbf{X}_0 + \int_0^t \left(\nabla A + A\frac{\nabla\varrho}{\varrho} \right) (\mathbf{X}_s)\, ds - \int_0^t (\varrho A\eta)\, (\mathbf{X}_s) d\ell_s$$

$$+ \int_0^t \left(\sqrt{2}\, \sigma(\mathbf{X}_s) \right) dW_s, \quad t \geq 0.$$

The law of $(\mathbf{X}_t)_{t\geq 0}$ *is given by the law* $\mathbb{P}_{\mu_0}$ *of the* $\mathcal{L}^p$-*strong Feller process* $\mathbf{M}^1$ *with initial distribution* μ_0.

Proof. Define a probability measure $\mathbb{P}_{\mu_0}$ on $(\mathbf{\Omega}^1, \mathcal{F}^1)$ by

$$\mathbb{P}_{\mu_0}(\cdot) := \int_{E_1} \mathbb{P}_x(\cdot)\, d\mu_0(x),$$

see (7.5) in Section 7.3 for details on the definition. Obviously, $\mathcal{L}(\mathbf{X}_0) = \mu_0$ under $\mathbb{P}_{\mu_0}$. By construction of $\mathbf{M}^1$ we have that $(\mathbf{X}_t)_{t\geq 0}$ has continuous paths on $[0, \infty)$. Furthermore, all paths stay in $E_1 \cup \{\Delta\}$. Since $\mathbb{P}_{\mu_0}(\mathcal{X} = \infty) = 1$, they do not hit Δ. So (i) is fulfilled. To prove property (ii) of the definition note that $((1 - 1_{\Gamma_2} 1_{\{\varrho > 0\}}) \cdot \ell)_t = 0$ for every $t \geq 0$, $\mathbb{P}_x^1$-a.s., thus also $\mathbb{P}_{\mu_0}$-a.s.

Let $N^{(i)}$ and $M^{(i)}$, $1 \leq i \leq d$, as defined before Theorem 6.3.2. Set $M := (M^{(1)}, ..., M^{(d)})$.

Let $\hat{\Lambda} \subset \Omega^1$ as in Theorem 6.3.2. Then the definition of $N^{(i)}$, $1 \leq i \leq d$, implies that on $\omega \in \hat{\Lambda}$ the integrals $\int_0^t |b_i(\mathbf{X}_s)| ds$, $1 \leq i \leq d$, and $\int_0^t |g_i(\mathbf{X}_s)| d\ell_s$, $1 \leq i \leq d$, exist and

$$\mathbf{X}_t = \mathbf{X}_0 + \int_0^t (\nabla A + A \frac{\nabla \varrho}{\varrho}) \, (\mathbf{X}_s) \, ds - \int_0^t (\varrho A \eta) \, (\mathbf{X}_s) d\ell_s + M_t$$

for all $0 \leq t < \infty$ since $\mathcal{X} = \infty$ by assumption.

Since $\mathbb{P}_x^1(\hat{\Lambda}) = 1$ for every $x \in E_1$, we have $\mathbb{P}_{\mu_0}(\hat{\Lambda}) = 1$. Thus this equality holds $\mathbb{P}_{\mu_0}$-a.s.

If $(C_t)_{t \geq 0}$ is an $\mathcal{F}_t^1$-martingale under $\mathbb{P}_x$ for every $x \in E_1$, then it is also one under $\mathbb{P}_{\mu_0}$. For the reducing sequence $(\tilde{\tau}_n)_{n \in \mathbb{N}}$ of $M^{(i)}$, $1 \leq i \leq d$, as in Theorem 6.3.2 we have $\tilde{\tau}_n \uparrow \infty$ $\mathbb{P}_x$-a.s. for every $x \in E_1$ hence also $\mathbb{P}_{\mu_0}$-a.s. So $M^{(i)}$ is again a local martingale with the same quadratic variation as in Theorem 6.3.2 for $1 \leq i \leq d$.

So it is left to construct a Brownian motion $(W_t)_{t \geq 0}$ such that $M_t = \int_0^t \sqrt{2}\sigma \, (\mathbf{X}_s) \, dW_s$. Note that $M^{(i)}$, $1 \leq i \leq d$, are continuous local martingales with

$$\langle M^{(i)}, M^{(j)} \rangle_t = 2 \int_0^t a_{ij}(\mathbf{X}_s) \, ds = \int_0^t \left(\sqrt{2}\sigma(\sqrt{2}\sigma)^\top \right) (\mathbf{X}_s) \, ds \text{ for } t \geq 0.$$

So we can adapt the proof of [KS91, Ch. 5, Prop. 4.6]. Starting from (4.12) therein we conclude the existence of an r-dimensional Brownian motion (possibly on an extension of the probability space of $\mathbf{M}^1$) such that $\int_0^t \sigma_{ij}^2(\mathbf{X}_s) \, ds < \infty$ $\mathbb{P}_{\mu_0}$-a.s. for $t \geq 0$, $1 \leq i \leq d$ and $1 \leq j \leq r$ and

$$M_t = \int_0^t \sqrt{2}\sigma \, (\mathbf{X}_s) \, dW_s \text{ for } t \geq 0.$$

Property (iv) follows immediately from the definition of $N^{(i)}$ and the relation between $M^{(i)}$ and $(W_t)_{t \geq 0}$. $\square$

6.4 Application to Interacting Particle Systems

We apply the results of the previous section to the stochastic dynamics for interacting particle systems of Section 5.1. So for $N \in \mathbb{N}$, let $\Omega = \Omega_0^N$, $E_1 = \Lambda_{ad}^N$, $\tilde{\Gamma}_2 = \Lambda_{ad}^N \cap \partial(\Lambda^N)$, $\hat{A}$, p and ϱ as in Section 5.1. We consider the gradient Dirichlet form of (5.3). Let $\mathbf{M}^1$ be the restriction to $E_1 \cup \{\Delta\}$ of the process from Section 5.1. Assume that $\mathbf{M}^1$ is conservative. Denote the coordinate maps just by $(\mathbf{X}_t)_{t \geq 0}$.

From Theorem 6.2.3 we get that the restricted surface measure $1_{\tilde{\Gamma}_2} 1_{\{\varrho > 0\}} \sigma$ is smooth in the strict sense. Denote the corresponding additive functional by $(\ell_t)_{t \geq 0}$.

Note that

$$\tilde{\Gamma}_2 \cap \{\varrho > 0\} = \{\varrho > 0\} \cap \bigcup_{k=1}^{N} \Omega_0^{k-1} \times \Gamma_2 \times \Omega_0^{N-k}.$$

Define $T^{(k)} := \Omega_0^{k-1} \times \Gamma_2 \times \Omega_0^{N-k}$, $1 \leq k \leq N$. For $x = (x^{(1)}, ..., x^{(N)}) \in T^{(k)}$ we have for the outward unit normal $\eta = (0, ..., 0, \eta_{\Gamma_2}(x^{(k)}), 0, ..., 0)$, i.e., in the k-th coordinate we have the outward unit normal of the boundary of the state space of the k-th particle. Set $\ell_t^{(k)} := (1_{T^{(k)}} \cdot \ell)_{t \geq 0}$, $1 \leq k \leq N$. Then $(\ell_t^{(k)})_{t \geq 0}$ grows only if the k-th particle is at the boundary. We want to apply the results of Section 6.3 to show that $(\mathbf{X}_t)_{t \geq 0}$ is a semimartingale. Let $\mathbf{N}$ and M as in (6.19) and (6.21), respectively. The coefficients b as in (6.20) take the following form: Write $b := (b^{(1)}, ..., b^{(N)})$ with $b^{(k)} : \Lambda_{\mathrm{ad}}^N \to \mathbb{R}^d$, $1 \leq k \leq d$. Then $b^{(k)}(x) = \left(\nabla_k A(x^{(k)}) - \sum_{l=1}^{N} A(x^{(k)})(\nabla \Psi)(x^{(k)} - x^{(l)}) \right)$, $1 \leq k \leq N$, $x = (x^{(1)}, ..., x^{(N)}) \in \Lambda_{\mathrm{ad}}^N$.

For $g(x) := \varrho(x) \hat{A} \eta(x)$, $x \in \tilde{\Gamma}_2$, we get: Write $g = (g^{(1)}, ..., g^{(N)})$, $g^{(k)} : \tilde{\Gamma}_2 \to \mathbb{R}^d$, $1 \leq k \leq d$. Then we have for $x \in T^{(k)}$, $g^{(i)}(x) = 0$, if $i \neq k$, and $g^{(k)} = \varrho(x) A(x^{(k)}) \eta_{\Gamma_2}(x^{(k)})$. So we can write $\mathbf{N}$ as $\mathbf{N} = (N^{(1)}, ..., N^{(N)})$ with $N^{(k)} : \Omega^1 \times [0, \infty) \to \mathbb{R}^d$, $1 \leq k \leq N$. Then we get:

$$N_t^{(k)} = \int_0^t \left(\nabla A(\mathbf{X}_s^{(k)}) - \sum_{l=1}^{N} A(\mathbf{X}_s^{(k)})(\nabla \Psi)(\mathbf{X}_s^{(k)} - \mathbf{X}_s^{(l)}) \right) ds$$

$$- \int_0^t \varrho(\mathbf{X}_s) A(\mathbf{X}_s^{(k)}) \eta_{\Gamma_2}(\mathbf{X}_s^{(k)}) d\ell_s^{(k)} \text{ for } 0 \leq t < \infty \text{ and } 1 \leq k \leq d.$$

Similarly we write the martingale part M as $M = (M^{(1)}, ..., M^{(N)})$ with $M^{(k)} = \mathbf{X}_t^{(k)} - \mathbf{X}_0^{(k)} - \mathbf{N}_t^{(k)}$. The quadratic covariation is given by $\hat{A}$ now. Thus for $1 \leq i, j, k, l \leq d$

$$\langle M_i^{(k)}, M_j^{(l)} \rangle_t = \delta_{k,l} 2 \int_0^t a_{ij}(\mathbf{X}_s^{(k)}) \, ds, \quad 0 \leq t < \infty.$$

So writing $\mathbf{X} := (\mathbf{X}^{(1)}, ..., \mathbf{X}^{(N)})$ we get

$$\mathbf{X}_t^{(k)} - \mathbf{X}_0^{(k)} = \int_0^t \left(\nabla A(\mathbf{X}_s^{(k)}) - \sum_{l=1}^N A(\mathbf{X}_s^{(k)})(\nabla\Psi)(\mathbf{X}_s^{(k)} - \mathbf{X}_s^{(l)}) \right) ds$$

$$- \int_0^t \varrho(\mathbf{X}_s) A(\mathbf{X}_s^{(k)}) \eta_{\Gamma_2}(\mathbf{X}_s^{(k)}) d\ell_s^{(k)} + M_t^{(k)} \text{ for } 0 \leq t < \infty \text{ and } 1 \leq k \leq N.$$

6.5 Stochastic Dynamics for an Interface Model with Area Conservation

We continue the discussion of interface models of Section 5.2. We apply the results of Section 6.3 to construct weak solutions to an SDE describing the stochastic dynamics for a wetting model with conservation. This model has been studied by Zambotti, see [Zam08].

With conservation we mean that the area between the interface and the hard wall is conserved. We take the SDE from [Zam08] and use (with some modifications) the notion therein.

We present the setting of [Zam08]. Let $N \in \mathbb{N}$ and define the $N \times N$ matrices

$$\sigma = \begin{pmatrix} -1 & & & & \\ 1 & . & & & \\ & . & . & & \\ & & 1 & -1 & \\ & & & 1 & 0 \end{pmatrix}, \quad \sigma^\top = \begin{pmatrix} -1 & 1 & & & \\ & . & . & & \\ & & . & . & \\ & & & -1 & 1 \\ & & & 0 & 0 \end{pmatrix}$$

and the $N \times (N-1)$ and $(N-1) \times N$ matrices

$$\widetilde{\sigma} = \begin{pmatrix} -1 & & & \\ 1 & . & & \\ & . & . & \\ & & 1 & -1 \\ & & & 1 \end{pmatrix}, \quad \widetilde{\sigma}^\top = \begin{pmatrix} -1 & 1 & & \\ & . & . & \\ & & . & . \\ & & & -1 & 1 \end{pmatrix}.$$

Let $\Gamma_N := \{1, ..., N\}$. The location of the interface at time t is described by the height variables $\phi_t = (\phi_t(1), ..., \phi_t(N)) \in [0, \infty)^{\Gamma_N}$. Let $V : \mathbb{R} \to \mathbb{R}$, C^2-smooth and convex. Define $\hat{V}' : \mathbb{R}^N \to \mathbb{R}^N$ by

$$y = (y_1, ..., y_N) \mapsto (V'(y_1), ..., V'(y_{N-1}), 0) =: \hat{V}'(y).$$

We assume furthermore

$$V(x) \overset{|x| \to \infty}{\longrightarrow} \infty.$$

Consider the following SDE: Let $(W_t)_{t \geq 0}$ be an $\mathbb{R}^N$-valued Brownian motion. Then the height variables $(\phi_t)_{t \geq 0}$ should fulfill

$$d\phi_t = -\sigma\sigma^\top (\sigma \hat{V}'(\sigma^\top \phi_t) dt - d\hat{\ell}_t) + \sqrt{2}\sigma dW_t, \qquad (6.26)$$

with initial condition $\phi_0 \in [0, \infty)^{\Gamma_N}$,

$$\phi_t(k) \geq 0 \quad \text{for } 1 \leq k \leq N,$$

$(\hat{\ell}_t)_{t \geq 0}$ is a $[0, \infty)^{\Gamma_N}$-valued functional, i.e., $\hat{\ell}_t = (\hat{\ell}_t^1, ..., \hat{\ell}_t^N)$, fulfilling

$$\hat{\ell}_0 = 0, \quad (\hat{\ell}_t)_{t \geq 0} \text{ is continuous and non-decreasing}$$

and

$$\int_0^\infty \phi_t d\hat{\ell}_t := \sum_{k=1}^N \int_0^\infty \phi_t(k) d\hat{\ell}_t^k = 0.$$

The last condition means that the functional ℓ^k grows only if the k-th coordinate is zero.

The variable $(\phi_t)_{t \geq 0}$ describes the height of a discretized interface with N positions. The surface area is defined by $\sum_{k=1}^N \phi_t(k)$, this corresponds to a piecewise constant interpolation of the interface. For $0 \leq c < \infty$ define

$$\mathbb{V}_N^{c,+} := \big\{ \phi \in [0, \infty)^{\Gamma_N} \mid \sum_{k=1}^N \phi(k) = c \big\}.$$

We take a different approach than [Zam08] to construct the dynamics. Observe that $\mathbb{V}_N^{c,+}$ defines an $(N-1)$-dimensional manifold. We construct first dynamics on a certain convex subset of $\mathbb{R}^{N-1}$ and then map the dynamics onto $\mathbb{V}_N^{c,+}$. In [Zam08] solutions are directly constructed on $\mathbb{V}_N^{c,+}$. Define

$$\widetilde{\mathbb{V}}_{N-1} := \Big\{ \widetilde{\phi} \in \mathbb{R}^{N-1} \mid \widetilde{\phi}(1) \leq c, \widetilde{\phi}(k) - \widetilde{\phi}(k-1) \leq c \text{ for } 2 \leq k \leq N-1,$$

$$\widetilde{\phi}(N-1) \geq -c \Big\}.$$

Observe that

$$\tilde{\sigma}\,\tilde{\mathbb{V}}_{N-1} = \big\{\phi \in [0,\infty)^{\Gamma_N} \mid \sum_{i=1}^{N} \phi(i) = 0,\ \phi(k) \geq -c \text{ for } 1 \leq k \leq N\big\}.$$

Hence with $\mathbf{1} = (1,...,1) \in [0,\infty)^{\Gamma_N}$ we have

$$\mathbb{V}_N^{c,+} = c\mathbf{1} + \tilde{\sigma}\,\tilde{\mathbb{V}}_{N-1}.$$

Define the Hamiltonian $\mathcal{H}_N$ on $\mathbb{V}_N^{c,+}$ by

$$\mathbb{V}_N^{c,+} \ni \phi \mapsto \mathcal{H}_N(\phi) = \sum_{k=1}^{N-1} V(\sigma_{k*}^{\top}\phi) = \sum_{k=1}^{N-1} \hat{V}_k(\sigma^{\top}\phi)$$

with $\hat{V} = (\hat{V}_1,...,\hat{V}_N)$, $\hat{V}_k((x_1,...,x_N)) := V(x_k)$ for $1 \leq k \leq N-1$, $\hat{V}_N := 0$. Define the corresponding Hamiltonian $\tilde{\mathcal{H}}_{N-1}$ on $\tilde{\mathbb{V}}_{N-1}$ by

$$\tilde{\mathbb{V}}_{N-1} \ni \tilde{\phi} \mapsto \tilde{\mathcal{H}}_{N-1}(\tilde{\phi}) := \mathcal{H}_N(\tilde{\sigma}\tilde{\phi}) = \sum_{k=1}^{N-1} \tilde{V}_k(\tilde{\sigma}^{\top}\tilde{\sigma}\tilde{\phi}),$$

with $\tilde{V} = (\tilde{V}_1,...,\tilde{V}_{N-1}) : \mathbb{R}^{N-1} \to \mathbb{R}^{N-1}$ and $\tilde{V}_k((x_1,...,x_{N-1})) := V(x_k)$, $1 \leq k \leq N-1$. Define

$$\tilde{\varrho} : \tilde{\mathbb{V}}_{N-1} \to \mathbb{R}_0^+, \tilde{\phi} \mapsto \frac{1}{Z}\exp(-\tilde{\mathcal{H}}_{N-1}(\tilde{\phi})),$$

$\mu := \tilde{\varrho}\,dx$, dx being the $(N-1)$-dimensional Lebesgue measure, $0 < Z < \infty$ some normalization constant.

Define the Dirichlet form $(\mathcal{E}, D(\mathcal{E}))$ as closure of

$$\mathcal{E}(u,v) = \int_{\tilde{\mathbb{V}}_{N-1}} (\nabla u, \nabla v)\,d\mu,$$

$$u, v \in \mathcal{D} := \{u \in C_c(\tilde{\mathbb{V}}_{N-1}) \mid u \in H_{\text{loc}}^{1,1}(\overset{\circ}{\tilde{\mathbb{V}}}_{N-1}), \mathcal{E}(u,u) < \infty\}.$$

The assumptions on the potential imply that $\tilde{\varrho}(\mathbb{R}^{N-1}) < \infty$, see [Zam08]. Thus the Dirichlet form is conservative, see Remark 4.1.16.

Let us describe the boundary of $\widetilde{\mathbb{V}}_{N-1}$ in terms of the conditions posed on $\widetilde{\phi}$. We have

$$\partial\widetilde{\mathbb{V}}_{N-1} := \left\{ \widetilde{\phi} \in \widetilde{\mathbb{V}}_{N-1} \,\middle|\, \exists\, 1 \leq k \leq N \text{ such that: } \widetilde{\phi}(k) = c \,(\text{if } k = 1),\right.$$

$$\widetilde{\phi}(k) - \widetilde{\phi}(k-1) = c \,(\text{if } 2 \leq k \leq N-1),$$

$$\left. \widetilde{\phi}(k-1) = -c \,(\text{if } k = N)\right\}.$$

Define

$$T^{(1)} := \left\{ \widetilde{\phi} \in \partial\widetilde{\mathbb{V}}_{N-1} \,\middle|\, \widetilde{\phi}(1) = c, \widetilde{\phi}(k) - \widetilde{\phi}(k-1) < c, 2 \leq k \leq N-1,\right.$$

$$\left. \widetilde{\phi}(N-1) > -c\right\},$$

$$T^{(i)} := \left\{ \widetilde{\phi} \in \partial\widetilde{\mathbb{V}}_{N-1} \,\middle|\, \widetilde{\phi}(i) - \widetilde{\phi}(i-1) = c,\right.$$

$$\widetilde{\phi}(1) < c, \widetilde{\phi}(k) - \widetilde{\phi}(k-1) < c, 2 \leq k \leq N-1, k \neq i,$$

$$\left. \widetilde{\phi}(N-1) > -c\right\}, 2 \leq i \leq N-1$$

and

$$T^{(N)} := \left\{ \widetilde{\phi} \in \partial\widetilde{\mathbb{V}}_{N-1} \,\middle|\, \widetilde{\phi}(1) < c, \widetilde{\phi}(k) - \widetilde{\phi}(k-1) < c, 2 \leq k \leq N-1,\right.$$

$$\left. \widetilde{\phi}(N-1) = -c\right\}.$$

Note that these boundary parts are contained in 1-codimensional subsets of $\mathbb{R}^{N-1}$. Furthermore, $\partial\widetilde{\mathbb{V}}_{N-1} \setminus \bigcup_{k=1}^{N} T^{(k)}$ is contained in a finite union of subsets of 2-codimensional subsets and hence has zero $\mathcal{H}^{N-1}$-Hausdorff measure and zero capacity, see Section 7.6. All boundary parts $T^{(k)}$, $1 \leq k \leq N$, are C^{∞}-smooth and the unit outward normal at $T^{(k)}$ is given (up to normalization) by the k-th column of $-\widetilde{\sigma}^{\top}$, i.e.,

$$\eta(\widetilde{\phi}) = -\frac{\widetilde{\sigma}^{\top}_{*k}}{\|\widetilde{\sigma}^{\top}_{*k}\|} \text{ for } \widetilde{\phi} \in T^{(k)}.$$

Define $\Gamma_2 := \bigcup_{k=1}^{N} T^{(k)}$, $E_1 := \overset{\circ}{\widetilde{\mathbb{V}}}_{N-1} \cup \Gamma_2$. For $u, \varphi \in C_c^2(E_1)$ the following partial integration holds:

$$\mathcal{E}(u, \varphi) = -\int_{\overset{\circ}{\widetilde{\mathbb{V}}}_{N-1}} \nabla \cdot \widetilde{\varrho} \nabla u \, \varphi \, dx + \sum_{k=1}^{N} \int_{T^{(k)}} \widetilde{\varrho} \varphi \eta \nabla u \, d\Sigma$$

$$= -\int_{\overset{\circ}{\widetilde{\mathbb{V}}}_{N-1}} \left(\Delta u + \left(\frac{\nabla \widetilde{\varrho}}{\widetilde{\varrho}}, \nabla u \right) \right) \varphi \, d\mu + \sum_{k=1}^{N} \int_{T^{(k)}} \widetilde{\varrho} \varphi \left(-\frac{\widetilde{\sigma}_{*k}^\top}{\|\widetilde{\sigma}_{*k}^\top\|}, \nabla u \right) d\Sigma$$

where we denote by Σ the surface measure. Note that for $\widetilde{\phi} \in \widetilde{\mathbb{V}}_{N-1}$ it holds

$$\frac{\nabla \widetilde{\varrho}}{\widetilde{\varrho}}(\widetilde{\phi}) = -\nabla \widetilde{\mathcal{H}}_{N-1}(\widetilde{\phi}) = -\nabla \left(\sum_{k=1}^{N-1} \widetilde{V}_k(\widetilde{\sigma}^\top \widetilde{\sigma} \widetilde{\phi}) \right) = -\widetilde{\sigma}^\top \widetilde{\sigma} \widetilde{V}'(\widetilde{\sigma}^\top \widetilde{\sigma} \widetilde{\phi})$$

where we define (under some abuse of notion) $\widetilde{V}'(y) := (V'(y_1), ..., V'(y_{N-1}))$ for $y = (y_1, ..., y_{N-1}) \in \mathbb{R}^{N-1}$.

So we may apply the results from the previous sections. Choose $p > \frac{N-1}{2}$. From Theorem 4.1.14 we obtain an $\mathcal{L}^p$-strong Feller diffusion $\mathbf{M}$ with coordinate maps $(\widetilde{\phi}_t)_{t \geq 0}$. Since $(\mathcal{E}, D(\mathcal{E}))$ is conservative, $\mathbf{M}$ is conservative for every starting point in E_1. From Theorem 6.2.3 we get a boundary local time $(\ell_t)_{t \geq 0}$ that is in Revuz correspondence to $1_{\Gamma_2} \Sigma$. By Theorem 6.3.2 we get a semimartingale decomposition for $(\widetilde{\phi}_t)_{t \geq 0}$. Now choose $\widetilde{\phi}_0 \in \widetilde{\mathbb{V}}_{N-1}$. Then we get by Theorem 6.3.5 a weak solution on a probability space $(\Omega, \mathcal{F}, \mathbb{P}_{\widetilde{\phi}_0})$ with an $(N-1)$-dimensional Brownian motion $(\widetilde{W}_t)_{t \geq 0}$ such that for $t \geq 0$

$$\widetilde{\phi}_t - \widetilde{\phi}_0 = -\int_0^t \widetilde{\sigma}^\top \widetilde{\sigma} \widetilde{V}'(\widetilde{\sigma}^\top \widetilde{\sigma} \widetilde{\phi}_s) \, ds$$

$$+ \sum_{k=1}^{N} \int_0^t \frac{\widetilde{\varrho}(\widetilde{\phi}_s)}{\|\widetilde{\sigma}_{*k}^\top\|} \widetilde{\sigma}_{*k}^\top 1_{T^{(k)}} (\widetilde{\phi}_s) \, d\ell_s + \sqrt{2} \, \widetilde{W}_t. \quad (6.27)$$

Define

$$\phi_t := c\mathbf{1} + \widetilde{\sigma} \widetilde{\phi}_t \in \mathbb{V}_N^{c,+}.$$

Let us identify the SDE that is solved by $(\phi_t)_{t\geq 0}$. Applying $\tilde{\sigma}$ on both sides of (6.27) we obtain

$$\phi_t - \phi_0 = -\tilde{\sigma} \int_0^t \tilde{\sigma}^\top \tilde{\sigma} \tilde{V}'(\tilde{\sigma}^\top \tilde{\sigma} \tilde{\phi}_s) \, ds$$

$$+ \sum_{k=1}^N \tilde{\sigma} \int_0^t \frac{\tilde{\varrho}(\tilde{\phi}_s)}{\|\tilde{\sigma}_{*k}^\top\|} \tilde{\sigma}_{*k}^\top 1_{T^{(k)}} (\tilde{\phi}_s) \, d\ell_s + \sqrt{2}\tilde{\sigma}\tilde{W}_t.$$

Let us identify the right-hand side by objects related to $(\phi_t)_{t\geq 0}$ instead of $(\tilde{\phi}_t)_{t\geq 0}$. First of all, note that $\tilde{\sigma}\tilde{\sigma}^\top = \sigma\sigma^\top$. We have $\tilde{\sigma}^\top(c1) = 0$, hence $\tilde{\sigma}^\top(\tilde{\sigma}\tilde{\phi}_s) = \tilde{\sigma}^\top(\phi_s)$, $0 \leq s < \infty$. The definition of $\tilde{V}'$ and $\hat{V}'$ together with the fact that the first $N-1$ rows of $\tilde{\sigma}^\top$ and $\hat{\sigma}^\top$ coincide yields then

$$\tilde{V}_k'(\tilde{\sigma}^\top \tilde{\sigma} \tilde{\phi}_s) = \hat{V}_k'(\sigma^\top \phi_s), 1 \leq k \leq N-1, 0 \leq s < \infty.$$

Thus

$$\tilde{\sigma}\tilde{V}'(\tilde{\sigma}^\top \tilde{\sigma} \tilde{\phi}_s) = \sigma\hat{V}'(\sigma^\top \phi_s), 0 \leq s < \infty.$$

So altogether, we get

$$-\tilde{\sigma} \int_0^t \tilde{\sigma}^\top \tilde{\sigma} \tilde{V}'(\tilde{\sigma}^\top \tilde{\sigma} \tilde{\phi}_s) \, ds = -\sigma\sigma^\top \int_0^t \sigma\hat{V}'(\sigma^\top \phi_s) \, ds.$$

Next we consider the integral w.r.t. the local time. Define

$$\hat{T}^{(k)} := \left\{ \phi \in \partial\mathbb{V}_N^{c,+} \mid \phi(k) = 0, \phi(i) > 0 \text{ for } 1 \leq i \leq N, i \neq k \right\}.$$

Note that $\hat{T}^{(k)} = c1 + \tilde{\sigma}T^{(k)}$. Define $\hat{\ell} : \mathbf{\Omega} \times [0,\infty) \to \mathbb{R}^N$, $\hat{\ell} = (\hat{\ell}^{(1)}, ..., \hat{\ell}^{(N)})$,

$$\hat{\ell}^{(k)} = \frac{\tilde{\varrho}}{\|\tilde{\sigma}_{*k}^\top\|} 1_{\hat{T}^{(k)}} \cdot \ell,$$

i.e.,

$$\hat{\ell}_t^{(k)} = \int_0^t \frac{\tilde{\varrho}(\tilde{\phi}_s)}{\|\tilde{\sigma}_{*k}^\top\|} 1_{T^{(k)}} (\tilde{\phi}_s) \, d\ell_s = \int_0^t \frac{\tilde{\varrho}(\tilde{\phi}_s)}{\|\tilde{\sigma}_{*k}^\top\|} 1_{\hat{T}^{(k)}} (\phi_s) \, d\ell_s.$$

Then

$$\sum_{k=1}^N \tilde{\sigma} \int_0^t \frac{\tilde{\varrho}(\tilde{\phi}_s)}{\|\tilde{\sigma}_{*k}^\top\|} \tilde{\sigma}_{*k}^\top 1_{T^{(k)}} (\tilde{\phi}_s) \, d\ell_s = \sum_{k=1}^N \tilde{\sigma} \int_0^t \tilde{\sigma}_{*k}^\top \, d\hat{\ell}_s^{(k)}$$

$$= \tilde{\sigma} \int_0^t \tilde{\sigma}^\top \, d\hat{\ell}_s = \tilde{\sigma}\tilde{\sigma}^\top \hat{\ell}_t = \sigma\sigma^\top \hat{\ell}_t.$$

Furthermore,

$$\int_0^\infty \phi_t \, d\hat{\ell}_t = \sum_{k=1}^N \int_0^\infty \phi_t(k) 1_{\hat{T}^{(k)}}(\phi_t) \, d\hat{\ell}_t^k = 0.$$

Finally we may extend the probability space to obtain an N-dimensional Brownian motion $(W_t)_{t \geq 0}$ such that $W^{(k)} = \widetilde{W}^{(k)}$ for $1 \leq k \leq N - 1$. Then for $0 \leq t < \infty$

$$\tilde{\sigma}\widetilde{W}_t = \sigma W_t.$$

Summarizing we get

$$\phi_t - \phi_0 = -\sigma\sigma^\top \int_0^t \sigma \hat{V}'(\sigma^\top \phi_s) ds + \sigma\sigma^\top \hat{\ell}_t + \sqrt{2}\sigma W_t \text{ for } t \geq 0.$$

Thus $(\phi_t)_{t \geq 0}$ is a solution to (6.26).

Let us now connect this result with results from [Zam08]. As in [Zam08] we assume that V is convex and C^2, hence the gradient is non-decreasing. So with the same proof as in [Zam08, Lem. 3.2] we get pathwise uniqueness of weak solutions. With the Yamada-Watanabe principle we can conclude uniqueness in law, see [KS91, Prop. 3.20]. We get the following result. Define $J : \widetilde{\mathbb{V}}_{N-1} \to \mathbb{V}_N^{c,+}$ by $\widetilde{\mathbb{V}}_{N-1} \ni \phi \mapsto c\mathbf{1} + \tilde{\sigma}\phi \in \mathbb{V}_N^{c,+}$. Then J is an isomorphism. Define by $\mathbf{J}$ the corresponding mapping from $C([0, \infty), \widetilde{\mathbb{V}}_{N-1})$ to $C([0, \infty), \mathbb{V}_N^{c,+})$, i.e., $\mathbf{J}((\tilde{\phi}_t)_{t \geq 0}) = (J\tilde{\phi}_t)_{t \geq 0}$.

Corollary 6.5.1. *Let* $(\mathbb{P}_{\phi_0})_{\phi_0 \in \widetilde{\mathbb{V}}_{N-1}}$ *the path measure of the* $\mathcal{L}^p$-*strong Feller process associated with* $(\mathcal{E}, D(\mathcal{E}))$ *due to Theorem 4.1.14. For* $\mu_0 \in \mathcal{P}(\widetilde{\mathbb{V}}_N^{c,+})$ *set* $\tilde{\mu}_0 := \mu_0 \circ J \in \mathcal{P}(\widetilde{\mathbb{V}}_{N-1})$ *(image measure under* J^{-1}*). Define* $\mathbb{P}_{\mu_0}^{\mathbb{V}_N^{c,+}} := \mathbb{P}_{\tilde{\mu}_0} \circ \mathbf{J}^{-1}$. *There exists a weak solution to* (6.26) *with initial distribution given by* μ_0. *The law is given by* $\mathbb{P}_{\mu_0}^{\mathbb{V}_N^{c,+}}$. *Solutions to* (6.26) *are pathwise unique and hence also unique in law. In particular, for every weak solution (with initial distribution* μ_0*) the law is given by* $\mathbb{P}_{\mu_0}^{\mathbb{V}_N^{c,+}}$.

Note that we allow general probability measures on $\widetilde{\mathbb{V}}_N^{c,+}$ as initial distribution. So in particular, one can construct weak solutions with deterministic starting configuration $\phi_0 \in \widetilde{\mathbb{V}}_N^{c,+}$.

In [Zam08] existence and pathwise uniqueness of weak solutions for every initial condition in $[0, \infty)^{\Gamma_N}$ is shown. Moreover, a family of path measures $(\mathbb{P}_x^*)_{x \in \mathbb{V}_N^{c,+}}$ is constructed from a strong Feller semigroup associated with a gradient Dirichlet form on $\mathbb{V}_N^{c,+}$. However, the identification of the law of the weak solution with the path measure $(\mathbb{P}_{\phi_0}^*)_{\phi_0 \in \mathbb{V}_N^{c,+}}$ is established for quasi-every configuration $\phi_0 \in \mathbb{V}_N^{c,+}$ only.

7 Appendix

7.1 Basics on Metric Spaces

In this section we consider some useful results concerning locally compact separable metric spaces with a locally finite Borel measure μ. Let $(E, \mathbf{d})$ be a metric space. We say that E is *locally compact* if for every $x \in E$ there exists an open neighborhood U of x such that $\overline{U}$ is compact.

We call $(E, \mathbf{d})$ *separable* if there exists a countable dense set $D \subset E$, i.e., $\overline{D} = E$.

Lemma 7.1.1. *Let $(E, \mathbf{d})$ be a locally compact separable metric space. Then there exists a sequence $B_n \subset E$, $n \in \mathbb{N}$, of open balls with compact closure covering E. Moreover, there exists an increasing sequence of compact sets K_n with $K_n \subset \overset{\circ}{K}_{n+1}$, $n \in \mathbb{N}$, covering E.*

See also [CB06, Cor. 2.77].

Proof. Let Q be a countable dense set in E. Define

$$E_k := \left\{ x \in E \,\middle|\, \overline{B_{\frac{1}{k}}(x)} \text{ is compact} \right\}.$$

Moreover, define

$$Q_k := \left\{ y \in Q \,\middle|\, \overline{B_{\frac{1}{2k}}(y)} \text{ is compact} \right\}.$$

Set

$$\widetilde{Q} := \left\{ (k, y) \in \mathbb{N} \times Q \,\middle|\, y \in Q_k \right\}.$$

Then $\widetilde{Q} \subset \mathbb{N} \times Q$ is also countable. Let $\mathbb{N} \ni n \mapsto (k_n, y_n) \in \widetilde{Q}$ be a bijection onto $\widetilde{Q}$. Set $B_n := B_{\frac{1}{2k_n}}(y_n)$. Then each B_n is open and has compact closure in E by definition of Q_k. We claim that it is an covering, i.e.,

$$E = \bigcup_{n \in \mathbb{N}} B_n = \bigcup_{(k,y) \in \widetilde{Q}} B_{\frac{1}{2k}}(y).$$

Let $x \in E$. Then there exists an $\varepsilon > 0$ such that $\overline{B_\varepsilon(x)}$ is compact. Choose $k \in \mathbb{N}$ such that $1/\varepsilon > 1/k$. Then also $\overline{B_{1/k}(x)}$ is compact. Thus $E = \bigcup_{k \in \mathbb{N}} E_k$. Furthermore, there exists $x' \in Q$ such that $\mathrm{d}(x, x') < \frac{1}{2k}$. Then $B_{\frac{1}{2k}}(x') \subset B_{\frac{1}{k}}(x)$ and $\overline{B_{\frac{1}{2k}}(x')}$ is compact. Thus $x' \in Q_k$ and $x \in B_{\frac{1}{2k}}(x')$.

So we get an open covering of sets with compact closure. However, these sets need not to be increasing. We construct such sets now. Denote for an open set $U \subset E$ by U^δ the set of all points in U with distance to the boundary greater or equal δ, i.e.,

$$U^\delta := \left\{ x \in U \mid \mathrm{dist}(x, \partial U) \geq \delta \right\}. \tag{7.1}$$

Note that this set is closed for $\delta > 0$ and since U is open we have $U = \bigcup_{n \in \mathbb{N}} U^{1/n}$. Define

$$K_n := \bigcup_{l=1}^{n} B_l^{1/n}, \, n \in \mathbb{N}.$$

Since $B_l^{1/n}$ is closed and a subset of $\overline{B_l}$, we have that $B_l^{1/n}$ is compact and thus also K_n is compact. And by construction we have that $K_n \subset \overset{\circ}{K}_{n+1}$. Clearly

$$\bigcup_{n \in \mathbb{N}} K_n = \bigcup_{n \in \mathbb{N}} \bigcup_{l=1}^{n} B_l^{1/n} = \bigcup_{k \in \mathbb{N}} B_k = E.$$

$\square$

Corollary 7.1.2. *Let* (E, d) *as in the previous theorem and assume that* μ *is a locally finite measure on* $(E, \mathcal{B}(E))$. *Then* $\mathcal{B}(E)$ *is generated by the open sets with compact closure and finite* μ-*measure.*

We state the following simple but technical lemma concerning partitions for finite sets.

Lemma 7.1.3. *(i). Let* A *be a set,* $A_i \subset A$, $1 \leq i \leq M$, $M \in \mathbb{N}$, *such that*

$$A = \bigcup_{i=1}^{M} A_i.$$

Then there exists a partition $(\tilde{A}_i)_{1 \leq i \leq \tilde{M}}$ *of* A, *i.e., the* $\tilde{A}_i$ *are pairwise disjoint and*

$$A = \bigcup_{i=1}^{\tilde{M}} \tilde{A}_i,$$

such that for each A_i, $1 \leq i \leq M$, there exists $I_i \subset \{1, ..., \tilde{M}\}$ such that

$$A_i = \bigcup_{i \in I_i} \tilde{A}_i.$$

Such a partition is called a refinement of $(A_i)_{1 \leq i \leq M}$.

(ii) Let $(C_i)_{1 \leq i \leq M}$, $M \in \mathbb{N}$, be a partition for some set C, $(D_i)_{1 \leq i \leq N}$, $N \in \mathbb{N}$, a partition for some set D. Then there exists a partition $(F_i)_{1 \leq i \leq \tilde{M}}$, $\tilde{M} \in \mathbb{N}$, such that

$$C \cup D = \bigcup_{i=1}^{\tilde{M}} F_i$$

and for each set G in $(C_i)_{1 \leq i \leq M}$ or $(D_i)_{1 \leq i \leq N}$ there exists $I_G \subset \{1, ..., \tilde{M}\}$ such that

$$G = \bigcup_{i \in I_G} F_i.$$

Proof. (i): We prove by induction. Clearly for $M = 1$ the claim is fulfilled. Now let $(A_i)_{1 \leq i \leq M+1}$ and let $(\tilde{A}_i)_{1 \leq i \leq \tilde{M}}$ be the corresponding partition for $\cup_{i=1}^{M} A_i$ and refinement of $(A_i)_{1 \leq i \leq M}$. Define

$$\mathcal{F} = \left\{ \tilde{A}_i \cap A_{M+1} \,\middle|\, 1 \leq i \leq \tilde{M} \right\} \cup \left\{ \tilde{A}_i \setminus A_{M+1} \,\middle|\, 1 \leq i \leq \tilde{M} \right\}$$
$$\cup \left\{ A_{M+1} \setminus \cup_{i=1}^{M} \tilde{A}_i \right\}.$$

Obviously, every $\tilde{A}_i$, $1 \leq i \leq \tilde{M}$, and A_{M+1} can be obtained as a finite union of sets in $\mathcal{F}$. By construction every A_i, $1 \leq i \leq \tilde{M}$ can be obtained as a finite union of sets in $(\tilde{A}_i)_{1 \leq i \leq M}$. Now choose a numbering for the sets in $\mathcal{F}$ to obtain the desired partition.

(ii): Define $F'_{i,j} = C_i \cap D_j$, for $1 \leq i \leq M$, $1 \leq j \leq N$. Clearly the $F'_{i,j}$ are pairwise disjoint from each other. Moreover,

$$C_i = \bigcup_{j=1}^{N} F'_{i,j},$$

the analogous property holds for every D_j, $1 \leq j \leq N$. Now drop all empty $F'_{i,j}$ and choose a numbering to get $(F_i)_{1 \leq i \leq \tilde{M}}$ for some $\tilde{M} \in \mathbb{N}$. □

Lemma 7.1.4. *Let $(K, \mathbf{d})$ be a compact metric space. Then there exists a family $\mathcal{Z} = (Z_i^n)_{1 \leq i \leq M_n, n \in \mathbb{N}}$ with $M_n \in \mathbb{N}$ of subsets of K with the property*

1. $K = \bigcup_{i=1}^{M_n} Z_i^n$ for every $n \in \mathbb{N}$.

2. $\sigma\left(\{Z_i^n \mid 1 \leq i \leq M_n, \ n \in \mathbb{N}\right) = \mathcal{B}(K)$, $M_n \in \mathbb{N}$,

Proof. For $n \in \mathbb{N}$ choose $x_1^{(n)}, ..., x_{M_n}^{(n)}$ such that

$$K = \bigcup_{i=1}^{M_n} B_{1/n}(x_i^{(n)}).$$

Such a covering exists due to compactness of K. Set $Z_i^n := B_{1/n}(x_i)$, $1 \leq i \leq M_n$, $n \in \mathbb{N}$.

Define

$$\mathcal{F} := \sigma\left(\left\{Z_i^{(n)} \mid 1 \leq i \leq M_n, \ n \in \mathbb{N}\right\}\right).$$

We prove $\mathcal{F} = \mathcal{B}(K)$. Clearly, $\mathcal{F} \subset \mathcal{B}(K)$. Let us check the converse equality. Let $U \subset K$ open. Define U^δ as in (7.1) for $\delta > 0$. These U^δ are closed and hence compact. Clearly, $U = \cup_{k=1}^\infty U^{1/k}$. For each $U^{1/k}$, $k \in \mathbb{N}$, we get a finite covering by balls with radius $1/3k$ and middle-point $x_i^{(3k)}$, i.e., by $Z_i^{(3k)}$, $1 \leq i \leq M_{3k}$. Fix $k \in \mathbb{N}$. Choose only those balls such that $U^{1/k} \cap B_{\frac{1}{3k}}(x_i) \neq \emptyset$. Denote by $\tilde{I}^k$ the corresponding index set. So $U^{1/k} \subset \cup_{i \in \tilde{I}^k} B_{\frac{1}{3k}}(x_i^{(3k)})$. For $y, z \in B_{\frac{1}{3k}}(x_i^{(3k)})$ it holds $\mathbf{d}(y,z) \leq \frac{2}{3k}$. There exists at least one $y \in B_{\frac{1}{3k}}(x_i^{(3k)}) \cap U^{1/k}$. Since $\mathrm{dist}(y, \partial U) \geq \frac{1}{k}$, we get for every $z \in B_{\frac{1}{3k}}(x_i^{(3k)})$ that $\mathrm{dist}(z, \partial U) \geq \frac{1}{3k}$. In particular, $B_{\frac{1}{3k}}(x_i) \subset U$. So

$$U = \bigcup_{k \in \mathbb{N}} U^{\frac{1}{k}} \subset \bigcup_{k \in \mathbb{N}} \bigcup_{i \in \tilde{I}^k} B_{\frac{1}{3k}}(x_i^{(3k)}) \subset U.$$

The middle expression is in $\mathcal{F}$ by assumption, hence $U \in \mathcal{F}$. So $\mathcal{B}(K) = \mathcal{F}$. $\square$

Corollary 7.1.5. *Let $(K, \mathbf{d})$ be a compact metric space. Then there exists a family of sets $(Z_i^{(n)})_{1 \leq i \leq M_n, n \in \mathbb{N}}$ with $M_n \in \mathbb{N}$, $n \in \mathbb{N}$, such that for every $n \in \mathbb{N}$ the family $(Z_i^{(n)})_{1 \leq i \leq M_n}$ is a partition of K and for each $1 \leq i \leq M_n$, there exists $I_n' \subset I_{n+1}$ such that*

$$Z_i^{(n)} = \bigcup_{j \in I_n'} Z_j^{(n+1)}$$

and

$$\mathcal{B}(K) = \sigma\left(\left\{Z_i^{(n)} \mid 1 \leq i \leq M_n, n \in \mathbb{N}\right\}\right).$$

Proof. Denote by $Z_i'^{(n)}$, $1 \leq i \leq M_n'$, $M_n', n \in \mathbb{N}$, the covering of Lemma 7.1.4. Define $Z^{(1)}$ to be the refinement of $Z'^{(1)}$ according to Lemma 7.1.3(i). Assume that $Z^{(k)}$, $1 \leq k \leq N$, are constructed. Then we obtain $Z^{(N+1)}$ by combining the refinement of $Z'^{(N+1)}$ (acc. to Lemma 7.1.3(i)) with $Z^{(N)}$ according to Lemma 7.1.3(ii).

Thus every set in $Z^{(n)}$, $n \in \mathbb{N}$, can be obtained as finite union of sets in $Z^{(k)}$ for every $k \geq n$. Thus

$$\mathcal{B}(K) = \sigma\left(\left\{Z_i'^{(n)} \mid 1 \leq i \leq M_n', n \in \mathbb{N}\right\}\right)$$
$$\subset \sigma\left(\left\{Z_i^{(n)} \mid 1 \leq i \leq M_n, n \in \mathbb{N}\right\}\right) \subset \mathcal{B}(K).$$

$\square$

We call a mapping $P : (X, \mathcal{G}) \times \mathcal{B}(K) \to \mathbb{R}_0^+$ a *transition kernel* if for every $x \in X$, $P(x, \cdot)$ is a measure on $\mathcal{B}(K)$ and for every $A \in \mathcal{B}(K)$ the mapping $X \ni x \mapsto P(x, A) \in \mathbb{R}$ is $\mathcal{G}$-measurable. A transition kernel is called (locally) finite if for every $x \in X$, $P(x, \cdot)$ is a (locally) finite measure.

For two measurable spaces $(S_1, \mathcal{A})$ and $(S_2, \mathcal{B})$ we denote by $\mathcal{A} \otimes \mathcal{B}$ the *product σ-algebra* on $S_1 \times S_2$ generated by the sets of the form $A \times B$, $A \in \mathcal{A}$, $B \in \mathcal{B}$.

Lemma 7.1.6. *Let $(X, \mathcal{G})$ be a measurable space, $(K, \mathbf{d})$ a compact metric space and μ a probability Borel measure on K. Assume that $P : (X, \mathcal{G}) \times \mathcal{B}(K) \to \mathbb{R}_0^+$ is a finite transition kernel such that for every $x \in X$ the measure $P(x, \cdot)$ is absolutely continuous w.r.t. μ. Then P has a $\mathcal{G} \otimes \mathcal{B}(K)$-measurable density.*

Proof. Applying for each $x \in X$ the Radon-Nikodym theorem to $P(x, \cdot)$ we obtain a mapping $p : X \times K \to \mathbb{R}_0^+$ such that $p(x, \cdot)$ is a density for $P(x, \cdot)$. Furthermore, $p(x, \cdot)$ is $\mathcal{B}(K)$-measurable for every $x \in X$. From this we construct a sequence of $\mathcal{G} \otimes \mathcal{B}(K)$-measurable functions and prove that they converge to a $\mathcal{G} \otimes \mathcal{B}(K)$-measurable mapping $\tilde{p}(x, y)$ which yields for every $x \in X$ a μ-version of $p(x, \cdot)$. Choose the family of sets $Z^{(n)} := (Z_i^{(n)})_{1 \leq i \leq M_n, n \in \mathbb{N}}$ with $M_n, n \in \mathbb{N}$, provided by Corollary 7.1.5. Define

$$p_n(x, y) = \sum_{1 \leq i \leq M_n} \frac{P(x, Z_i^n)}{\mu(Z_i^n)} 1_{Z_i^n}(y).$$

Clearly p_n is $\mathcal{G} \otimes \mathcal{B}(K)$-measurable. Set $\mathcal{F}_n := \sigma(Z^{(n)})$, $n \in \mathbb{N}$. Note that $\mathcal{F}_n \subset \mathcal{F}_{n+1}$, $n \in \mathbb{N}$, by construction. Moreover, $\sigma(\mathcal{F}_n \mid n \in \mathbb{N}) = \mathcal{B}(K)$

by Corollary 7.1.5. Denote by $\mathbb{E}[\,\cdot\,]$ the expectation and by $\mathbb{E}[\,\cdot\mid\cdot\,]$ the conditional expectation w.r.t. μ. We claim that $p_n(x,\cdot) = \mathbb{E}[p(x,\cdot)|\mathcal{F}_n]$. Let $n \in \mathbb{N}$ and $i \in I_n$. We show that

$$\mathbb{E}[p(x,\cdot)1_A(\cdot)] = \mathbb{E}[p_n(x,\cdot)1_A(\cdot)] \quad \text{for all } A \in \mathcal{F}_n. \tag{7.2}$$

Since $\mathcal{F}_n$ consists of the family of finite (disjoint) unions of Z_i^n, $1 \le i \le M_n$, it is enough to check (7.2) on these sets by linearity of the expression. We have

$$\mathbb{E}[p(x,\cdot)1_{Z_i^n}(\cdot)] = \int_K p(x,y)1_{Z_i^n}(y)d\mu(y) = P(x, Z_i^n),$$

and

$$\mathbb{E}[p_n(x,\cdot)1_{Z_i^n}(\cdot)] = \mathbb{E}\left[\frac{P(x, Z_i^n)}{\mu(Z_i^n)}1_{Z_i^n}(\cdot)\right] = \frac{P(x, Z_i^n)}{\mu(Z_i^n)}\mathbb{E}[1_{Z_i^n}(\cdot)] = P(x, Z_i^n).$$

This shows (7.2). Thus $(p_n(x,\cdot))_{n\in\mathbb{N}}$ is an $\mathcal{F}_n$-martingale with $p_n(x,\cdot) = \mathbb{E}[p(x,\cdot)|\mathcal{F}_n]$ and by martingale convergence it follows that μ-a.s. it holds

$$p_n(x,y) \overset{n\to\infty}{\longrightarrow} p(x,y).$$

Moreover, we have that the set

$$M := \big\{(x,y) \in X \times K \,\big|\, \text{The limit of } p_n(x,y) \text{ exists.}\big\},$$

is $\mathcal{G} \otimes \mathcal{B}(K)$-measurable since the functions p_n are $\mathcal{G} \otimes \mathcal{B}(K)$-measurable. Set $\tilde{p}(x,y) := 1_M \lim_{n\to\infty} p_n(x,y)$. Then $\tilde{p}(x,y)$ is well-defined and $\mathcal{G} \otimes \mathcal{B}(K)$-measurable. Moreover, for every $x \in X$ it holds

$$\tilde{p}(x,y) = 1_M \lim_{n\to\infty} p_n(x,y) = p(x,y) \quad \text{for } \mu\text{-a.e. } y \in K.$$

Thus $\tilde{p}(x,\cdot)$ is a μ-version of $p(x,\cdot)$ and hence also a density for $P(x,\cdot)$, $x \in X$. $\qquad\square$

The essential proof steps are taken from [Doo53, Theo. 2.5] and [Doo53, Exa. 2.7]. They can be also found in [DM82, Ch. 5, Theo. 58].

Next we drop the compactness assumption. Let $(E, \mathbf{d})$ be a metric space. Denote by $\mathcal{B}(E)|_K$, $K \in \mathcal{B}(E)$, $(E, \mathbf{d})$ a metric space, the trace σ-algebra of $\mathcal{B}(E)$ on K, i.e., $\mathcal{B}(E)|_K = \{F \cap K \in \mathcal{B}(E) \,|\, F \in \mathcal{B}(E)\}$. Note that $\mathcal{B}(E)|_K = \mathcal{B}(K)$ if K is endowed with the trace topology on K, i.e., $U \subset K$ is open iff there exists $\widetilde{U} \subset E$ open such that $U = K \cap \widetilde{U}$.

Theorem 7.1.7. *Let $(X, \mathcal{G})$ be a measurable space, $(E, \mathbf{d})$ a locally compact separable metric space and μ a locally finite Borel measure on E. Assume that $P : (X, \mathcal{G}) \times \mathcal{B}(E) \to \mathbb{R}_0^+$ is a transition kernel of probabilities such that for every $x \in X$, the measure $P(x, \cdot)$ is absolutely continuous w.r.t. μ. Then p has a $\mathcal{G} \otimes \mathcal{B}(E)$-measurable density.*

Proof. By Lemma 7.1.1 we get an increasing sequence of compact sets $(K_n)_{n \in \mathbb{N}}$ such that $E = \bigcup_{n \in \mathbb{N}} K_n$. Set $K_0 := \emptyset$. Set $I := \{n \in \mathbb{N} \,|\, \mu(K_n) > 0\}$. Let $p : X \times \mathcal{B}(E) \to \mathbb{R}_0^+$ be the density of P due to Radon-Nikodym theorem. Define $q_n := \frac{1}{\mu(K_n)} p|_{X \times K_n}$. Then for every $n \in \mathbb{N}$, $q_n(x, \cdot)$ is the density of the probability measure $q_n \mu|_{\mathcal{B}(K_n)}$ on $\mathcal{B}(K_n)$. So by Lemma 7.1.6 there exists a $\mathcal{G} \otimes \mathcal{B}(K_n)$-measurable density $\tilde{q}_n(x, \cdot)$. Since $\mathcal{B}(E)|_{K_n} = \mathcal{B}(K_n)$, the mapping $X \otimes E \ni (x, y) \mapsto 1_{K_n \setminus K_{n-1}}(y) \tilde{q}_n(x, \cdot)$ is $\mathcal{G} \otimes \mathcal{B}(E)$-measurable for $n \in I$. Set

$$\tilde{p}(x, y) := \sum_{n \in I} \mu(K_n \setminus K_{n-1}) 1_{K_n \setminus K_{n-1}}(y) \tilde{q}_n(x, \cdot).$$

Note that by absolute continuity of $P(x, \cdot)$ w.r.t. μ we have that $P(x, A) = P(x, A \cap (\bigcup_{n \in I} K_n))$. So $\tilde{p}$ is a $\mathcal{G} \otimes \mathcal{B}(E)$-measurable density for P. $\square$

7.2 Sub-Markovian Semigroups and Resolvents

In this section we provide important facts concerning sub-Markovian semigroups and resolvents on L^p-spaces. The most important statement is the Beurling-Deny theorem which allows to extend sub-Markovian analytic L^2-semigroups to sub-Markovian analytic L^p-semigroups for $1 < p < \infty$.

We assume the reader to be familiar with the notion of Dirichlet forms, strongly continuous contraction semigroups/resolvents and generators as well as the relation between these objects. See [FOT11, Ch. I] or [MR92, Ch. I]. For the semigroup theory without connection to Dirichlet forms the reader can consult [Paz83] or [EN00].

Let $(X, \|\cdot\|)$ be a Banach space and $(T_t)_{t \geq 0}$ a strongly continuous contraction semigroup (s.c.c.s) on X. Let $(G_\lambda)_{\lambda > 0}$ be a family of bounded linear operators on X. We say that $(G_\lambda)_{\lambda > 0}$ is the resolvent of $(T_t)_{t \geq 0}$ if for all $u \in X$ it holds

$$G_\lambda u = \int_0^\infty \exp(-\lambda s) T_s u \, ds \quad \text{for all } \lambda > 0. \tag{7.3}$$

The integral is constructed as a Bochner integral. Note that $(G_\lambda)_{\lambda>0}$ is then a strongly continuous contraction resolvent (s.c.c.r) on X, cf. [MR92, Ch. I, Prop. 1.10]. The resolvent G_λ, $\lambda > 0$, is the inverse to $\lambda - L$. Here $(L, D(L))$ is the infinitesimal generator of $(T_t)_{t\geq 0}$.

If $X = L^p(E, \mu)$ for a measure space $(E, \mathcal{B}, \mu)$ and some $1 \leq p \leq \infty$, we say that $(T_t)_{t\geq 0}$ is an L^p-s.c.c.s. and $(G_\lambda)_{\lambda>0}$ is an L^p-s.c.c.r.

Let X be a Hilbert space with scalar product $(\cdot, \cdot)_X$. We say that a resolvent family $(G_\lambda)_{\lambda>0}$ is associated with a symmetric Dirichlet form $(\mathcal{E}, D(\mathcal{E}))$ if for $\lambda > 0$, $u \in X$ it holds

$$\mathcal{E}_\lambda(G_\lambda u, v) = (u, v)_X \quad \text{for all } v \in D(\mathcal{E}).$$

In this case $(G_\lambda)_{\lambda>0}$ is a strongly continuous contraction resolvent, see [MR92, Ch. I, Theo. 2.8]. We say that a s.c.c.s. $(T_t)_{t\geq 0}$ is associated with $(\mathcal{E}, D(\mathcal{E}))$ if the resolvent of the semigroup is associated with $(\mathcal{E}, D(\mathcal{E}))$.

So let us assume that $(T_t)_{t\geq 0}$ is a s.c.c.s with resolvent $(G_\lambda)_{\lambda>0}$ and generator $(L, D(L))$. Then we get the following lemma.

Lemma 7.2.1. *Both the semigroup and resolvent are contractive and strongly continuous on $D(L)$, i.e.,*

$\|T_t u\|_{D(L)} \leq \|u\|_{D(L)}$ *for $t > 0$ and* $\|\lambda G_\lambda u\|_{D(L)} \leq \|u\|_{D(L)}$ *for $\lambda > 0$ and all $u \in D(L)$.*

For $u \in D(L)$ it holds

$$\|T_t u - u\|_{D(L)} \overset{t\to 0}{\longrightarrow} 0 \text{ and } \|\lambda G_\lambda u - u\|_{D(L)} \overset{\lambda\to\infty}{\longrightarrow} 0.$$

The statements follow immediately from the fact that L and $(G_\lambda)_{\lambda>0}$ as-well as L and $(T_t)_{t\geq 0}$ commute on $D(L)$.

For the notion of an analytic semigroup we refer to [MR92, Ch. I, Def. 2.19], [Paz83, Ch. 2, Def. 5.1] or [EN00, Ch. II, Def. 4.5].

We state the following two important properties of analytic semigroups. Denote by $D(L^n)$ the domain of the n-th power of the generator L, $n \in \mathbb{N}$. By $\|\cdot\|_{D(L)}$ we denote the graph norm of the operator L on $D(L)$, i.e., $\|u\|_{D(L)} = \|u\| + \|Lu\|$, $u \in D(L)$.

Proposition 7.2.2. *Let $(T_t)_{t\geq 0}$ be a strongly continuous contraction semigroup on the Banach space $(X, \|\cdot\|_X)$. Assume that $(T_t)_{t>0}$ admits an analytic continuation to a sector of the complex plane. Then it holds:*

(i) $T_t u \in D(L^n)$ for every $n \in \mathbb{N}$ and $t > 0$.

(ii) There exists a constant $C < \infty$ such that

$$\|T_t u\|_{D(L)} \leq \left(1 + \frac{C}{t}\right) \|u\|_X.$$

For the proof see [EN00, Ch. II, Theo. 4.6c] or [Paz83, Ch. 2, Theo. 5.2d].
We turn now to sub-Markovian operators and need more structure on the space X. So we assume that $(E, \mathcal{B}, \mu)$ is a measure space and $X = L^2(E, \mu)$. A semigroup $(T_t)_{t \geq 0}$ is called sub-Markovian if it holds

$$0 \leq u \leq 1 \ \mu - \text{a.e. implies } 0 \leq T_t u \leq 1 \ \mu - \text{a.e. for } t \geq 0.$$

A resolvent family $(G_\lambda)_{\lambda > 0}$ is called sub-Markovian if it holds

$$0 \leq u \leq 1 \ \mu - \text{a.e. implies } 0 \leq \lambda G_\lambda u \leq 1 \ \mu - \text{a.e. for } \lambda > 0.$$

If $(T_t)_{t \geq 0}$ and $(G_\lambda)_{\lambda > 0}$ are associated with a coercive closed form, then these properties are equivalent to the *Dirichlet property* of the associated coercive bilinear form, i.e., for $u \in D(\mathcal{E})$ it holds $u^+ \wedge 1 \in D(\mathcal{E})$ and

$$\mathcal{E}(u + u^+ \wedge 1, u - u^+ \wedge 1) \geq 0$$
$$\mathcal{E}(u - u^+ \wedge 1, u + u^+ \wedge 1) \geq 0.$$

If $(\mathcal{E}, D(\mathcal{E}))$ is symmetric this condition is equivalent to

$$\mathcal{E}(u^+ \wedge 1, u^+ \wedge 1) \leq \mathcal{E}(u, u).$$

Now we can formulate the Beurling-Deny theorem.

Theorem 7.2.3. *Let $(T_t^2)_{t \geq 0}$ be a sub-Markovian L^2-s.c.c.s. on some measure space $(E, \mathcal{B}, \mu)$.*

(i) $T_t^2|_{L^1(E,\mu) \cap L^\infty(E,\mu)}$ extends to a contractive operator T_t^p on $L^p(E, \mu)$ for all $t \geq 0$ and $p \in [1, \infty)$. The family $(T_t^p)_{t \geq 0}$ forms a sub-Markovian L^p-s.c.c.s.

(ii) For $1 < p < \infty$, $(T_t^p)_{t \geq 0}$ is the restriction of an analytic semigroup.

See [LS96, Prop. 1.8] and [LS96, Rem. 1.2].
We denote the generator of $(T_t^p)_{t \geq 0}$ by $(L_p, D(L_p))$. The resolvent family we denote by $(G_\lambda^p)_{\lambda > 0}$.
We get the following lemma.

Lemma 7.2.4. *Let $(T_t^p)_{t \geq 0}$, $1 \leq p < \infty$, as in the previous theorem.*
(i) For $f \in L^1(E, \mu) \cap L^\infty(E, \mu)$ it holds $G_\lambda^2 f = G_\lambda^p f$ for all $\lambda > 0$.
(ii) Let $u \in D(L_2) \cap L^p(E, \mu)$. If $L_2 u \in L^p(E, \mu)$, then $u \in D(L_p)$.

Proof. (i): follows from (7.3).

(ii): Let $p \in [1, \infty)$ fixed. From (i) we can conclude that for $f \in L^2(E, \mu) \cap L^p(E, \mu)$ it holds $G_\lambda^p f = G_\lambda^2 f$. Indeed, define $f_n := 1_{\{\frac{1}{n} < |f| < n\}} f$. Then $f_n \overset{n \to \infty}{\longrightarrow} f$ both in L^2 and L^p. Hence $G_\lambda^2 f = G_\lambda^p f$ for $\lambda > 0$. Since $(1 - L_2)u \in L^p(E, \mu)$, we can define $\hat{u} := G_\lambda^p(1 - L_2)u \in D(L_p)$. Then we get

$$\hat{u} = G_\lambda^p(1 - L_2)u = G_\lambda^2(1 - L_2)u = u.$$

Thus $u = \hat{u} \in D(L_p)$. Applying $1 - L_p$ we get

$$(1 - L_p)\hat{u} = (1 - L_p)G_\lambda^p(1 - L_2)u = (1 - L_2)u.$$

So altogether, $L_p u = L_2 u$. □

7.3 Markov Processes

In this section we state basic definitions of stochastic processes and provide several useful lemmata. The definitions are taken from [MR92, Ch. IV, Sec. I]. First we introduce the notion of a *completion* of a σ-algebra. Let $(\Omega, \mathcal{B})$ be a measurable space and $\mathbb{P}$ a probability measure on $(\Omega, \mathcal{B})$. Define the *completion* $\mathcal{B}^\mathbb{P}$ of $\mathcal{B}$ w.r.t. $\mathbb{P}$ by

$$\mathcal{B}^\mathbb{P} := \{A \subset \Omega \,|\, A \Delta B \subset N \text{ for some } B, N \in \mathcal{B} \text{ with } \mathbb{P}(N) = 0\}.$$

Here Δ denotes the symmetric difference of A and B, i.e., $A \Delta B = (A \cup B) \setminus (A \cap B)$. The measure $\mathbb{P}$ is extended to $\mathcal{B}^\mathbb{P}$ in the natural way. Let $\mathcal{N}_\mathbb{P} := \{N \in \mathcal{B}^\mathbb{P} \,|\, \mathbb{P}(N) = 0\}$. For a sub-$\sigma$-algebra $\mathcal{C} \subset \mathcal{B}$ we define the completion of $\mathcal{C}$ in $\mathcal{B}$, denoted also by $\mathcal{C}^\mathbb{P}$, by

$$\mathcal{C}^\mathbb{P} := \{A \subset \Omega \,|\, A \Delta B \subset N \text{ for some } B \in \mathcal{C}, N \in \mathcal{B} \text{ with } \mathbb{P}(N) = 0\}.$$

Note that we assume $N \in \mathcal{B}$ here. We have $\mathcal{C}^\mathbb{P} = \sigma(\mathcal{C}, \mathcal{N}_\mathbb{P})$.

Let $(\Omega, \mathcal{M})$ be a measurable space. A family $(\mathcal{M}_t)_{t \geq 0}$ with $\mathcal{M}_t \subset \mathcal{M}$, $t \geq 0$, is called a *filtration* if $\mathcal{M}_s \subset \mathcal{M}_t$, $0 \leq s \leq t$. Define

$$\mathcal{M}_\infty = \sigma \Big(\bigcup_{0 \leq t < \infty} \mathcal{M}_t \Big)$$

and

$$\mathcal{M}_{t+} = \bigcap_{s > t} \mathcal{M}_s.$$

A filtration $(\mathcal{M}_t)_{t\geq 0}$ is called *right-continuous* if $\mathcal{M}_{t+} = \mathcal{M}_t$ for $t \geq 0$. One can prove that $(\mathcal{M}_{t+})^{\mathbb{P}} = (\mathcal{M}_t^{\mathbb{P}})_+$ for every $t \geq 0$ and probability measure $\mathbb{P}$ on $(\Omega, \mathcal{M})$.

To a given state space E we adjoin an extra point Δ. In the case of a locally compact metric space $(E, \mathbf{d})$ we take the Alexandrov compactification, i.e., the neighborhoods of Δ are the complements of compact sets in E.

Definition 7.3.1. We call $(\mathbf{X}_t)_{t\geq 0}$ a *stochastic process* with state space E and life time $\mathcal{X}$ (on a measurable space $(\Omega, \mathcal{M})$) if

(i) $\mathbf{X}_t : \Omega \mapsto E^{\Delta}$ is $\mathcal{M}/\mathcal{B}(E^{\Delta})$-measurable for all $t \in [0, \infty)$,

(ii) $\mathcal{X} : \Omega \mapsto [0, \infty]$ is $\mathcal{M}$-measurable.

(iii) For $\omega \in \Omega$, $\mathbf{X}_t(\omega) \in E$ for $t < \mathcal{X}$ and $\mathbf{X}_t(\omega) = \Delta$ for $t \geq \mathcal{X}$.

Item (iii) means that $\mathcal{X}(\omega)$ is the time at which the process hits the extra point Δ and the process stays at Δ forever. So Δ can be considered as a trap, cemetery or heaven for the process.

A stochastic process $(\mathbf{X}_t)_{t\geq 0}$ is called *measurable* if $(t, \omega) \mapsto \mathbf{X}_t(\omega)$ is $(\mathcal{B}([0, \infty)) \otimes \mathcal{M})/\mathcal{B}(E_{\Delta})$-measurable. $(\mathbf{X}_t)_{t\geq 0}$ is called $\mathcal{M}_t$-adapted if $\mathbf{X}_t$ is measurable w.r.t. $\mathcal{M}_t$, $t \geq 0$.

Definition 7.3.2. A tuple $\mathbf{M} := (\Omega, \mathcal{M}, (\mathcal{M}_t)_{t\geq 0}, (\mathbf{X}_t)_{t\geq 0}, (\mathbb{P}_x)_{x\in E^{\Delta}})$ is called a time-homogeneous *Markov process* with state space E and life time $\mathcal{X}$ if:

(i) The mapping $(\mathbf{X}_t)_{t\geq 0}$ is an $\mathcal{M}_t$-adapted stochastic process with life time $\mathcal{X}$ and state space E.

(ii) For each $t \geq 0$ there exists a *shift operator* $\theta_t : \Omega \to \Omega$ such that $\mathbf{X}_s \circ \theta_t = \mathbf{X}_{s+t}$ for all $t, s \geq 0$.

(iii) For each $x \in E^{\Delta}$, $\mathbb{P}_x$ is a probability measure on $\mathcal{M}$ and the mapping $x \mapsto \mathbb{P}_x(\Gamma)$ is $\mathcal{B}^*(E^{\Delta})$-measurable for $\Gamma \in \mathcal{M}$ and $\mathcal{B}(E^{\Delta})$-measurable for $\Gamma \in \sigma(\{\mathbf{X}_s \mid 0 \leq s < \infty\})$. Furthermore, $\mathbb{P}_{\Delta}(\mathbf{X}_0 = \Delta) = 1$.

(iv) For every $x \in E^{\Delta}$, $A \in \mathcal{B}(E^{\Delta})$ and $t, s \geq 0$

$$\mathbb{P}_x(\mathbf{X}_{s+t} \in A \mid \mathcal{M}_t) = \mathbb{P}_{\mathbf{X}_t}(\mathbf{X}_s \in A) \quad \mathbb{P}_x - \text{a.s..} \tag{7.4}$$

Property (iv) is called the *Markov property* of a process.

Analogously we define a Markov process on dyadic time-parameters, by restricting all times to $s, t \in S$. See Lemma 7.3.7 below for strong consequences of the Markov property.

Let $(\mathcal{G}_t)_{t \geq 0}$ be another filtration on $(\Omega, \mathcal{M})$. We say that $\mathbf{M}$ is *Markov w.r.t.* $(\mathcal{G}_t)_{t \geq 0}$ if (i) and (iv) still hold after replacing $(\mathcal{M}_t)_{t \geq 0}$ with $(\mathcal{G}_t)_{t \geq 0}$.

Denote by $\mathcal{P}(E^\Delta)$ the set of all probability measures on $(E^\Delta, \mathcal{B}(E^\Delta))$. We sometimes shortly write *a probability measure* ν *on* E^Δ if ν is defined on the Borel σ-algebra $\mathcal{B}(E^\Delta)$. An important σ-algebra containing $\mathcal{B}(E^\Delta)$ is $\mathcal{B}^*(E^\Delta)$ the family of *universally measurable sets* defined by

$$\mathcal{B}^*(E^\Delta) := \bigcap_{\nu \in \mathcal{P}(E^\Delta)} (\mathcal{B}(E^\Delta))^\nu.$$

Analogously we define $\mathcal{B}^*(E)$. Note that every $\nu \in \mathcal{P}(E^\Delta)$ canonically extends to a probability measure on $\mathcal{B}^*(E^\Delta)$. Let $\mathbf{M}$ be a Markov process with path measure $(\mathbb{P}_x)_{x \in E^\Delta}$. Let $\nu \in \mathcal{P}(E^\Delta)$. We define the probability measure $\mathbb{P}_\nu$ in the following way: Extend ν to a probability measure on $\mathcal{B}^*(E^\Delta)$. Define

$$\mathcal{M} \ni \Gamma \mapsto \mathbb{P}_\nu(\Gamma) := \int_{E^\Delta} \mathbb{P}_x(\Gamma) \, d\nu(x). \tag{7.5}$$

By Definition 7.3.2(iii) this is well-defined. Using monotone convergence one easily checks that this yields a measure and since $\mathbb{P}_x(\Omega) = 1$ for every $x \in E^\Delta$, $\mathbb{P}_\nu$ is also a probability measure.

If (7.4) holds with $\mathbb{P}_x$ replaced by $\mathbb{P}_\nu$ for a probability measure $\nu \in \mathcal{P}(E^\Delta)$, then we say that $\mathbf{M}$ is Markov under $\mathbb{P}_\nu$.

Define

$$\mathcal{F}'_t := \sigma(\mathbf{X}_s \mid 0 \leq s \leq t), 0 \leq t < \infty, \tag{7.6}$$

the *filtration generated by* $(\mathbf{X}_t)_{t \geq 0}$ and $\mathcal{F}' := \sigma(\mathbf{X}_t \mid 0 \leq t < \infty)$. Define

$$\mathcal{F} := \bigcap_{\nu \in \mathcal{P}(E^\Delta)} (\mathcal{F}')^{\mathbb{P}_\nu} \tag{7.7}$$

and

$$\mathcal{F}_t := \bigcap_{\nu \in \mathcal{P}(E^\Delta)} (\mathcal{F}'_t)^{\mathbb{P}_\nu}, \, 0 \leq t < \infty, \tag{7.8}$$

here $(\mathcal{F}'_t)^{\mathbb{P}_\nu}$ is the completion in $\mathcal{F}'$. Then $(\mathcal{F}_t)_{t \geq 0}$ is called *the natural filtration* of $\mathbf{M}$.

Lemma 7.3.3. *Let* **M** *be a Markov process (as in Definition 7.3.2) on the measurable space* $(\Omega, \mathcal{M})$. *Then the following measurability statements hold.*
(i) Let $F : \Omega \to \mathbb{R}$ *be* $\mathcal{F}'$-*measurable such that* F *is positive or* $\mathbb{E}_x[|F|] < \infty$ *for every* $x \in E^\Delta$. *Then the mapping* $E^\Delta \ni x \mapsto \mathbb{E}_x[F]$ *is* $\mathcal{B}(E^\Delta)$-*measurable.*
(ii) Let F *as in (i) but assume that* F *is only* $\mathcal{M}$-*measurable. Then the mapping* $E^\Delta \ni x \mapsto \mathbb{E}_x[F]$ *is* $\mathcal{B}^*(E^\Delta)$-*measurable.*

Proof. Using property (iii) of Definition 7.3.2 we get that for $\Gamma \in \mathcal{F}'$ the mapping $x \mapsto \mathbb{E}_x[1_\Gamma]$ is $\mathcal{B}(E^\Delta)$-measurable. So we may apply the functional monotone class theorem (Theorem 2.2.3) on the measurable space $(\Omega, \mathcal{F}')$ to conclude that $x \mapsto \mathbb{E}_x[F]$ is $\mathcal{B}(E^\Delta)$-measurable for $F \in \mathcal{B}_b(E^\Delta)$. Using the usual arguments we can conclude the measurability for the functions stated in (i).
(ii): We get again by Definition 7.3.2(iii) that for $\Gamma \in \mathcal{M}$ the mapping $x \mapsto \mathbb{E}_x[1_\Gamma]$ is $\mathcal{B}^*(E^\Delta)$-measurable. Then the proof works as in (i) but on $(\Omega, \mathcal{M})$. □

The following lemma is useful when dealing with completions.

Lemma 7.3.4. *Let* $(\mathbf{X}_t)_{t \geq 0}$ *be a stochastic process on* $(\Omega, \mathcal{M}')$ *adapted to the filtration* $(\mathcal{M}'_t)_{t \geq 0}$. *Let* E *be the state space of* $(\mathbf{X}_t)_{t \geq 0}$, *extended by* Δ. *Denote by* $\mathcal{M}$ *the completion of* $\mathcal{M}'$ *and by* $\mathcal{M}_t$ *the completion of* $\mathcal{M}'_t$ *in* $\mathcal{M}'$ *under* $\mathcal{P}(E^\Delta)$ *as in (7.8). Then the following statements hold.*
(i) $\mathbf{X}_t$ *is* $\mathcal{M}_t/\mathcal{B}^*(E^\Delta)$-*measurable.*
(ii) Assume that for $\Gamma \in \mathcal{M}'$ *it holds that the mapping*

$$E^\Delta \ni x \mapsto \mathbb{P}_x(\Gamma)$$

is $\mathcal{B}(E^\Delta)$-*measurable. Then for* $\Gamma \in \mathcal{M}$ *the mapping* $E^\Delta \ni x \mapsto \mathbb{P}_x(\Gamma)$ *is* $\mathcal{B}^*(E^\Delta)$-*measurable. Of course,* $\mathbb{P}_x$, $x \in E^\Delta$, *is extended to* $\mathcal{M}$ *in the canonical way.*

See [BG68, Ch. I, Prop. 5.10] and [BG68, Ch. I, Prop. 5.8].

Remark 7.3.5. Combining item (i) and (ii) of the previous lemma we see that $\Omega : \omega \mapsto \mathbb{P}_{\mathbf{X}_t(\omega)}(\Gamma)$ is $\mathcal{F}_t$-measurable for $\Gamma \in \mathcal{F}$.

Lemma 7.3.6. *Let* **M** *be a Markov process with filtration* $(\mathcal{M}_t)_{t \geq 0}$. *Then*

(i) **M** *is also Markov w.r.t.* $(\mathcal{F}'_t)_{t \geq 0}$. *If* $\mathcal{F}'_{t+} \subset \mathcal{M}_t$, $t \geq 0$, *then* **M** *is also Markov w.r.t.* $(\mathcal{F}'_{t+})_{t \geq 0}$.

(ii) **M** *is Markov under* $\mathbb{P}_\nu$ *w.r.t. the completion of* $(\mathcal{F}'_t)_{t \geq 0}$ *in* $\mathcal{F}$ *under* $\mathbb{P}_\nu$ *for every probability measure* ν *on* E^Δ.

(iii) **M** *is Markov w.r.t.* $(\mathcal{F}_t)_{t\geq 0}$ *under* $\mathbb{P}_\nu$ *for every probability measure* ν *on* E^Δ.

(iv) *If* **M** *is Markov w.r.t.* $(\mathcal{F}'_{t+})_{t\geq 0}$, *then* $(\mathcal{F}'_t)^{\mathbb{P}_\nu}$ *is right-continuous for every probability measure* ν *on* E^Δ.

Proof. (i): By definition, $(\mathbf{X}_t)_{t\geq 0}$ is $\mathcal{M}_t$-adapted. Thus $\mathcal{F}'_t \subset \mathcal{M}_t$ for every $t \geq 0$. Since $\mathbb{E}_{\mathbf{X}_t}[1_A(\mathbf{X}_s)]$ is already $\mathcal{F}'_t$-measurable for $s \geq 0$, $A \in \mathcal{B}(E^\Delta)$, we get the Markov property under $(\mathcal{F}'_t)_{t\geq 0}$.

The statement for $(\mathcal{F}'_{t+})_{t\geq 0}$ follows analogously.

(ii): Let $\Gamma \in (\mathcal{F}'_t)^{\mathbb{P}_\nu}$, then there exists $\tilde\Gamma \in \mathcal{F}'_t$, $N_1 \subset \tilde{N}_1 \in \mathcal{F}'$ and $N_2 \subset \tilde{N}_2 \in \mathcal{F}'$ with $\mathbb{P}_\nu(\tilde{N}_1) = 0$ and $\mathbb{P}_\nu(\tilde{N}_2) = 0$ such that $\Gamma = (\tilde\Gamma \cup N_2) \setminus N_1$. W.l.o.g. we may choose N_2 disjoint from $\tilde\Gamma$ and $N_1 \subset \tilde\Gamma$. Let $A \in \mathcal{B}(E)$. Then

$$\mathbb{E}_\nu\big[1_\Gamma 1_A(\mathbf{X}_{t+s})\big] = \mathbb{E}_\nu\big[(1_{\tilde\Gamma} + 1_{N_2} - 1_{N_1})1_A(\mathbf{X}_{t+s})\big] = \mathbb{E}_\nu\big[1_{\tilde\Gamma}1_A(\mathbf{X}_{t+s})\big]$$
$$= \mathbb{E}_\nu\big[1_{\tilde\Gamma}\,\mathbb{E}_{\mathbf{X}_t}[1_A(\mathbf{X}_s)]\big] = \mathbb{E}_\nu\big[1_\Gamma\,\mathbb{E}_{\mathbf{X}_t}[1_A(\mathbf{X}_s)]\big].$$

(iii): Let $\nu \in \mathcal{P}(E^\Delta)$. Let $A \in \mathcal{F}_t$, $B \in \mathcal{B}(E^\Delta)$. Then $A \in (\mathcal{F}'_t)^{\mathbb{P}_\nu}$. By (ii) we get

$$\mathbb{E}_\nu\big[1_A 1_B(\mathbf{X}_{t+s})\big] = \mathbb{E}_\nu\big[1_A \mathbb{E}_{\mathbf{X}_t}[1_B(\mathbf{X}_s)]\big].$$

Since $\mathbb{E}_{\mathbf{X}_t}[1_B(\mathbf{X}_s)]$ is $\mathcal{F}_t$-measurable, this yields the conditional expectation. (iv): We give the essential ideas of the proof of [BG68, Ch. I, Prop. 8.12]. If **M** is Markov both w.r.t. $\mathcal{F}'_t$ and $\mathcal{F}'_{t+}$, we get

$$\mathbb{E}_\nu[Y \mid \mathcal{F}'_{t+}] = \mathbb{E}_\nu[Y \mid \mathcal{F}'_t],$$

for every $Y \in \mathcal{B}_b(\mathcal{F}')$. Setting $Y = 1_\Gamma$ for $\Gamma \in \mathcal{F}'_{t+}$ this yields that Γ coincides up to a subset of a $\mathbb{P}_\nu$ nullset with an element in $\mathcal{F}'_t$. Thus $\mathcal{F}'_{t+} \subset (\mathcal{F}'_t)^{\mathbb{P}_\nu}$. Since $(\mathcal{F}'_{t+})^{\mathbb{P}_\nu} = (\mathcal{F}'_t)^{\mathbb{P}_\nu}_+$, we have

$$(\mathcal{F}'_t)^{\mathbb{P}_\nu}_+ \subset (\mathcal{F}'_t)^{\mathbb{P}_\nu}.$$

$\square$

This lemma means that one can replace the filtration $(\mathcal{M}_t)_{t\geq 0}$ of **M** by the filtration $(\mathcal{F}_t)_{t\geq 0}$, keeping all nice properties of **M**. The natural filtration is suitable for the process since it is (modulo the augmentation procedure) generated by the process. This allows to apply monotone class theorems for proving several properties. Observe that if the conclusion in Lemma 7.3.6(iv) holds, then $(\mathcal{F}_t)_{t \geq 0}$ is also right-continuous. The following lemma states important consequences of the Markov property.

Lemma 7.3.7. *The Markov property in Definition 7.3.2(iv) is equivalent to either of the following properties:*

(i) $\mathbb{E}_x[f(\mathbf{X}_{s+t}) \mid \mathcal{M}_t] = \mathbb{E}_{\mathbf{X}_t}[f(\mathbf{X}_s)]$ *for* $0 \leq s, t < \infty$, $x \in E^{\Delta}$, $f \in \mathcal{B}_b(E^{\Delta})$.

(ii) $\mathbb{E}_x[F \circ \theta_t \mid \mathcal{M}_t] = \mathbb{E}_{\mathbf{X}_t}[F]$ *for* $0 \leq t < \infty$, $x \in E^{\Delta}$, *for all* $F : (\Omega, \mathcal{F}') \to \mathbb{R}$ *measurable and bounded functions.*

Furthermore, it holds
(iii) $\mathbb{E}_x[F \circ \theta_t \mid \mathcal{F}_t] = \mathbb{E}_{\mathbf{X}_t}[F]$ *for* $0 \leq t < \infty$, $x \in E^{\Delta}$, *for all* $F : (\Omega, \mathcal{F}) \to \mathbb{R}$ *measurable and bounded functions.*
(iv) Assume that $F : (\Omega, \mathcal{F}) \to \mathbb{R}$ *is measurable,* $\mathbb{E}_x[|F|] < \infty$ *and* $\mathbb{E}_x[\mathbb{E}_{\mathbf{X}_t}[|F|]] < \infty$ *for every* $0 \leq t < \infty$ *and* $x \in E^{\Delta}$. *Then* $\mathbb{E}_x[F \circ \theta_t \mid \mathcal{F}_t] = \mathbb{E}_{\mathbf{X}_t}[F]$ *for* $0 \leq t < \infty$, $x \in E^{\Delta}$.

If $(E, \mathbf{d})$ *is a locally compact separable metric space and* μ *a locally finite Borel measure on* E, *then (i) is equivalent to:*

(i') $\mathbb{E}_x[f(\mathbf{X}_{s+t}) \mid \mathcal{M}_t] = \mathbb{E}_{\mathbf{X}_t}[f(\mathbf{X}_s)]$ *for* $0 \leq s, t < \infty$, $x \in E$, $f \in C_b(E) \cap \mathcal{L}^p(E, \mu)$ *and some* $1 \leq p < \infty$.

Proof. For the proof of the equivalence of (i) and (ii) to the Markov property, see [BG68, Ch. I, Theo. 3.6]. The equivalence to (i') can be proven using Corollary 2.2.5.

For the proof of (iii), see [BG68, Ch. I, Prop. 5.12].

(iv): From $\mathbb{E}_x[|F|] < \infty$ for all $x \in E^{\Delta}$ we conclude that the mapping $E^{\Delta} \ni x \mapsto \mathbb{E}_x[F]$ is $\mathcal{B}^*(E^{\Delta})$-measurable. Hence $\mathbb{E}_{\mathbf{X}_t}[F]$ is $\mathcal{F}_t$-adapted. Let $x \in E^{\Delta}$, $0 \leq t < \infty$. Assume first that F is positive. Then using monotone convergence and applying (iii) we get for $\Gamma \in \mathcal{F}_t$ and $x \in E^{\Delta}$

$$\mathbb{E}_x\big[\mathbf{1}_\Gamma \, (F \circ \theta_t)\big] = \lim_{n \to \infty} \mathbb{E}_x\big[\mathbf{1}_\Gamma \, (F \wedge n) \circ \theta_t\big]$$
$$= \lim_{n \to \infty} \mathbb{E}_x\big[\mathbf{1}_\Gamma \, \mathbb{E}_{\mathbf{X}_t}\big[F \wedge n\big]\big] = \mathbb{E}_x\big[\mathbf{1}_\Gamma \, \mathbb{E}_{\mathbf{X}_t}\big[F\big]\big] < \infty. \quad (7.9)$$

Thus $F \circ \theta_t$ is integrable and $\mathbb{E}_x[F \circ \theta_t \mid \mathcal{F}_t] = \mathbb{E}_{\mathbf{X}_t}[F]$. For the general case apply (7.9) first to $|F|$ to conclude integrability of $F \circ \theta_t$. Then apply the equation to F^+ and F^- and use linearity to conclude the expression for the conditional expectation. $\square$

Next we introduce the notion of a semigroup associated with a Markov process **M**. Let $(P_t^\Delta)_{t\geq 0}$ be a family of linear operators mapping from $\mathcal{B}_b(E^\Delta)$ into itself. We say that $(P_t^\Delta)_{t\geq 0}$ is the *transition semigroup* of **M** if for all $u \in \mathcal{B}_b(E^\Delta)$ it holds

$$P_t^\Delta u(x) = \mathbb{E}_x[u(\mathbf{X}_t)] \quad \text{for all } x \in E^\Delta, t \geq 0.$$

Observe that due to the Markov property, $(P_t^\Delta)_{t\geq 0}$ is indeed a semigroup of linear operators. This semigroup naturally extends to a semigroup of linear operators on $\mathcal{B}_b^*(E^\Delta)$. Furthermore, for $\mathcal{B}^*(E^\Delta)$-measurable u such that $P_t^\Delta |u|(x) < \infty$ for $0 < t < \infty$ the mapping $x \mapsto P_t^\Delta u(x)$ is well-defined and $\mathcal{B}^*(E^\Delta)$-measurable. We say that $(P_t^\Delta)_{t>0}$ is *absolutely continuous* w.r.t. some measure μ on $\mathcal{B}(E)$ if for $\mathcal{B}(E^\Delta)$-measurable f with $f(\Delta) = 0$ and $f(x) = 0$ for μ-a.e. $x \in E$, it holds $P_t^\Delta f(x) = 0$ for every $x \in E$ and $t > 0$. In this case, $(P_t^\Delta)_{t>0}$ is also absolutely continuous w.r.t. μ on $\mathcal{B}^*(E)$. We continue with the introduction of further processes.

Definition 7.3.8. Let **M** $:= (\Omega, \mathcal{M}, (\mathcal{M}_t)_{t\geq 0}, (\mathbf{X}_t)_{t\geq 0}, (\mathbb{P}_x)_{x\in E^\Delta})$ be a Markov process with life time $\mathcal{X}$ and state space E. We say that **M** is a right-process if the following additional properties hold:

(i) $\mathbb{P}_x[\mathbf{X}_0 = x] = 1$ for all $x \in E^\Delta$.

(ii) For each $\omega \in \Omega$, $t \mapsto \mathbf{X}_t(\omega)$ is right-continuous on $[0, \infty)$.

(iii) $(\mathcal{M}_t)_{t\geq 0}$ is right-continuous and for every $(\mathcal{M}_t)$-stopping time σ it holds

$$\mathbb{P}_x[\mathbf{X}_{\sigma+t} \in A \mid \mathcal{M}_\sigma] = \mathbb{P}_{\mathbf{X}_\sigma}[\mathbf{X}_t \in A] \quad \mathbb{P}_x - \text{a.s. for } A \in \mathcal{B}(E^\Delta) \text{ and } x \in E^\Delta.$$

Property (iii) is called the *strong Markov property*. For any other filtration $(\mathcal{G}_t)_{t\geq 0}$ we say that **M** is strong Markov w.r.t. $(\mathcal{G}_t)_{t\geq 0}$ if (iii) holds for this filtration.

Definition 7.3.9. Let **M** be a right-process with the same objects as in Definition 7.3.8. We say that **M** is a *Hunt process* if additionally it holds:

(i) $\mathbf{X}_{t-} := \lim_{s\uparrow t, s<t} \mathbf{X}_s$ exists in E for all $t \in [0, \infty)$, $\mathbb{P}_x$-a.s., $x \in E_1$.

(ii) If $\tau, \tau_n, n \in \mathbb{N}$, are $\mathcal{F}_t$-stopping times such that $\tau_n \uparrow \tau$, then $\mathbf{X}_{\tau_n} \to \mathbf{X}_\tau$ as $n \to \infty$, $\mathbb{P}_x$-a.s. for $x \in E_1$.

A right-process with left limits is also called *càdlàg* (continue à droit, limites à droite).

Let us introduce the potential operators $(U^\alpha)_{\alpha>0}$. For $f \in \mathcal{B}_b(E)$ define

$$U^\alpha f(x) := \mathbb{E}_x \left[\int_0^\infty e^{-\alpha s} f(\mathbf{X}_s)\, ds \right]. \tag{7.10}$$

From Definition 7.3.2(iii) one concludes using the usual arguments that $U^\alpha f \in \mathcal{B}_b(E^\Delta)$.

The following theorem is a slight modification of [BG68, Ch. I, Theo. 8.11]. We formulate the theorem for locally compact separable metric spaces and μ a locally finite measure.

Theorem 7.3.10. *Let $(E, \mathbf{d})$ be a locally compact metric space and μ a locally finite Borel measure. Let $\mathbf{M} := (\Omega, \mathcal{M}, (\mathcal{M}_t)_{t \geq 0}, (\mathbf{X}_t)_{t \geq 0}, (\mathbb{P}_x)_{x \in E^\Delta})$ be a Markov process with right-continuous paths and state space E. Assume that $(\mathcal{M}_t)_{t \geq 0}$ is right-continuous. Suppose there exists a space $\mathbf{L} \subset C_b(E)$ such that*

(i) The mapping $t \mapsto U^\alpha f(\mathbf{X}_t)$ is $\mathbb{P}_x$-a.s. right-continuous on $[0, \mathcal{X})$ for $f \in \mathbf{L}$, $\alpha > 0$.

(ii) For every open set $G \subset E$ with $\mu(G) < \infty$ there exists a sequence $(f_n)_{n \in \mathbb{N}}$ with $f_n \uparrow 1_G$.

Then $\mathbf{M}$ is strong Markov.

Remark 7.3.11. Clearly one can choose as $\mathbf{L}$ the set $C_b(E) \cap L^p(E, \mu)$ for every $1 \leq p \leq \infty$. We have to introduce this modified version of the theorem since in our application we know the nice properties of $U^\alpha f(\mathbf{X}_t)$ only if $f \in L^p(E, \mu)$ for some $1 \leq p < \infty$.

We sketch the proof of the theorem.

Let τ be a $\mathcal{M}_t$-stopping time and $f \in \mathbf{L}$. Following the proof of [BG68, Ch. I, Theo. 8.11] we get

$$\mathbb{E}_x\big[f(\mathbf{X}_{t+\tau})\big] = \mathbb{E}_x\big[\mathbb{E}_{\mathbf{X}_\tau}\big[f(\mathbf{X}_t)\big]\big] \text{ for every } x \in E.$$

Due to (ii) we may apply Corollary 2.2.5 to conclude that this equality holds for every $f \in \mathcal{B}_b(E)$. Now let $\Gamma \in \mathcal{M}_\tau$. Then by [BG68, Ch. I, Prop. 8.2] we get

$$\mathbb{E}_x\big[f(\mathbf{X}_{t+\tau})1_\Gamma\big] = \mathbb{E}_x\big[\mathbb{E}_{\mathbf{X}_\tau}\big[f(\mathbf{X}_t)\big]1_\Gamma\big] \text{ for every } f \in \mathcal{B}_b(E^\Delta), \ x \in E^\Delta.$$

So $\mathbb{E}_{\mathbf{X}_\tau}[f(\mathbf{X}_t)]$ satisfies the equation for the conditional expectation. Furthermore, $y \mapsto \mathbb{E}_y[f(\mathbf{X}_t)]$ is $\mathcal{B}(E^\Delta)$-measurable and $\mathbf{X}_\tau$ is $\mathcal{M}_\tau/\mathcal{B}(E^\Delta)$-measurable. Thus $E_{\mathbf{X}_\tau}[f(\mathbf{X}_t)]$ is $\mathcal{M}_\tau$-measurable and hence

$$\mathbb{E}_x\big[f(\mathbf{X}_{t+\tau})|\mathcal{M}_\tau\big] = \mathbb{E}_{\mathbf{X}_\tau}\big[f(\mathbf{X}_t)\big] \ \mathbb{P}_x\text{-a.s. for every } x \in E^\Delta.$$

Next we consider processes with continuous paths.

Definition 7.3.12. Let $(E, \mathbf{d})$ be a locally compact separable metric space, $(E^\Delta, \mathbf{d})$ the corresponding metrization of the Alexandrov one-point compactification. A right-process $\mathbf{M} = (\Omega, \mathcal{M}, (\mathcal{M}_t)_{t \geq 0}, (\mathbf{X}_t)_{t \geq 0}, (\mathbb{P}_x)_{x \in E^\Delta})$ is called a *diffusion* if

$$\mathbb{P}_x(\mathbf{X}. : [0, \mathcal{X}) \to E \text{ is continuous}) = 1 \quad \text{for every } x \in E.$$

We call it a *diffusion* on $[0, \infty)$ if

$$\mathbb{P}_x(\mathbf{X}. : [0, \infty) \to E^\Delta \text{ is continuous}) = 1 \quad \text{for every } x \in E^\Delta.$$

Remark 7.3.13. This means that for every $\omega \in \Omega$ the mapping $\mathbf{X}_s(\omega)$: $[0, \mathcal{X}(\omega)) \to E$ is continuous with respect to the original topology on E and for $t_n \overset{n \to \infty}{\longrightarrow} \mathcal{X}(\omega)$, $0 \leq t_n < \mathcal{X}$, $n \in \mathbb{N}$, it holds $\mathbf{X}_{t_n}(\omega) \overset{n \to \infty}{\longrightarrow} \Delta$. The latter means that $\mathbf{X}_{t_n}(\omega)$ leaves *continuously* every compact set for $n \to \infty$. If $\mathbf{M}$ has only continuous paths on $[0, \mathcal{X})$, then it may „jump" to the cemetery.

We come now to the essential results connecting Dirichlet forms and stochastic processes. We restrict to the setting covered in this work, i.e., the regular symmetric case, but give also reference to the more general results in [MR92].

Definition 7.3.14. Let $(\mathcal{E}, D(\mathcal{E}))$ be a Dirichlet form on $L^2(E, \mu)$, $\mathbf{M}$ a right-process with state space E and transition semigroup $(P_t)_{t \geq 0}$. We say $\mathbf{M}$ is (properly) associated with $(\mathcal{E}, D(\mathcal{E}))$ if $P_t f$ is a $\mathcal{E}$-quasi-continuous μ-version of $T_t^{(2)} f$ for $t \geq 0$, $f \in \mathcal{B}_b(E) \cap L^2(E, \mu)$.

Theorem 7.3.15. *Let $(\mathcal{E}, D(\mathcal{E}))$ be a regular symmetric Dirichlet form on a locally compact separable metric space $(E, \mathbf{d})$. Then there exists a Hunt process $\mathbf{M}$ that is properly associated with $(\mathcal{E}, D(\mathcal{E}))$.*

See [FOT11, Theo. 7.2.1]. For the corresponding generalization to the quasi-regular (non-symmetric) case see [MR92, Ch. IV, Theo. 3.5].

If the Dirichlet form satisfies additional assumptions, we get more path regularity.

Definition 7.3.16. We say that $(\mathcal{E}, D(\mathcal{E}))$ is *local* if for every $u, v \in D(\mathcal{E})$ it holds:

If $\mathrm{supp}[u]$ and $\mathrm{supp}[v]$ are disjoint compact sets, then $\mathcal{E}(u, v) = 0$.

The Dirichlet form is said to be *strongly local* if the above conclusion already holds if the compact supports are not assumed to be disjoint, but only that v is constant on a neighborhood of $\mathrm{supp}[u]$.

Compare [FOT11, Ch. 1, p. 6] or [MR92, Ch. V, Def. 1.1].

Proposition 7.3.17. *Let $(\mathcal{E}, D(\mathcal{E}))$ be a symmetric regular Dirichlet form on a locally compact separable metric space $(E, \mathbf{d})$. If $(\mathcal{E}, D(\mathcal{E}))$ has the local property, then there exists a right-process* **M** *such that*

$$\mathbb{P}_x(\mathbf{X}. : [0, \mathcal{X}) \to E \text{ is continuous}) = 1 \quad \text{for every } x \in E.$$

If $(\mathcal{E}, D(\mathcal{E}))$ is strongly local, then there exists a right-process **M** *such that*

$$\mathbb{P}_x(\mathbf{X}. : [0, \infty) \to E^{\Delta} \text{ is continuous}) = 1 \quad \text{for every } x \in E^{\Delta}.$$

The continuity holds with respect to topology of the Alexandrov one-point compactification E^{Δ} of E.

For the proof see [FOT11, Theo. 4.5.1] and [FOT11, Theo. 4.5.3]. For the generalization to the quasi-regular case see [MR92, Ch. V, Theo. 1.11].

We call a symmetric Dirichlet form *conservative* if for the associated sub-Markovian L^2-s.c.c.s $(T_t^2)_{t \geq 0}$ it holds

$$f_n \in L^2(E, \mu), \ n \in \mathbb{N}, \text{ with } f_n \uparrow 1_E \quad \text{implies} \quad T_t^2 f_n \uparrow 1 \ \mu - \text{a.e.}$$

A stochastic process $(\mathbf{X}_t)_{t \geq 0}$ is called *conservative* under $\mathbb{P}_x$, $x \in E$, if $\mathbb{P}_x(\mathcal{X} = \infty) = 1$, i.e., the process does not hit the cemetery $\mathbb{P}_x$-a.s. From [FOT11, Exer. 4.5.1] we conclude that if $(\mathcal{E}, D(\mathcal{E}))$ is conservative then the associated process **M** is conservative for $\mathcal{E}$-quasi-every starting point.

Next we consider restrictions of Hunt processes. This means that we construct from a given process a process whose paths stay in a certain subset. This is important for the construction of additive functionals.

Let us first introduce the trace of a σ-algebra and restriction of a measure. For $(\Omega, \mathcal{B})$ a measurable space and $\tilde{E} \subset \Omega$ define $\mathcal{B}|_{\tilde{E}}$ the *trace of $\mathcal{B}$ on $\tilde{E}$* by

$$\mathcal{B}|_{\tilde{E}} := \{A \cap \tilde{E} \mid A \in \mathcal{B}\}.$$

For $\tilde{E} \in \mathcal{B}$ and μ a measure on $\mathcal{B}$ define the *restriction of μ to $\mathcal{B}|_{\tilde{E}}$* by

$$\mu|_{\mathcal{B}|_{\tilde{E}}}(A) := \mu(A), A \in \mathcal{B}|_{\tilde{E}}.$$

Note that this is well-defined since $\mathcal{B}|_{\tilde{E}} \subset \mathcal{B}$ if $\tilde{E} \in \mathcal{B}$.

Let $\mathbf{M} := (\Omega, \mathcal{F}, (\mathcal{F}_t)_{t \geq 0}, (\mathbf{X}_t)_{t \geq 0}, (\mathbb{P}_x)_{x \in E^{\Delta}})$ be a Hunt process where $\mathcal{F}$ and $(\mathcal{F}_t)_{t \geq 0}$ are defined as in (7.7) and (7.8). Let $\mathcal{F}'$, $(\mathcal{F}_t')_{t \geq 0}$ as in (7.6).

Let $\widetilde{E} \in \mathcal{B}^n(E^\Delta)$ nearly Borel, see (7.18), such that $E^\Delta \setminus \widetilde{E}$ is properly exceptional. Define

$$\Omega^{\widetilde{E}} := \{\omega \in \Omega \mid X_t(\omega) \in \widetilde{E}, X_{t-}(\omega) \in \widetilde{E} \text{ for all } 0 \leq t < \infty\}.$$

From [FOT11, p. 398] it follows $\Omega^{\widetilde{E}} \in \mathcal{F}$. Define $\mathbf{X}_t^{\widetilde{E}} := \mathbf{X}_t|_{\Omega^{\widetilde{E}}}$, $t \geq 0$. We consider measurability w.r.t. the Borel σ-algebra $\mathcal{B}(\widetilde{E}) = \mathcal{B}(E^\Delta)|_{\widetilde{E}}$ on $\widetilde{E}$. Let $\mathcal{F}'|_{\Omega^{\widetilde{E}}}$ and $(\mathcal{F}'_t|_{\Omega^{\widetilde{E}}})_{t \geq 0}$ be the trace-σ-algebra of $\mathcal{F}'$ and $(\mathcal{F}'_t)_{t \geq 0}$, respectively, on $\Omega^{\widetilde{E}}$. Note that $\mathcal{F}'|_{\Omega^{\widetilde{E}}} = \sigma(\mathbf{X}_t^{\widetilde{E}} \mid 0 \leq t \leq \infty)$ and $\mathcal{F}'_t|_{\Omega^{\widetilde{E}}} = \sigma(\mathbf{X}_s^{\widetilde{E}} \mid 0 \leq s \leq t)$. For $x \in \widetilde{E}$ define $\mathbb{P}_x^{\widetilde{E}} := \mathbb{P}_x|_{\mathcal{F}'|_{\Omega^{\widetilde{E}}}}$, $x \in \widetilde{E}$. This is well-defined since $\mathcal{F}'|_{\Omega^{\widetilde{E}}} \subset \mathcal{F}$.
Define

$$\mathcal{F}^{\Omega^{\widetilde{E}}} := \bigcap_{\nu \in \mathcal{P}(\widetilde{E})} (\mathcal{F}'|_{\Omega^{\widetilde{E}}})^{\mathbb{P}_\nu^{\widetilde{E}}}$$

and

$$\mathcal{F}_t^{\Omega^{\widetilde{E}}} := \bigcap_{\nu \in \mathcal{P}(\widetilde{E})} (\mathcal{F}'_t|_{\Omega^{\widetilde{E}}})^{\mathbb{P}_\nu^{\widetilde{E}}}, \quad 0 \leq t < \infty.$$

Here $(\mathcal{F}'_t|_{\Omega^{\widetilde{E}}})^{\mathbb{P}_\nu^{\widetilde{E}}}$ is the completion in $\mathcal{F}'|_{\Omega^{\widetilde{E}}}$. Note that $\mathbb{P}_x^{\widetilde{E}}$, $x \in \widetilde{E}$, canonically extends to $\mathcal{F}^{\Omega^{\widetilde{E}}}$.

Definition 7.3.18. For the Hunt process $\mathbf{M}$ and $\widetilde{E}$ as above, define the *restriction* of $\mathbf{M}$ by

$$\mathbf{M}^{\widetilde{E}} := (\Omega^{\widetilde{E}}, \mathcal{F}^{\Omega^{\widetilde{E}}}, (\mathcal{F}_t^{\Omega^{\widetilde{E}}})_{t \geq 0}, (\mathbf{X}_t^{\widetilde{E}})_{t \geq 0}, (\mathbb{P}_x^{\widetilde{E}})_{x \in \widetilde{E}}).$$

We denote the corresponding semigroup by $(P_t^{\widetilde{E}})_{t \geq 0}$. See [FOT11, Appendix, (A.2.23)] for this definition. As in the mentioned reference we get the following theorem.

Theorem 7.3.19. *The restricted process $\mathbf{M}^{\widetilde{E}}$ is a Hunt process with state space $\widetilde{E}$ and cemetery point Δ.*

Lemma 7.3.20. *Let $\widetilde{E}_2 \subset \widetilde{E}_1 \subset E^\Delta$ Borel-measurable such that $E^\Delta \setminus \widetilde{E}_1$ and $E^\Delta \setminus \widetilde{E}_2$ are properly exceptional. Let $\mathbf{M}^{\widetilde{E}_1}$ and $\mathbf{M}^{\widetilde{E}_2}$ be the corresponding restricted processes. Then $\mathcal{F}^{\Omega^{\widetilde{E}_1}}|_{\Omega^{\widetilde{E}_2}} \subset \mathcal{F}^{\Omega^{\widetilde{E}_2}}$ and $\mathcal{F}_t^{\Omega^{\widetilde{E}_1}}|_{\Omega^{\widetilde{E}_2}} \subset \mathcal{F}_t^{\Omega^{\widetilde{E}_2}}$ for*

$0 \le t < \infty$. In particular, for $Y_0 : \Omega^{\tilde{E}_1} \to \mathbb{R}$, $\mathcal{F}^{\Omega^{\tilde{E}_1}}$-measurable, it holds that $Y := Y_0|_{\Omega^{\tilde{E}_2}}$ is $\mathcal{F}^{\Omega^{\tilde{E}_2}}$-measurable.

Let $Y : \Omega^{\tilde{E}_2} \to \mathbb{R}$, $\mathcal{F}^{\Omega^{\tilde{E}_2}}$-measurable, be given. Then there exists $Y_0 :$ $\Omega^{\tilde{E}_1} \to \mathbb{R}$ such that $Y = Y_0|_{\Omega^{\tilde{E}_2}}$ and Y_0 is $(\mathcal{F}'|_{\Omega^{\tilde{E}_1}})^{\mathbb{P}_\nu^{\tilde{E}_1}}$-measurable for every $\nu \in \mathcal{P}(\tilde{E}_1)$ such that $\nu(\tilde{E}_1 \setminus \tilde{E}_2) = 0$. For every $x \in \tilde{E}_2$ it holds for the corresponding expectations:

$$\mathbb{E}_x^{\tilde{E}_1}[Y_0] = \mathbb{E}_x^{\tilde{E}_1}[1_{\Omega^{\tilde{E}_2}} Y] = \mathbb{E}_x^{\tilde{E}_2}[Y]. \tag{7.11}$$

Proof. We first prove that $\mathcal{F}^{\Omega^{\tilde{E}_1}}|_{\Omega^{\tilde{E}_2}} \subset \mathcal{F}^{\Omega^{\tilde{E}_2}}$ and $\mathcal{F}_t^{\Omega^{\tilde{E}_1}}|_{\Omega^{\tilde{E}_2}} \subset \mathcal{F}_t^{\Omega^{\tilde{E}_2}}$ for $0 \le t < \infty$. Let $\hat{\Gamma} \in \mathcal{F}^{\Omega^{\tilde{E}_1}}|_{\Omega^{\tilde{E}_2}}$ $(\mathcal{F}_t^{\Omega^{\tilde{E}_1}}|_{\Omega^{\tilde{E}_2}})$. Then there exists $\Gamma \in \mathcal{F}^{\Omega^{\tilde{E}_1}}$ $(\mathcal{F}_t^{\Omega^{\tilde{E}_1}})$ such that $\hat{\Gamma} = \Gamma \cap \Omega^{\tilde{E}_2}$.

Let $\nu \in \mathcal{P}(\tilde{E}_2)$, define $\hat{\nu} \in \mathcal{P}(\tilde{E}_1)$ by $\hat{\nu}(\cdot) := \nu(\cdot \cap \tilde{E}_2)$. Since $\Gamma \in \mathcal{F}^{\Omega^{\tilde{E}_1}}$ $(\mathcal{F}_t^{\Omega^{\tilde{E}_1}})$, there exists $\tilde{\Gamma}, N \in \mathcal{F}'|_{\Omega^{\tilde{E}_1}}$ $(\tilde{\Gamma} \in \mathcal{F}_t'|_{\Omega^{\tilde{E}_1}}$, $N \in \mathcal{F}'|_{\Omega^{\tilde{E}_1}})$ such that $\mathbb{P}_{\hat{\nu}}^{\tilde{E}_1}(N) = 0$ and $\Gamma \triangle \tilde{\Gamma} \subset N$. Thus

$$\Gamma \cap \Omega^{\tilde{E}_2} \triangle \tilde{\Gamma} \cap \Omega^{\tilde{E}_2} \subset N \cap \Omega^{\tilde{E}_2}.$$

Note that $\tilde{\Gamma} \cap \Omega^{\tilde{E}_2}, N \cap \Omega^{\tilde{E}_2} \in \mathcal{F}'|_{\Omega^{\tilde{E}_2}}$ $(\tilde{\Gamma} \cap \Omega^{\tilde{E}_2} \in \mathcal{F}_t'|_{\Omega^{\tilde{E}_2}}$, $N \cap \Omega^{\tilde{E}_2} \in \mathcal{F}'|_{\Omega^{\tilde{E}_2}})$. Furthermore,

$$\mathbb{P}_\nu^{\tilde{E}_2}(N \cap \Omega^{\tilde{E}_2}) = \int_{\tilde{E}_2} \mathbb{P}_x^{\tilde{E}_2}(N \cap \Omega^{\tilde{E}_2}) \, d\nu(x)$$

$$= \int_{\tilde{E}_2} \mathbb{P}_x^{\tilde{E}_1}(N \cap \Omega^{\tilde{E}_2}) d\nu(x) = \int_{\tilde{E}_2} \mathbb{P}_x^{\tilde{E}_1}(N) \, d\nu(x) = \int_{\tilde{E}_2} \mathbb{P}_x^{\tilde{E}_1}(N) \, d\hat{\nu}(x)$$

$$= \int_{\tilde{E}_1} \mathbb{P}_x^{\tilde{E}_1}(N) \, d\hat{\nu}(x) = \mathbb{P}_{\hat{\nu}}^{\tilde{E}_1}(N) = 0.$$

Here we used that $\mathbb{P}_x^{\tilde{E}_1}(\Omega^{\tilde{E}_2}) = 1$ for every $x \in \tilde{E}_2$ and $\hat{\nu}(\tilde{E}_1 \setminus \tilde{E}_2) = 0$.

Since $\nu \in \mathcal{P}(\tilde{E}_2)$ was chosen arbitrarily, we get $\Gamma \cap \Omega^{\tilde{E}_2} \in \mathcal{F}^{\Omega^{\tilde{E}_2}}$ $(\Gamma \cap \Omega^{\tilde{E}_2} \in \mathcal{F}_t^{\Omega^{\tilde{E}_2}})$.

Now let $Y : \Omega^{\tilde{E}_2} \to \mathbb{R}$ be $\mathcal{F}^{\Omega^{\tilde{E}_2}}$-measurable. Define $Y_0 : \Omega^{\tilde{E}_1} \to \mathbb{R}$ by $Y_0 := 1_{\Omega^{\tilde{E}_2}} Y$. Clearly, $Y_0|_{\Omega^{\tilde{E}_2}} = Y$.

For $B \in \mathcal{B}(\mathbb{R})$ we have

$$Y_0^{-1}(B) = \begin{cases} \Omega^{\tilde{E}_2} \cap Y^{-1}(B) & \text{if } 0 \notin B \\ \Omega^{\tilde{E}_2} \cap Y^{-1}(B) \cup (\Omega^{\tilde{E}_2})^c & \text{else.} \end{cases}$$

Let $\nu \in \mathcal{P}(\widetilde{E}_1)$ such that $\nu(\widetilde{E}_1 \setminus \widetilde{E}_2) = 0$. Then $\widetilde{\nu}(\cdot) := \nu(\cdot \cap \widetilde{E}_2)$ defines a probability measure on $\mathcal{B}(\widetilde{E}_2)$.

Since $\Omega^{\widetilde{E}_2} \in \mathcal{F}$ and $\Omega^{\widetilde{E}_2} \subset \Omega^{\widetilde{E}_1}$, we have $\Omega^{\widetilde{E}_2} \in \mathcal{F}|_{\Omega^{\widetilde{E}_1}} \subset \mathcal{F}^{\Omega^{\widetilde{E}_1}} \subset (\mathcal{F}'|_{\Omega^{\widetilde{E}_1}})^{\mathbb{P}_\nu^{\widetilde{E}_1}}$.

Let $\Gamma := Y^{-1}(B) \in \mathcal{F}^{\Omega^{\widetilde{E}_2}}$. Since $\mathcal{F}^{\Omega^{\widetilde{E}_2}} \subset (\mathcal{F}'|_{\Omega^{\widetilde{E}_2}})^{\mathbb{P}_{\widetilde{\nu}}^{\widetilde{E}_2}}$, we have $\Gamma \in (\mathcal{F}'|_{\Omega^{\widetilde{E}_2}})^{\mathbb{P}_{\widetilde{\nu}}^{\widetilde{E}_2}}$. Thus $\Gamma \cap \Omega^{\widetilde{E}_2} \in (\mathcal{F}'|_{\Omega^{\widetilde{E}_1}})^{\mathbb{P}_\nu^{\widetilde{E}_1}}$.

It is left to prove (7.11). Let $x \in \widetilde{E}_2$ arbitrary. If Y is the indicator function of some set in $\mathcal{F}'|_{\Omega^{\widetilde{E}_2}}$ the claim follows by definition of $\mathbb{P}_x^{\widetilde{E}_2}$. For general $\mathcal{F}'|_{\Omega^{\widetilde{E}_2}}$-measurable functions the equation follows by a monotone class argument. For general $\mathcal{F}^{\Omega^{\widetilde{E}_2}}$-measurable Y the claim follows since $\mathbb{P}_x^{\widetilde{E}_2}$ is extended to $(\mathcal{F}'|_{\Omega^{\widetilde{E}_2}})^{\mathbb{P}_x^{\widetilde{E}_2}}$ in the same way as $\mathbb{P}_x^{\widetilde{E}_1}$ is extended to $(\mathcal{F}'|_{\Omega^{\widetilde{E}_1}})^{\mathbb{P}_x^{\widetilde{E}_1}}$. $\qquad\square$

Lemma 7.3.21. *Let $N \subset E^\Delta$ be properly exceptional. Let $(A_t)_{t \geq 0}$ be an CAF of $\mathbf{M}^{E^\Delta \setminus N}$ with additivity set Λ. Let $\widetilde{E}_1 \subset E^\Delta$ Borel measurable such that $E^\Delta \setminus \widetilde{E}_1$ is properly exceptional. Define $\widetilde{E}_2 := \widetilde{E}_1 \setminus N$. Then $(A_t|_{\Omega^{\widetilde{E}_2}})_{t \geq 0}$ is a CAF of the restricted process $\mathbf{M}^{\widetilde{E}_2}$ with additivity set $\Lambda \cap \Omega^{\widetilde{E}_2}$.*

Proof. Except for the $\mathcal{F}_t^{\Omega^{\widetilde{E}_2}}$-adaptedness all properties are immediate. From Lemma 7.3.20 we get $\mathcal{F}_t^{\Omega^{E^\Delta \setminus N}}|_{\Omega^{\widetilde{E}_2}} \subset \mathcal{F}_t^{\Omega^{\widetilde{E}_2}}$, $t \geq 0$. Thus also adaptedness holds. $\qquad\square$

The next lemma is important for the construction of strict additive functionals.

Lemma 7.3.22. *Let $\widetilde{E}_1 \in \mathcal{B}(E^\Delta)$, $\widetilde{E}_2 \in \mathcal{B}(E^\Delta)$ with $\widetilde{E}_2 \subset \widetilde{E}_1$ such that $E^\Delta \setminus \widetilde{E}_1$ and $E^\Delta \setminus \widetilde{E}_2$ are properly exceptional. Assume that for $(P_t^{\widetilde{E}_1})_{t > 0}$ it holds $P_t^{\widetilde{E}_1}(x, \widetilde{E}_2) = 1$ for every $x \in \widetilde{E}_1$. Let $\Lambda \in \mathcal{F}^{\Omega^{\widetilde{E}_2}}$. Then for every $\varepsilon > 0$, $\theta_\varepsilon^{-1}(\Lambda) \cap \Omega^{\widetilde{E}_1} \in \mathcal{F}^{\Omega^{\widetilde{E}_1}}$.*

Now assume $\mathbb{P}_x^{\widetilde{E}_2}(\Lambda) = 1$ for $x \in \widetilde{E}_2$. Then

$$\mathbb{P}_x^{\widetilde{E}_1}\left(\bigcap_{n \in \mathbb{N}} \theta_{1/n}^{-1}(\Lambda) \cap \Omega^{\widetilde{E}_1} \right) = 1 \quad \text{for } x \in \widetilde{E}_1. \tag{7.12}$$

Define $\Lambda_0 := \bigcap_{n \in \mathbb{N}} \theta_{1/n}^{-1}(\Lambda) \cap \Omega^{\widetilde{E}_1}$. Let $(A_t)_{t \geq 0}$ be a family of mappings from $\Omega^{\widetilde{E}_2}$ to $\mathbb{R}$ that are $\mathcal{F}_t^{\Omega^{\widetilde{E}_2}}$-adapted. Let $\varepsilon > 0$. Define $A_t^\varepsilon : \Omega^{\widetilde{E}_1} \to \mathbb{R}$

by $\omega \mapsto 1_{\Lambda_0}(\omega) A_{t-\varepsilon}(\theta_\varepsilon \omega)$ *for* $t \geq \varepsilon$ *and* 0 *for* $0 \leq t < \varepsilon$. *Then* $(A_t^\varepsilon)_{t \geq 0}$ *is* $\mathcal{F}_t^{\Omega^{\widetilde{E}_1}}$-*adapted.*

Proof. Let $\varepsilon > 0$ and $\nu \in \mathcal{P}(\widetilde{E}_1)$. Define $\nu_1(\cdot) := \int_{\widetilde{E}_1} P_\varepsilon^{\widetilde{E}_1}(x, \cdot) \, d\nu(x)$. So ν_1 is the image measure of $\mathbf{X}_\varepsilon^{\widetilde{E}_1}$ under $\mathbb{P}_\nu^{\widetilde{E}_1}$. Thus ν_1 is a probability measure on $\mathcal{B}^*(\widetilde{E}_1)$ with $\nu_1(\widetilde{E}_2) = 1$. Hence $\mathbb{P}_{\nu_1}^{\widetilde{E}_1}(\Omega^{\widetilde{E}_2}) = 1$. Let $\widetilde{\nu}_1$ be the corresponding probability measure on $\mathcal{B}^*(\widetilde{E}_2)$, obtained by first restricting ν_1 to $\mathcal{B}(\widetilde{E}_2)$ and then extending it in the canonical way.

Since $\Lambda \in \mathcal{F}^{\Omega^{\widetilde{E}_2}}$, there exist $B \in \mathcal{F}'|_{\Omega^{\widetilde{E}_2}}$ and $C \in \mathcal{F}'|_{\Omega^{\widetilde{E}_2}}$ with $\Lambda \Delta B \subset C$ such that $\mathbb{P}_{\widetilde{\nu}_1}^{\widetilde{E}_2}(B) = \mathbb{P}_{\widetilde{\nu}_1}^{\widetilde{E}_2}(\Lambda)$ and $\mathbb{P}_{\widetilde{\nu}_1}^{\widetilde{E}_2}(C) = 0$. Choose $\widetilde{B} \in \mathcal{F}'$ and $\widetilde{C} \in \mathcal{F}'$ such that $B = \widetilde{B} \cap \Omega^{\widetilde{E}_2}$ and $C = \widetilde{C} \cap \Omega^{\widetilde{E}_2}$. Note that $\Lambda \subset \Omega^{\widetilde{E}_2}$. So $\Lambda \cap \Omega^{\widetilde{E}_2} \Delta \widetilde{B} \cap \Omega^{\widetilde{E}_2} \subset \widetilde{C} \cap \Omega^{\widetilde{E}_2}$ and $\Lambda \Delta \widetilde{B} \subset \widetilde{C} \cap \Omega^{\widetilde{E}_2} \cup (\Omega^{\widetilde{E}_2})^c$. Thus

$$\theta_\varepsilon^{-1}(\Lambda) \cap \Omega^{\widetilde{E}_1} \Delta \theta_\varepsilon^{-1}(\widetilde{B}) \cap \Omega^{\widetilde{E}_1} \subset \theta_\varepsilon^{-1}\big(\widetilde{C} \cap \Omega^{\widetilde{E}_2} \cup (\Omega^{\widetilde{E}_2})^c\big) \cap \Omega^{\widetilde{E}_1}.$$

Using Lemma 7.3.20 we get

$$\mathbb{P}_{\nu_1}^{\widetilde{E}_1}\big((\widetilde{C} \cap \Omega^{\widetilde{E}_2} \cup (\Omega^{\widetilde{E}_2})^c) \cap \Omega^{\widetilde{E}_1}\big)$$
$$= \mathbb{P}_{\nu_1}^{\widetilde{E}_1}\big(\widetilde{C} \cap \Omega^{\widetilde{E}_1} \cap \Omega^{\widetilde{E}_2}\big) + \mathbb{P}_{\nu_1}^{\widetilde{E}_1}\big((\Omega^{\widetilde{E}_2})^c \cap \Omega^{\widetilde{E}_1}\big) = \mathbb{P}_{\widetilde{\nu}_1}^{\widetilde{E}_2}(C) = 0$$

and

$$\mathbb{P}_{\nu_1}^{\widetilde{E}_1}\big(\widetilde{B} \cap \Omega^{\widetilde{E}_1}\big) = \mathbb{P}_{\nu_1}^{\widetilde{E}_1}\big(\widetilde{B} \cap \Omega^{\widetilde{E}_1} \cap \Omega^{\widetilde{E}_2}\big) = \mathbb{P}_{\widetilde{\nu}_1}^{\widetilde{E}_2}(B) = \mathbb{P}_{\widetilde{\nu}_1}^{\widetilde{E}_2}(\Lambda). \tag{7.13}$$

Note that shift-invariance of $\Omega^{\widetilde{E}_1}$ implies $\theta_\varepsilon^{-1}(\,\cdot\,) \cap \Omega^{\widetilde{E}_1} = \theta_\varepsilon^{-1}(\,\cdot \cap \Omega^{\widetilde{E}_1}) \cap \Omega^{\widetilde{E}_1}$.

Using the Markov property and the definition of ν_1 we get

$$\mathbb{P}_\nu^{\widetilde{E}_1}\big(\theta_\varepsilon^{-1}(\widetilde{C} \cap \Omega^{\widetilde{E}_2} \cup (\Omega^{\widetilde{E}_2})^c) \cap \Omega^{\widetilde{E}_1}\big)$$
$$= \mathbb{P}_\nu^{\widetilde{E}_1}\big(\theta_\varepsilon^{-1}(\widetilde{C} \cap \Omega^{\widetilde{E}_2} \cup (\Omega^{\widetilde{E}_2})^c \cap \Omega^{\widetilde{E}_1}) \cap \Omega^{\widetilde{E}_1}\big)$$
$$= \int_{\widetilde{E}_1} P_\varepsilon^{\widetilde{E}_1}\big(\mathbb{E}^{(\widetilde{E}_1)}\big[1_{\widetilde{C} \cap \Omega^{\widetilde{E}_2} \cup (\Omega^{\widetilde{E}_2})^c \cap \Omega^{\widetilde{E}_1}}\big]\big)(x) \, d\nu(x)$$
$$= \int_{\widetilde{E}_1} \mathbb{E}_x^{\widetilde{E}_1}\big[1_{\widetilde{C} \cap \Omega^{\widetilde{E}_2} \cup (\Omega^{\widetilde{E}_2})^c \cap \Omega^{\widetilde{E}_1}}\big] \, d\nu_1(x)$$
$$= \mathbb{P}_{\nu_1}^{\widetilde{E}_1}\big(\widetilde{C} \cap \Omega^{\widetilde{E}_2} \cup (\Omega^{\widetilde{E}_2})^c \cap \Omega^{\widetilde{E}_1}\big) = 0.$$

Since $\theta_\varepsilon^{-1}\big(\widetilde{C} \cap \Omega^{\widetilde{E}_2} \cup (\Omega^{\widetilde{E}_2})^c\big) \cap \Omega^{\widetilde{E}_1} \in \mathcal{F}^{\Omega^{\widetilde{E}_1}} \subset (\mathcal{F}'|_{\Omega^{\widetilde{E}_1}})^{\mathbb{P}_\nu^{\widetilde{E}_1}}$, there exists $N' \in \mathcal{F}'|_{\Omega^{\widetilde{E}_1}}$ with $\mathbb{P}_\nu^{\widetilde{E}_1}(N') = 0$ and $\theta_\varepsilon^{-1}\big(\widetilde{C} \cap \Omega^{\widetilde{E}_2} \cup (\Omega^{\widetilde{E}_2})^c\big) \cap \Omega^{\widetilde{E}_1} \subset N'$. Clearly, $\theta_\varepsilon^{-1}(\widetilde{B}) \cap \Omega^{\widetilde{E}_1} \in \mathcal{F}'|_{\Omega^{\widetilde{E}_1}}$. So altogether,

$$\theta_\varepsilon^{-1}(\Lambda) \cap \Omega^{\widetilde{E}_1} \Delta \theta_\varepsilon^{-1}(\widetilde{B}) \cap \Omega^{\widetilde{E}_1} \subset N'.$$

Thus $\theta_\varepsilon^{-1}(\Lambda) \cap \Omega^{\widetilde{E}_1} \in (\mathcal{F}'|_{\widetilde{E}_1})^{\mathbb{P}_\nu^{\widetilde{E}_1}}$ and, together with (7.13), we get

$$\mathbb{P}_\nu^{\widetilde{E}_1}\big(\theta_\varepsilon^{-1}(\Lambda) \cap \Omega^{\widetilde{E}_1}\big) = \mathbb{P}_\nu^{\widetilde{E}_1}\big(\theta_\varepsilon^{-1}(\widetilde{B}) \cap \Omega^{\widetilde{E}_1}\big)$$
$$= \mathbb{P}_{\nu_1}^{\widetilde{E}_1}\big(\widetilde{B} \cap \Omega^{\widetilde{E}_1}\big) = \mathbb{P}_{\widetilde{\nu}_1}^{\widetilde{E}_2}(B) = \mathbb{P}_{\widetilde{\nu}_1}^{\widetilde{E}_2}(\Lambda).$$

Since $\nu \in \mathcal{P}(\widetilde{E}_1)$ was chosen arbitrary, we get $\theta_\varepsilon^{-1}(\Lambda) \cap \Omega^{\widetilde{E}_1} \in \mathcal{F}^{\Omega^{\widetilde{E}_1}}$. Now assume $\mathbb{P}_x^{\widetilde{E}_2}(\Lambda) = 1$ for every $x \in \widetilde{E}_2$. If we choose $\nu := \varepsilon_x$ for $x \in \widetilde{E}_1$ arbitrary, we get (7.12).

Next we consider adaptedness of $(A_t^\varepsilon)_{t \geq 0}$. Since $\mathbb{P}_\nu^{\widetilde{E}_1}(\Lambda_0) = 1$ for all $\nu \in \mathcal{P}(\widetilde{E}_1)$, we get $\Lambda_0 \in \mathcal{F}_t^{\Omega^{\widetilde{E}_1}}$ for all $t \geq 0$. Let $\varepsilon > 0$ and $t > \varepsilon$.

First we prove that for $\Lambda \in \mathcal{F}_{t-\varepsilon}^{\Omega^{\widetilde{E}_2}}$ it holds $\theta_\varepsilon^{-1}(\Lambda) \cap \Lambda_0 \in \mathcal{F}_t^{\Omega^{\widetilde{E}_1}}$. Let $\nu \in \mathcal{P}(\widetilde{E}_1)$. Define $\widetilde{\nu}_1 \in \mathcal{P}(\widetilde{E}_2)$ as above.

Then there exists $\widetilde{B} \in \mathcal{F}_{t-\varepsilon}'$ and $\widetilde{C} \in \mathcal{F}'$ with $\mathbb{P}_{\widetilde{\nu}_1}^{\widetilde{E}_2}(\widetilde{C} \cap \Omega^{\widetilde{E}_2}) = 0$ such that

$$\Lambda \Delta \widetilde{B} \cap \Omega^{\widetilde{E}_2} \subset \widetilde{C} \cap \Omega^{\widetilde{E}_2}.$$

Note that for $\omega \in \Lambda_0$ it holds $\theta_\varepsilon(\omega) \in \Omega^{\widetilde{E}_2}$. So we get

$$\theta_\varepsilon^{-1}(\Lambda) \cap \Lambda_0 \Delta \theta_\varepsilon^{-1}(\widetilde{B}) \cap \Lambda_0 \subset \theta_\varepsilon^{-1}(\widetilde{C}) \cap \Lambda_0.$$

So

$$\theta_\varepsilon^{-1}(\Lambda) \cap \Lambda_0 \Delta \theta_\varepsilon^{-1}(\widetilde{B}) \cap \Omega^{\widetilde{E}_1} \subset \theta_\varepsilon^{-1}(\widetilde{C}) \cap \Lambda_0 \cup \Omega^{\widetilde{E}_1} \setminus \Lambda_0.$$

We have $\mathbb{P}_\nu^{\widetilde{E}_1}(\Omega^{\widetilde{E}_1} \setminus \Lambda_0) = 0$ and as above we get $\mathbb{P}_\nu^{\widetilde{E}_1}(\theta_\varepsilon^{-1}(\widetilde{C}) \cap \Lambda_0) = 0$. So we get $N' \in \mathcal{F}'|_{\Omega^{\widetilde{E}_1}}$ such that $\theta_\varepsilon^{-1}(\widetilde{C}) \cap \Lambda_0 \cup \Omega^{\widetilde{E}_1} \setminus \Lambda_0 \subset N'$ and $\mathbb{P}_\nu^{\widetilde{E}_1}(N') = 0$.

Since also $\theta_\varepsilon^{-1}(\widetilde{B}) \cap \Omega^{\widetilde{E}_1} \in \mathcal{F}_t'|_{\Omega^{\widetilde{E}_1}}$, we get $\theta_\varepsilon^{-1}(\Lambda) \cap \Lambda_0 \in (\mathcal{F}_t'|_{\Omega^{\widetilde{E}_1}})^{\mathbb{P}_\nu^{\widetilde{E}_1}}$. Because $\nu \in \mathcal{P}(\widetilde{E}_1)$ was arbitrary, we get $\theta_\varepsilon^{-1}(\Lambda) \cap \Lambda_0 \in \mathcal{F}_t^{\Omega^{\widetilde{E}_1}}$.

Now we have everything at hand to prove $\mathcal{F}_t^{\Omega^{\widetilde{E}_1}}$-adaptedness of $(A_t^\varepsilon)_{t\geq 0}$. Observe that for $t \leq \varepsilon$, A_t^ε is constantly zero, hence adapted. For $t > \varepsilon$ we have for $B \in \mathcal{B}(\mathbb{R})$:

$$(A_t^\varepsilon)^{-1}(B) = \begin{cases} \Lambda_0 \cap \theta_\varepsilon^{-1}((A_{t-\varepsilon})^{-1}(B)) & \text{if } 0 \notin B \\ \Lambda_0 \cap \theta_\varepsilon^{-1}((A_{t-\varepsilon})^{-1}(B)) \cup \Omega^{\widetilde{E}_1} \setminus \Lambda_0 & \text{else.} \end{cases}$$

Since $(A_{t-\varepsilon})^{-1}(B) \in \mathcal{F}_{t-\varepsilon}^{\Omega^{\widetilde{E}_2}}$ and $\Lambda_0 \in \mathcal{F}_t^{\Omega^{\widetilde{E}_1}}$, we get from the above considerations that $(A_t^\varepsilon)^{-1}(B) \in \mathcal{F}_t^{\Omega^{\widetilde{E}_1}}$.

$\square$

Lemma 7.3.23. *Let $\widetilde{E}_2 \subset \widetilde{E}_1 \subset E^\Delta$ be Borel such that $E^\Delta \setminus \widetilde{E}_1$ and $\widetilde{E}_1 \setminus \widetilde{E}_2$ are properly exceptional. Assume that $Y : \Omega^{\widetilde{E}_2} \to \mathbb{R}$ is $\mathcal{F}^{\Omega^{\widetilde{E}_2}}$-measurable and bounded. Define $\Lambda_0 := \Omega^{\widetilde{E}_1} \cap \bigcap_{n\in\mathbb{N}} \theta_{1/n}^{-1}(\Lambda)$ for some $\Lambda \in \mathcal{F}^{\Omega^{\widetilde{E}_2}}$ with $\mathbb{P}_x^{\widetilde{E}_2}(\Lambda) = 1$ for $x \in \widetilde{E}_2$. Assume that $P_t^{\widetilde{E}_1}(x, \widetilde{E}_1 \setminus \widetilde{E}_2) = 0$ for $x \in \widetilde{E}_1$. For $\varepsilon > 0$ define $\widetilde{Y}_\varepsilon : \Omega^{\widetilde{E}_1} \to \mathbb{R}$ by $\omega \mapsto 1_{\Lambda_0} Y(\theta_\varepsilon \omega)$. Then $\widetilde{Y}_\varepsilon$ is $\mathcal{F}^{\Omega^{\widetilde{E}_1}}$-measurable and for every $x \in \widetilde{E}_1$ it holds*

$$\mathbb{E}_x^{\widetilde{E}_1}[\widetilde{Y}_\varepsilon] = \mathbb{E}_x^{\widetilde{E}_1}\big[\mathbb{E}_{\mathbf{X}_\varepsilon}^{\widetilde{E}_2}[Y]\big].$$

The claim is also true if instead of boundedness of Y it holds $\mathbb{E}_x^{\widetilde{E}_1}\big[\mathbb{E}_{\mathbf{X}_\varepsilon}^{\widetilde{E}_2}[|Y|]\big] < \infty$ for every $x \in \widetilde{E}_2$.

Proof. From Lemma 7.3.22 we get measurability of $\widetilde{Y}_\varepsilon$ and that $\Lambda_0 \in \mathcal{F}_t^{\widetilde{E}_1}$ and $\mathbb{P}_x^{\widetilde{E}_1}(\Lambda_0) = 1$ for every $x \in \widetilde{E}_1$.

Assume there exist $0 \leq t_1 \leq ... \leq t_n < \infty$ and $f_1, ..., f_n : \mathbb{R} \to \mathbb{R}$, $n \in \mathbb{N}$, such that $Y = f_1(\mathbf{X}_{t_1}^{\widetilde{E}_2})...f_n(\mathbf{X}_{t_n}^{\widetilde{E}_2})$.

Then $1_{\Lambda_0} Y(\theta_\varepsilon \cdot) = 1_{\Lambda_0}(f_1(\mathbf{X}_{\varepsilon+t_1}^{\widetilde{E}_1})...f_n(\mathbf{X}_{\varepsilon+t_n}^{\widetilde{E}_1}))$. So the claim follows from the Markov property of $\mathbf{M}^{\widetilde{E}_1}$ and Lemma 7.3.20. Using a monotone class argument we get the claim for general Y that are $\mathcal{F}'|_{\Omega^{\widetilde{E}_2}}$-measurable.

Let Y be $\mathcal{F}^{\Omega^{\widetilde{E}_2}}$-measurable. Let ν be the restriction to $\mathcal{B}(\widetilde{E}_2)$ of the image measure of $\mathbf{X}_\varepsilon^{\widetilde{E}_1}$ under $\mathbb{P}_x^{\widetilde{E}_1}$. There exists $Y^{(1)}$ and $Y^{(2)}$ $\mathcal{F}'|_{\Omega^{\widetilde{E}_2}}$-measurable such that $Y^{(1)} \leq Y \leq Y^{(2)}$ and $\mathbb{P}_\nu^{\widetilde{E}_2}(Y_1 < Y_2) = 0$. Let $\widetilde{Y}_\varepsilon^{(1)}$ and $\widetilde{Y}_\varepsilon^{(2)}$ be analogously defined to $\widetilde{Y}_\varepsilon$. Then we get from the already proven facts:

$$\mathbb{E}_x^{\widetilde{E}_1}[\widetilde{Y}_\varepsilon^{(1)}] = \mathbb{E}_x^{\widetilde{E}_1}\big[\mathbb{E}_{\mathbf{X}_\varepsilon}^{\widetilde{E}_2}[Y^{(1)}]\big] = \mathbb{E}_x^{\widetilde{E}_1}\big[\mathbb{E}_{\mathbf{X}_\varepsilon}^{\widetilde{E}_2}[Y]\big]$$
$$= \mathbb{E}_x^{\widetilde{E}_1}\big[\mathbb{E}_{\mathbf{X}_\varepsilon}^{\widetilde{E}_2}[Y^{(2)}]\big] = \mathbb{E}_x^{\widetilde{E}_1}[\widetilde{Y}_\varepsilon^{(2)}].$$

Moreover, we have

$$\mathbb{E}_x^{\widetilde{E}_1}\big[\widetilde{Y}_\varepsilon^{(1)}\big] \leq \mathbb{E}_x^{\widetilde{E}_1}\big[\widetilde{Y}_\varepsilon\big] \leq \mathbb{E}_x^{\widetilde{E}_1}\big[\widetilde{Y}_\varepsilon^{(2)}\big]$$

and hence

$$\mathbb{E}_x^{\widetilde{E}_1}\big[\widetilde{Y}_\varepsilon\big] = \mathbb{E}_x^{\widetilde{E}_1}\big[\mathbb{E}_{\mathbf{X}_\varepsilon}^{\widetilde{E}_2}[Y]\big] \quad \text{for every } x \in \widetilde{E}_1.$$

If Y is not assumed to be bounded but just fulfills the integrability conditions, we can apply a similar approximation argument as in the proof of Lemma 7.3.7(iv).

$\square$

7.4 Point Separating at the Boundary

The proof of the following lemma is based on [Lun95, Prop. 0.3.1], some ideas were obtained from [Tri78, Sec. 2.9]. Denote by $C_b^m(\mathbb{R}^d)$ the set of m-times differentiable functions such that the function and derivatives up to order m are bounded on $\mathbb{R}^d$. By $C_b^m(\overline{\Omega})$, $\Omega \subset \mathbb{R}^d$ open, denote the space of m-times differentiable functions such that the function and the derivatives up to order m admit a bounded and continuous extension to the boundary.

Lemma 7.4.1. *Let* $u \in C_c^2(\mathbb{R}^{d-1})$, $v \in C_c^1(\mathbb{R}^{d-1})$. *Then there exists* $\widetilde{u} \in C_b^2(\mathbb{R}^d \cap \{x_d \geq 0\})$ *with* $\widetilde{u}(x',0) = u(x')$ *and* $\partial_d\widetilde{u}(x',0) = v(x')$ *for* $x' \in \mathbb{R}^{d-1}$.

Proof. Let $\varphi \in C_c^2(\mathbb{R}^{d-1})$ such that $\int_{\mathbb{R}^{d-1}} \varphi = 1$. Define for $g \in C_c^2(\mathbb{R}^{d-1})$

$$P_0 g(x',x_d) = \int_{\mathbb{R}^{d-1}} \varphi(\xi)g(x' + x_d\xi)d\xi$$

and for $h \in C_c^2(\mathbb{R}^{d-1})$ define

$$P_1 h(x',x_d) = x_d \int_{\mathbb{R}^{d-1}} \varphi(\xi)h(x' + x_d\xi)d\xi.$$

Then P_0 defines a linear operator mapping from $C_c^2(\mathbb{R}^{d-1})$ to $C_b^2(\mathbb{R}^d \cap \{x_d \geq 0\})$.
Moreover, $P_1 h \in C_b^2(\mathbb{R}^d \cap \{x_d \geq 0\})$. Using partial integration one sees that the $C_b^2(\mathbb{R}^d \cap \{x_d \geq 0\})$-norm of $P_1 g$ can be estimated by the $C_b^1(\mathbb{R}^{d-1})$-norm of h. Thus the mapping extends to linear bounded operator $P_1 : C_c^1(\mathbb{R}^{d-1}) \to C_b^2(\mathbb{R}^d \cap \{x_d \geq 0\})$.

Now set $\lambda_0 = 1$, $\lambda_1 = \frac{1}{2}$. Then for $a_0 = -1$, $a_1 = 2$ we have $a_0 + a_1 = 1$, $\lambda_0 a_0 + \lambda_1 a_1 = 0$. For $u \in C_c^2(\mathbb{R}^{d-1})$ and $v \in C_c^1(\mathbb{R}^{d-1})$ define

$$\tilde{u}(x', x_d) := a_0 P_0 u(x', \lambda_0 x_d) + a_1 P_0 u(x', \lambda_1 x_d) + P_1 v(x', x_d).$$

Then by the mapping properties of P_0 and P_1 we have $\tilde{u} \in C_b^2(\mathbb{R}^d \cap \{x_d \geq 0\})$ and by the choice of a_0, a_1 and λ_0, λ_1 the boundary conditions are fulfilled. □

Lemma 7.4.2. *Let U be a neighborhood of 0, $A : U \to \mathbb{R}^{d \times d}$, $A = (a_{ij})_{1 \leq i,j \leq d}$ a matrix-valued mapping of symmetric strictly elliptic matrices with $(a_{ij})_{1 \leq i,j \leq d} \in C^1(U \cap \{x_d \geq 0\}; \mathbb{R}^{d \times d})$. Then the set*

$$D := \left\{ u \in C_c^2(U \cap \{x_d \geq 0\}) \,\Big|\, (e_d, A\nabla u) = 0 \, on \, U \cap \{x_d = 0\} \right\}$$

is point separating in 0. For $x \in U \cap \{x_d = 0\}$, $\nabla u(x)$ denotes the continuous extension of the gradient of u to $U \cap \{x_d \geq 0\}$, e_d denotes the d-th unit vector. More precisely, there exists a sequence of functions $(f_k)_{k \in \mathbb{N}}$ in D such that for every $y \in U \cap \{x_d \geq 0\}$ with $y \neq 0$ there exists $k \in \mathbb{N}$ with $f_k(0) = 1$ and $f_k(y) = 0$.

Proof. Let $k \in \mathbb{N}$. Choose a function $g_k \in C_c^\infty(B_{1/k}^{(d-1)}(0))$ such that $g_k(0) = 1$. Here $B_{1/k}^{(d-1)}(0) \subset \mathbb{R}^{d-1}$ denotes the ball of radius $1/k$ in $\mathbb{R}^{d-1}$. Set $v_k(x) := -\sum_{j=1}^{d-1} \frac{1}{a_{dd}} a_{dj} \partial_j g_k(x)$ for $x \in \mathbb{R}^{d-1}$.

Since $a^{dd} > 0$ and the coefficients a_{ij} are C^1-smooth, $v_k \in C_c^1(\mathbb{R}^{d-1})$ with $\text{supp}[v_k] \subset B_{1/k}^{(d-1)}(0)$. Then by Lemma 7.4.1 there exists a function $h_k \in C_b^2(\mathbb{R}^d \cap \{x_d \geq 0\})$ such that $h_k(x', 0) = g_k(x')$ and $\partial_d h_k(x', 0) = v_k(x', 0)$. By the last equality we have $\sum_{j=1}^d a_{dj} \partial_j h_k(x', 0) = 0$. Choosing φ in the proof of Lemma 7.4.1 such that $\text{supp}\,\varphi \subset B_{\frac{1}{k}}^{(d-1)}(0)$ we get that for $x_d \leq 1$ it holds $\text{supp}\, h_k(\cdot, x_d) \subset [-2/k, 2/k]^{d-1}$.

Indeed. Note that the integral is taken only over the support of φ. Assume that $x' \notin [-2/k, 2/k]^{d-1}$. Then there exists $i < d$ with $|x_i'| > 2/k$. Let $\xi \in B_{1/k}^{(d-1)}(0)$. Since $x_d \leq 1$, we have

$$|x' + x_d \xi|_i > 2/k - |x_d \xi| \geq 2/k - 1/k = 1/k.$$

So $x' + x_d \xi$ is not in the support of g for $\xi \in B_{1/k}^{(d-1)}(0)$. So the integrand is zero for $\xi \in B_{1/k}^{(d-1)}(0)$. Since $\text{supp}[\xi] \subset B_{1/k}^{(d-1)}(0)$, it follows $P_0 g = 0$. The same holds for P_1. Choose a function $\phi_k \in C_c^\infty(\mathbb{R}_0^+)$ such that $\phi_k(z) = 1$ for $z < 1/2k$ and $\phi_k(z) = 0$ for $z \geq 1/k$.

Define $f_k(x) = h_k(x)\phi_k(x_d)$. For k_0 large enough, $[-2/k_0, 2/k_0]^d \cap \{x_d \geq 0\} \subset\subset U \cap \{x_d \geq 0\}$. For $k \geq k_0$ we have $f_k \in C_c^2(U \cap \{x_d \geq 0\})$ and since $\phi_k(x_d)$ is constant for $x_d < 1/2k$ the boundary condition is also fulfilled. Obviously the functions $(f_k)_{k \geq k_0}$ are point separating in 0. □

We are now ready to state the following lemma.

Lemma 7.4.3. *Let Ω be a domain, $x \in \partial\Omega$ with local C^2-smooth boundary in a neighborhood W of x. Furthermore, let $A = (a_{ij})_{1 \leq i,j \leq d} \in C^1(W \cap \partial\Omega)$. Then the set*

$$D := \left\{ u \in C_c^2(W \cap \overline{\Omega}) \;\middle|\; \sum_{i=1}^d \eta_i a_{ij} \partial_j u = 0 \text{ on } W \cap \partial\Omega \right\}$$

is point separating in x. More precisely, there exists a sequence $(f_k)_{k \in \mathbb{N}}$ in D such that for every $y \in \overline{\Omega}$ there exists f_k with $f_k(0) = 1$ and $f_k(y) = 0$.

Proof. By definition of C^2-smoothness there exists a neighborhood V of x, a neighborhood U of 0 and a C^2-diffeomorphism $\psi : V \to U$ with $\psi(V \cap \Omega) = U \cap \{x_d \geq 0\}$, $\psi(V \cap \partial\Omega) = U \cap \{x_d = 0\}$ and $\psi(x) = 0$. Moreover, for $y \in V \cap \partial\Omega$ the tangential space T_y is spanned by the $d - 1$ first column vectors of $(D\psi)^{-1}(y)$, i.e.,

$$T_y = \text{span}\left\{ (D\psi)^{-1}(y)_{*i} \;\middle|\; 1 \leq i \leq d - 1 \right\}.$$

In particular a vector $z \in \mathbb{R}^d$ is an element of the tangential space T_y iff the d-th coordinate of $(D\psi(y))z$ is zero.

For a function u on $V \cap \overline{\Omega}$, $\tilde{u} := u \circ \psi^{-1}$ defines a function on $U \cap \{x_d \geq 0\}$ and vice versa. The mapping properties of ψ and the C^2-smoothness of ψ and ψ^{-1} yield $u \in C_c^2(V \cap \overline{\Omega})$ iff $\tilde{u} \in C_c^2(U \cap \{x_d \geq 0\})$. Applying the chain rule we get $\nabla u = (D\psi)^\top \nabla \tilde{u}$. Additionally, $A\nabla u$ is an element of the tangent space at y iff $(e_d, (D\psi)A\nabla u) = 0$. Define

$$\tilde{A}(\tilde{y}) := (D\psi)A(D\psi)^\top(\psi^{-1}(\tilde{y})).$$

Using that η is orthogonal to the tangential space we get

$$(\eta, A\nabla u) = 0 \Leftrightarrow (e_d, (D\psi)A\nabla u) = 0$$

$$\Leftrightarrow (e_d, (D\psi)A(D\psi)^\top \nabla \tilde{u}) = (e_d, \tilde{A}\nabla \tilde{u}) = 0.$$

Summarizing we get

$$\left\{ u \in C_c^2(V \cap \overline{\Omega}) \;\middle|\; (\eta, A\nabla u) = 0 \text{ on } V \cap \partial\Omega \right\} = \left\{ \tilde{u} \circ \psi \;\middle|\; \tilde{u} \in \tilde{D} \right\} \quad (7.14)$$

with

$$\tilde{D} = \left\{ \tilde{u} \in C_c^2(U \cap \{x_d \geq 0\}) \;\middle|\; (e_d, \tilde{A}\nabla\tilde{u}) = 0 \text{ on } U \cap \{x_d = 0\} \right\}.$$

The set in the left-hand side of (7.14) can be embedded into D by continuing the functions by zero in $\overline{\Omega}$ outside $V \cap \overline{\Omega}$. By Lemma 7.4.2 the set $\tilde{D}$ is point separating in 0 thus D is point separating in x. $\qquad\square$

7.5 Results on Sobolev Spaces

In this section we state several well-known results concerning Sobolev spaces. However, we provide corresponding local versions of these results, i.e., we pose just local assumptions on the boundary smoothness. In most of the literature these results are only proven in the case of relatively compact sets what does not cover our setting.

Recall that an *open cuboid* is a set $Q := \times_{i=1}^d (a_i, b_i) \subset \mathbb{R}^d$, $d \in \mathbb{N}$, with $a_i < b_i$, $a_i, b_i \in \mathbb{R} \cup \{-\infty, \infty\}$, $1 \leq i \leq d$.

Lemma 7.5.1. *Let $U' \subset \mathbb{R}^{d-1}$ be an open cuboid, $0 < r < \infty$, $h : \overline{U'} \to \mathbb{R}$ Lipschitz continuous. Let $U := \{(x', x_d) \in \mathbb{R}^d \,|\, x' \in U', |x_d - h(x')| < r\}$. Define $\Psi^{(1)} : U \to U' \times (-r, r)$ by $(x', x_d) \mapsto (x', x_d - h(x'))$. Then $\Psi^{(1)}$ is bi-Lipschitz continuous with inverse $(\Psi^{(1)})^{-1} : U' \times (-r, r) \to U$ given by $(\Psi^{(1)})^{-1}(x', y) = (x', h(x') + y)$.*

Definition 7.5.2. Let $\Omega \subset \mathbb{R}^d$, $d \in \mathbb{N}$, open. We say that the boundary is $C^{m,\alpha}$-*smooth* at $x \in \partial\Omega$ for $m \in \mathbb{N}_0$, $\alpha \in [0, 1]$ if the following holds: There exists an open cuboid $U'_x \subset \mathbb{R}^{d-1}$, $0 < r_x < \infty$, a function $h_x \in C^{m,\alpha}(\overline{U'_x})$ and a mapping $\Psi_x^{(0)}$ composed of a translation and rotation such that $U_x := \{(x', x_d) \in \mathbb{R}^d \,|\, x' \in U'_x, |x_d - h(x')| < r_x\}$ is a neighborhood of $\Psi_x^{(0)} x$ and:

$$U_x \cap \Psi_x^{(0)}\Omega = \left\{ (x', x_d) \in U_x \,\middle|\, x_d > h_x(x') \right\}$$
$$U_x \cap \Psi_x^{(0)} \partial\Omega = \left\{ (x', x_d) \in U_x \,\middle|\, x_d = h_x(x') \right\}$$
$$U_x \cap \Psi_x^{(0)}\overline{\Omega}^c = \left\{ (x', x_d) \in U_x \,\middle|\, x_d < h_x(x') \right\}.$$

Define $\widetilde{U}_x := (\Psi_x^{(0)})^{-1} U_x$. Let $\Psi_x^{(1)} : U_x \to U' \times (-r, r)$ defined as in Lemma 7.5.1 corresponding to h_x, $\Psi_x := \Psi_x^{(1)} \circ \Psi_x^{(0)}$. For $\Gamma \subset \partial\Omega$ we say Γ is $C^{m,\alpha}$-smooth if the boundary is $C^{m,\alpha}$-smooth at every $x \in \Gamma$. We call

$$\mathcal{A} := \{(U_x', U_x, \widetilde{U}_x, h_x, r_x, \Psi_x^{(0)}, \Psi_x) \,|\, x \in \Gamma\}$$

a *parametrization* of Γ. If Γ is compact we can cover Γ by finitely many of the $\widetilde{U}_x$, $x \in \Gamma$, and obtain a *finite parametrization*

$$\mathcal{A} := \{(U_i', U_i, \widetilde{U}_i, h_i, r_i, \Psi_i^{(0)}, \Psi_i) \,|\, 1 \le i \le N\}$$

where $(U_i', U_i, \widetilde{U}_i, h_i, r_i, \Psi_i^{(0)}, \Psi_i) = (U_{x_i}', U_{x_i}, \widetilde{U}_{x_i}, h_{x_i}, r_{x_i}, \Psi_{x_i}^{(0)}, \Psi_{x_i})$ for some $x_i \in \Gamma$, $1 \le i \le N$ and $N \in \mathbb{N}$. For $m = 0$, $\alpha = 1$ we call Γ *locally Lipschitz smooth*.

Remark 7.5.3. We say *locally* Lipschitz smooth to emphasize that there need not to be one common Lipschitz function h for the whole boundary part Γ. So our definition is completely local. This definition is adapted from [Dob10, Def. 6.1].

In the case $m \ge 1$ we get a different characterization of boundary smoothness. We call a mapping $\Psi : U \to V$, $U, V \subset \mathbb{R}^d$ open, a $C^{m,\alpha}$-*diffeomorphism*, $m \in \mathbb{N}$, $\alpha \in (0, 1)$, if Ψ is bijective and both Ψ and Ψ^{-1} are $C^{m,\alpha}$-smooth on U and V, respectively. With Ψ being $C^{m,\alpha}$-*smooth* on U we mean that Ψ is m-times continuously differentiable and the m-th derivative is Hölder continuous of index α.

Theorem 7.5.4. *Let $m \in \mathbb{N}$, $\Omega \subset \mathbb{R}^d$, $d \in \mathbb{N}$, open and $\Gamma \subset \partial\Omega$. Then are equivalent:*
(i) Γ is $C^{m,\alpha}$-smooth.
(ii) For each $x \in \Gamma$ there exists an open neighborhood U and $C^{m,\alpha}$-diffeomorphism $\Psi : U \to B_1(0)$ with

$$\Psi(U \cap \Omega) = B_1^+(0),$$
$$\Psi(U \cap \partial\Omega) = B_1^0(0),$$
$$\Psi(U \cap \overline{\Omega}^c) = B_1^-(0).$$

The proof works as the one of [Dob10, Satz 6.3].
For several localization arguments we need a partition of unity.

Theorem 7.5.5 ([AD75],Theo. 3.14). *Let $A \subset \mathbb{R}^d$, $d \in \mathbb{N}$, be an arbitrary subset and let $\mathcal{O}$ be a family of open sets in $\mathbb{R}^d$ covering A. Then there exists*

a family **F** *of functions* $\phi \in C_c^\infty(\mathbb{R}^d)$ *having the following properties:*
(i) For every $\phi \in$ **F** *it holds* $0 \le \phi \le 1$.
(ii) If $K \subset A$ *is compact, then at most finitely many* ϕ *are non-zero on* K.
(iii) For every $\phi \in$ **F** *there exists* $U \in \mathcal{O}$ *such that* $supp[\phi] \subset U$.
(iv) For every $x \in A$, $\sum_{\phi \in \mathbf{F}} \phi(x) = 1$.
Such a family is called a C^∞-*partition of unity for* A *subordinate to* U. *If* A
is compact, then every covering $\mathcal{O}$ *can be reduced to a finite open covering.*
Then the corresponding family **F** *consists of only finitely many functions.*

See also [Dob10, Lem. 5.13] for the compact case.

Remark 7.5.6. Let $\Omega \subset \mathbb{R}^d$, $d \in \mathbb{N}$, open, $K \subset \overline{\Omega}$ compact. Define
$\Gamma := K \cap \partial\Omega$. Assume that Γ is locally Lipschitz smooth. Let $\mathcal{A} = \{(U_i', U_i, \widetilde{U}_i, h_i, r_i, \Psi_i^{(0)}) \,|\, 1 \le i \le N\}$ be the finite parametrization of Γ
according to Definition 7.5.2. Set $\mathcal{A}_0 := \{\widetilde{U}_i \,|\, 1 \le i \le N\}$. There exists a
finite family of open cuboids $\mathcal{B} := \{B_i \,|\, 1 \le i \le N_1\}$, $\overline{B}_i \subset \Omega$, $1 \le i \le N_1$,
$N_1 \in \mathbb{N}$, such that $\mathcal{A}_0 \cup \mathcal{B}$ is a finite covering of K. According to Theorem
7.5.5 there exists a finite partition of unity $(\phi_i)_{1 \le i \le N'}$, $N' \in \mathbb{N}$, subordinate
to the covering $\mathcal{A}_0 \cup \mathcal{B}$.

For some purposes the following weaker assumptions on the boundary are
sufficient.

Definition 7.5.7. A domain $\Omega \subset \mathbb{R}^d$, $d \in \mathbb{N}$, is said to have the *segment*
property if for every $x \in \partial\Omega$ there exists an open neighborhood U_x of x and
a nonzero vector $v_x \in \mathbb{R}^d$ such that for $y \in \overline{\Omega} \cap U_x$ it holds $y + tv_x \in \Omega$ for
$0 < t < 1$.

One can easily show that if Ω has C^0-smooth boundary, then Ω has the
segment property.

Theorem 7.5.8 ([AD75],Theo. 3.18). *If* $\Omega \subset \mathbb{R}^d$ *open*, $d \in \mathbb{N}$, *has the*
segment property, then $C_c^\infty(\mathbb{R}^d)|_\Omega$ *is dense in* $H^{m,p}(\Omega)$, $m \in \mathbb{N}$, $1 \le p < \infty$.
If $u \in C_c^0(\overline{\Omega}) \cap H^{m,p}(\Omega)$, *then there exists* $(u_n)_{n \in \mathbb{N}}$ *in* $C_c^\infty(\mathbb{R}^d)$ *such that*
$u_n \overset{n \to \infty}{\longrightarrow} u$ *both in* $H^{m,p}(\Omega)$ *and w.r.t. sup-norm on* $\overline{\Omega}$.

The first claim is proven in [AD75, Theo. 3.18]. The convergence in sup-
norm can be deduced by a careful analysis of the proof of [AD75, Theo. 3.18].
We need the following lemma which states that the neighborhood from
Definition 7.5.2 viewed as a domain has at least the segment property. This
is important for local approximation arguments.

Lemma 7.5.9. *Let $U' \subset \mathbb{R}^{d-1}$, $d \in \mathbb{N}$, be an open cuboid, $h : \overline{U'} \to \mathbb{R}$ Lipschitz continuous, $0 < r < \infty$. Define $\hat{U} := \{(x', x_d) \in \mathbb{R}^d \mid x' \in U', h(x') < x_d < h(x') + r\}$. Then $\hat{U}$ has the segment property.*

Proof. The boundary of $\hat{U}$ consists of three parts.

$$\Gamma_1 := \{(x', x_d) \mid x' \in U', x_d = h(x')\},$$
$$\Gamma_2 := \{(x', x_d) \mid x' \in \partial U', h(x') \leq x_d \leq h(x') + r\},$$
$$\Gamma_3 := \{(x', x_d) \mid x' \in U', x_d = h(x') + r\}$$

and $\partial \hat{U} = \Gamma_1 \cup \Gamma_2 \cup \Gamma_3$. For $x \in \Gamma_1$ we may take as $v_x := r(0, 1/2)$, $\widetilde{U} := \hat{U} \cap \{(x', x_d) \in \hat{U} \mid x_d < h(x') + \frac{r}{2}\}$. For $x \in \Gamma_3$ we may take as $v_x := -r(0, 1/2)$, $\widetilde{U} := \hat{U} \cap \{(x', x_d) \in \hat{U} \mid x_d > h(x') + \frac{r}{2}\}$. Let $x = (x', x_d) \in \Gamma_2$ with $h(x') < x_d < h(x') + r$. We can find an open cuboid $U_2 = U_2' \times (a, b) \subset \mathbb{R}^d$ containing x such that for $(y', y_d) \in U_2$ it holds $h(x') < y_d < h(x') + r$. Here $U_2' \subset \mathbb{R}^{d-1}$ is an open cuboid and $0 < a < b < \infty$. Set $\widetilde{U_2} := (U' \cap U_2') \times (a, b)$. Then $x \in \partial \widetilde{U_2}$. Since $\widetilde{U_2}$ is a cuboid, it has also the segment property. So we find an open neighborhood $\widetilde{U_3} \subset \mathbb{R}^d$ of x and a vector $v_x \in \mathbb{R}^d$ such that for every $y \in \widetilde{U_3} \cap \overline{\widetilde{U_2}}$ it holds $y + tv_x \in \widetilde{U_2}$ for $0 < t < 1$. Making $\widetilde{U_3}$ smaller we may assume $\widetilde{U_3} \cap \hat{U} \subset \overline{\widetilde{U_2}}$. So with $\widetilde{U_3} \cap \hat{U} = \widetilde{U_3} \cap \overline{\widetilde{U_2}}$ we get the segment property of $\hat{U}$ at x.

Now let $x = (x', x_d) \in \Gamma_2$ with $x' \in \partial U'$ and $x_d = h(x')$. Since U' has the segment property, we find a neighborhood $U_0' \subset \mathbb{R}^{d-1}$ containing x' and $v_{x'}' \subset \mathbb{R}^{d-1}$ such that for $y' \in U_0' \cap \overline{U'}$ and $0 < t < 1$ it holds $y' + tv_{x'}' \in U'$. Define $v_x := (v_{x'}', 2L\|v_{x'}'\|)$ with L being the Lipschitz constant of h on $\overline{U'}$. Define

$$U_1 := U_0' \times \left(h(x'), h(x') + \frac{r}{2}\right).$$

Let $y = (y', y_d) \in U_1 \cap \hat{U}$, then $h(y') \leq y_d < h(x') + \frac{r}{2}$ and $y + tv_x = \left(y' + tv_{x'}', y_d + t2L\|v_{x'}'\|\right)$ for $0 < t < 1$. We have

$$h(y' + tv_{x'}') \leq h(y') + tL\|v_{x'}'\| < y_d + 2Lt\|v_{x'}'\|$$

and

$$h(y' + tv_{x'}') + r \geq h(x') + r - (Lt\|v_{x'}'\| - L\|y' - x'\|)$$
$$> y_d + \frac{r}{2} - (Lt\|v_{x'}'\| - L\|y' - x'\|).$$

Thus

$$y_d + t2L\|v'_{x'}\| < h(y' + tv'_{x'}) + \frac{r}{2} + L\|y' - x'\| + t3L\|v'_{x'}\|.$$

So we can make U'_0 and U_1 suitable smaller and choose $0 < T \le 1$ such that for $y \in U_1$ and $z := y + tTv_x = (z', z_d)$ it holds $h(z') < z_d < h(z') + r$. So $z \in \hat{U}$. □

The following lemma concerning weakly differentiability of Lipschitz functions should be well-known. We state it here for completeness.

Lemma 7.5.10. *Let* $\Omega \subset \mathbb{R}^d$, $d \in \mathbb{N}$, *open with finite measure. If* u *is Lipschitz continuous on* $\overline{\Omega}$, *then* $u \in H^{1,p}(\Omega)$ *for every* $1 \le p \le \infty$. *Moreover,* $\partial_i u \le L$, *dx-a.e. with* L *being the Lipschitz constant of* u, $1 \le i \le d$.

See [GT77, Lem. 7.24] and [GT77, Problem 7.7].

We need the following lemma concerning the transformation of weakly differentiable functions and Hölder continuous functions.

Lemma 7.5.11. *Let* $U' \subset \mathbb{R}^{d-1}$ *be an open cuboid,* $0 < r < \infty$, $h : \overline{U'} \to \mathbb{R}$ *Lipschitz continuous. Let* $U := \{(x', x_d) \in \mathbb{R}^d \,|\, x' \in U', |x_d - h(x')| < r\}$. *Let* $\Psi^{(1)} : U \to U' \times (-r, r)$ *be the mapping* $(x', x_d) \mapsto (x', x_d - h(x'))$ *as defined in Lemma 7.5.1 with inverse* $\Psi_1^{-1} : U' \times (-r, r) \to U$ *given by* $\Psi_1^{-1}(x', y) = (x', h(x') + y)$. *Let* Ψ_0 *be a composition of a translation and rotation on* $\mathbb{R}^d$. *Set* $\hat{U} := \Psi_0^{-1}\Psi_1^{-1}(U' \times (0, r))$. *Define* $\Psi : \hat{U} \to U' \times (0, r)$ *by* $\Psi := \Psi_1 \circ \Psi_0$.
Define $T : \mathbb{R}^{U' \times (0,r)} \to \mathbb{R}^{\hat{U}}$ *by*

$$u \mapsto u \circ \Psi.$$

Then T *is a linear bijective operator with inverse* T^{-1} *given by* $u \mapsto u \circ \Psi^{-1}$. *Moreover,* T *induces a continuous and bijective operator from* $H^{1,p}(U' \times (0, r))$ *onto* $H^{1,p}(\hat{U})$ *for* $1 \le p < \infty$. *Furthermore,* T *induces a continuous and bijective operator from* $C^{0,\beta}(\overline{U' \times (0, r)})$ *to* $C^{0,\beta}(\hat{U})$ *for* $0 \le \beta \le 1$.

Proof. We follow the arguments in the proof of [Dob10, Lem. 6.6]. Note that for a weakly differentiable function u the corresponding function $u \circ \Psi_0$ is weakly differentiable as well. Since the Lebesgue measure is invariant under rotation and translation, we may assume $\Psi_0 = 1$. Let $u\ \mathcal{B}(U' \times (0,r))$-measurable then $u \circ \Psi$ is $\mathcal{B}(\hat{U})$-measurable. We calculate with $x = (x', x_d)$,

$$
\begin{aligned}
\int_{\hat{U}} u \circ \Psi\,(x)\,dx &= \int_{U'} \int_{h(x')}^{h(x')+r} u \circ \Psi\,(x', x_d)\,dx_d\,dx' \\
&= \int_{U'} \int_{h(x')}^{h(x')+r} u(x', x_d - h(x'))\,dx_d\,dx' \\
&= \int_{U'} \int_0^r u(x', x_d)\,dx_d\,dx'. \quad (7.15)
\end{aligned}
$$

Similarly we get for $\mathcal{B}(\hat{U})$-measurable u:

$$
\int_{U' \times (0,r)} u \circ \Psi^{-1}\,(x)\,dx = \int_{\hat{U}} u\,(x)\,dx.
$$

Observe that for $\mathcal{B}(U' \times (0,r))$-measurable u_1 and u_2 it holds

$$
1_{\{u_1 \neq u_2\}} \circ \Psi = 1_{\{u_1 \circ \Psi \neq u_2 \circ \Psi\}}.
$$

So plugging $1_{\{u_1 \neq u_2\}}$ into the right-hand side of (7.15) we get that u_1 is a dx-version of u_2 iff $u_1 \circ \Psi$ is a dx-version of $u_2 \circ \Psi$. So the operator T maps measurable functions to measurable functions and respects dx-equivalence classes. Furthermore, we can conclude that T is a linear continuous operator from $L^p(U' \times (0,r))$ to $L^p(\hat{U})$ with continuous inverse given by $T^{-1}u = u \circ \Psi^{-1}$.

Next we prove that the operator T induces a continuous linear operator mapping from $H^{1,p}(U' \times (0,r))$ to $H^{1,p}(\hat{U})$. Since U' is a cuboid and h is Lipschitz continuous on $\overline{U'}$, we get by Theorem 7.5.8 a sequence $(h_n)_{n \in \mathbb{N}}$ in $C_c^\infty(\mathbb{R}^d)$ converging to h both w.r.t. sup-norm on $\overline{U'}$ and in $H^{1,p}(U')$. Denote by Ψ_n, $n \in \mathbb{N}$, the transformation $\hat{U} \ni (x', x_d) \mapsto (x', x_d - h_n(x')) \in \mathbb{R}^d$.

Let $u \in C_c^\infty(\mathbb{R}^d)$. Using the chain rule we get

$$
\begin{aligned}
\partial_i\, u(x', x_d - h_n(x')) = {}&(\partial_i u)\,(x', x_d - h_n(x')) \\
&- (\partial_d u)\,(x', x_d - h_n(x'))\,(\partial_i h_n)\,(x') \quad (7.16)
\end{aligned}
$$

for $1 \leq i \leq d-1$ and

$$
\partial_d\, u(x', x_d - h_n(x')) = (\partial_d u)\,(x', x_d - h_n(x')) \text{ for } (x', x_d) \in \hat{\overline{U}} \quad (7.17)
$$

Since u has compact support in $\mathbb{R}^d$, the function and the derivatives are uniformly continuous. Together with uniform convergence of $(h_n)_{n\in\mathbb{N}}$ on $\overline{U'}$ we get

$$\lim_{n\to\infty} (\partial_i u)(x', x_d - h_n(x')) = (\partial_i u)(x', x_d - h(x'))$$

uniformly on $\overline{U}$ for $1 \leq i \leq d$. This convergence also holds in $L^p(\hat{U})$.

In the same way we get that $u \circ \Psi_n \overset{n\to\infty}{\longrightarrow} u \circ \Psi$ converges in $L^p(\hat{U})$. Furthermore, $(\partial_i h_n)_{n\in\mathbb{N}}$ converges in $L^p(U', dx')$ for $1 \leq i \leq d$. So altogether, $(u \circ \Psi_n)_{n\in\mathbb{N}}$ forms a Cauchy sequence in $H^{1,p}(\hat{U})$. But $(u \circ \Psi_n)_{n\in\mathbb{N}}$ converges to $u \circ \Psi$ in $L^p(\hat{U})$. Thus $Tu = u \circ \Psi \in H^{1,p}(\hat{U})$. For the weak derivatives of h we have $\partial_i h \leq L$, $1 \leq i \leq d$, dx-a.e. with L being the Lipschitz constant of h on $\overline{U'}$, see Lemma 7.5.10. Taking the limit in (7.16) and (7.17) we get together with (7.15) that the $H^{1,p}(\hat{U})$-norm of $u \circ \Psi$ can be estimated by the $H^{1,p}(U' \times (0,r))$-norm of u with a constant independent of u.

Since $H^{1,p}(U' \times (0,r)) \cap C_c^\infty(\mathbb{R}^d)|_{U'\times(0,r)}$ is dense in $H^{1,p}(U' \times (0,r))$, the operator T extends to a continuous linear operator mapping from $H^{1,p}(U' \times (0,r))$ to $H^{1,p}(\hat{U})$.

Similarly one can show that the operator $\widetilde{T}$ defined by $u \mapsto u \circ \Psi^{-1}$ induces a continuous linear operator mapping from $H^{1,p}(\hat{U})$ to $H^{1,p}(U' \times (0,r))$. The only difference in the proof is that we have to do an approximation argument with functions in $H^{1,p}(\hat{U}) \cap C_c^\infty(\mathbb{R}^d)|_{\hat{U}}$. Since $\hat{U}$ has the segment property, we have that this space is dense in $H^{1,p}(\hat{U})$. In particular, if $v \in H^{1,p}(\hat{U})$, then $v \circ \Psi^{-1} = \widetilde{T}v \in H^{1,p}(U' \times (0,r))$. Clearly $v = T(v \circ \Psi^{-1})$. Thus T is surjective and the inverse is given by $\widetilde{T}$.

If $u \in C^{0,\beta}(\overline{U' \times (0,r)})$, then $u \circ \Psi$ is in $C^{0,\beta}(\overline{\hat{U}})$ as a composition of a β-Hölder continuous and Lipschitz continuous function. Of course, the corresponding norm can be estimated. The analogous statement holds for Ψ^{-1} and $u \in C^{0,\beta}(\overline{\hat{U}})$. So we get the claimed properties for T also on Hölder function spaces. $\square$

Now we can formulate the Sobolev embedding theorem.

Theorem 7.5.12. Let $\Omega \subset \mathbb{R}^d$ open, $d \in \mathbb{N}$. Let $\Omega_0 \subset \Omega$ open, $\Gamma_0 := \partial\Omega_0 \cap \partial\Omega$. Assume that there exists a locally Lipschitz smooth boundary part $\Gamma \subset \partial\Omega$ such that $\Gamma_0 \subset \Gamma$ and $\Omega_0 \cup \Gamma_0 \subset\subset \Omega \cup \Gamma$. Let $p > d$ and $0 < \beta \leq 1 - \frac{d}{p}$. Then every $u \in H^{1,p}(\Omega)$ has a continuous version $\widetilde{u}$ on $\overline{\Omega}_0$. The mapping

$$\iota_{\Omega_0} : H^{1,p}(\Omega) \ni u \mapsto \widetilde{u} \in C^{0,\beta}(\overline{\Omega}_0)$$

is a continuous linear operator.

Proof. Let $u \in H^{1,p}(\Omega)$. Let $K := \overline{\Omega_0}$. Then K is compact in $\overline{\Omega}$ and $K \cap \partial\Omega = \Gamma_0$. Let $\mathcal{A}_0 \cup \mathcal{B}$ be the finite covering of K according to Remark 7.5.6. For $B_i \in \mathcal{B}$, $1 \leq i \leq N'$, we have that B_i is an open cuboid with $\overline{B_i} \subset \Omega$. Then $u|_{B_i} \in H^{1,p}(B_i)$. Thus as in the proof of [AD75, Ch. V, Lem. 5.17] we get that u has a continuous version $\widetilde{u} \in C^{0,\beta}(\overline{B_i})$ and the mapping $\iota_x : H^{1,p}(\Omega) \to C^{0,\beta}(\overline{B_i})$, $u \mapsto \widetilde{u}$ is continuous.

For $\widetilde{U}_i \in \mathcal{A}_0$, $1 \leq i \leq N'$, let $(U_i', U_i, \widetilde{U}_i, h_i, r_i, \Psi_i^{(0)}, \Psi_i)$ be the corresponding parametrization. Then $u|_{\widetilde{U}_i \cap \Omega} \in H^{1,p}(\widetilde{U}_i \cap \Omega)$. Let T_i be the transformation on function spaces corresponding to Ψ_i as in Lemma 7.5.11. Then $T_i^{-1} u|_{\widetilde{U}_i \cap \Omega} \in H^{1,p}(U_i' \times (0, r_i))$. So we can apply again the proof steps of [AD75, Ch. V, Lem. 5.17] to conclude that $T_i^{-1} u$ has a continuous version $\widetilde{T_i^{-1}u} \in C^{0,\beta}(\overline{U_i'} \times [0, r_i])$. Thus $T_i(\widetilde{T_i^{-1}u}) \in C^{0,\beta}(\widetilde{\overline{U}_i \cap \Omega}) \cap H^{1,p}(\widetilde{U}_i \cap \Omega)$. Since $\widetilde{T_i^{-1}u}$ is a version of $u \circ \Psi_i^{-1}$, $T_i \widetilde{T_i^{-1}u}$ is a Hölder continuous version of u. Moreover, the continuity of T_i and T_i^{-1} on the corresponding spaces and the estimates in the proof of [AD75, Ch. V, Lem. 5.17] yield that the mapping $\iota_i : H^{1,p}(\Omega) \to C^{0,\beta}(\widetilde{U}_i \cap \overline{\Omega})$, $u \mapsto \widetilde{u|_{\widetilde{U}_i \cap \Omega}}$ is continuous. The existence of a Hölder continuous version $\widetilde{u}$ of u on $\overline{\Omega_0} \cup \Gamma_0$ follows by a localization using a partition of unity $(\phi_i)_{1 \leq i \leq N'}$ subordinate to $\mathcal{A}_0 \cup \mathcal{B}$. The corresponding local estimate of the Hölder norm yields continuity of the mapping $\iota : H^{1,p}(\Omega) \to C^{0,\beta}(\overline{\Omega_0})$, $u \mapsto \widetilde{u}$. □

We say that a domain $\Omega \subset \mathbb{R}^d$ possesses the *cone property* if there exists a finite cone $C \subset \Omega$ and for every $x \in \Omega$, C can be translated and rotated to a cone $C_x \subset \Omega$ such that x is the vertex of C_x, compare [AD75, Ch. IV, Def. 4.3].

Theorem 7.5.13. *Assume that $\Omega \subset \mathbb{R}^d$ open, $d \in \mathbb{N}$, has the cone property. Then for $1 \leq p_0 < d$, $p_0 \leq p_1 \leq \frac{dp_0}{d-p_0}$ the embedding $H^{1,p_0}(\Omega) \to L^{p_1}(\Omega)$ exists and is continuous.*

Next we consider traces of Sobolev functions at Lipschitz smooth boundary parts. We first introduce the Hausdorff measure and give a local parametrization at boundary parts.

Definition 7.5.14.
(i) Let $A \subset \mathbb{R}^d$, $d \in \mathbb{N}$, $0 \le s < \infty$, $0 < \delta \le \infty$. Define

$$\mathcal{H}^s_\delta(A) := \inf \left\{ \sum_{i=1}^{\infty} \alpha(s) \left(\frac{\operatorname{diam} C_i}{2} \right)^s \right.$$

$$\left| A \subset \bigcup_{i=1}^{\infty} C_i, \operatorname{diam} C_i \le \delta, C_i \subset \mathbb{R}^d, i \in \mathbb{N} \right\}$$

where

$$\alpha(s) := \frac{\pi^{s/2}}{\Gamma(\frac{s}{2} + 1)}$$

and $\operatorname{diam} C = \sup \{ |x - y| \,|\, x, y \in C \}$. Here $\Gamma(s) := \int_0^\infty e^{-x} x^{s-1} \, dx$, $0 < s < \infty$, is the Gamma function.
(ii) For A and s as in (i), define

$$\mathcal{H}^s(A) := \lim_{\delta \to 0} \mathcal{H}^s_\delta(A) = \sup_{\delta > 0} \mathcal{H}^s_\delta(A).$$

We call $\mathcal{H}^s$ the *s-dimensional Hausdorff measure on* $\mathbb{R}^d$.

This definition is taken from [EG09, Ch. 2]. There it is shown that $\mathcal{H}^s$ defines a Borel regular measure for $0 \le s < \infty$. The constant $\alpha(s)$ is chosen in such a way that $\mathcal{H}^d$ is just the Lebesgue measure on $\mathbb{R}^d$.

Definition 7.5.15. For $\Omega \subset \mathbb{R}^d$ open, $d \in \mathbb{N}$, define the *surface measure* on $\partial\Omega$ by

$$\sigma := 1_{\partial\Omega} \, \mathcal{H}^{d-1}.$$

For Lipschitz smooth boundary parts we obtain locally a representation in terms of the parametrization of the boundary.

Lemma 7.5.16. *Let* $\Omega \subset \mathbb{R}^d$ *open,* $d \in \mathbb{N}$, *and* $x \in \partial\Omega$. *Assume that* $\partial\Omega$ *is locally Lipschitz smooth at* x *with* $(U', U, \widetilde{U}, h, r, \Psi^{(0)}, \Psi)$ *the local parametrization at* x *as in Definition 7.5.2. Let* $u \in \mathcal{B}^+(\mathbb{R}^d)$. *Then it holds*

$$\int_{\partial\Omega \cap \widetilde{U}} u(y) \, d\sigma(y) = \int_{U'} (u \circ \Psi_0^{(-1)})(x', h(x')) \sqrt{1 + |\nabla h(x')|^2} \, dx'.$$

Proof. Define $f : U' \to \tilde{U} \cap \partial\Omega$ by

$$x' \mapsto \Psi_0^{-1}(x', h(x')).$$

Since Ψ_0^{-1} is the composition of a rotation and translation, we have with Jf as in [EG09, Sec. 3.2.2] that $Jf(x') = \sqrt{1 + |\nabla h(x')|^2}$. Define $g : U' \to \mathbb{R}$ by $g := u \circ f$, i.e., $g(x') = (u \circ \Psi_0^{-1})(x', h(x'))$ for $x' \in U'$. From the change of variable formula, see [EG09, Sec. 3.3.3, Theo. 2], we get

$$\int_{U'} (u \circ \Psi_0^{-1})(x', h(x')) \sqrt{1 + |\nabla h(x')|^2}\, dx' = \int_{U'} g(x') \sqrt{1 + |\nabla h(x')|^2}\, dx'$$

$$= \int_{\mathbb{R}^d} \left[\sum_{x' \in f^{-1}(\{y\})} g(x') \right] d\mathcal{H}^{d-1}(y).$$

If $y \notin \partial\Omega \cap \tilde{U}$, then $f^{-1}(\{y\}) = \emptyset$ hence the integrand is zero in this case. For $y \in \partial\Omega \cap \tilde{U}$ there exists a unique $x' \in U'$ with $y = \Psi_0^{-1}(x', h(x'))$. Thus $f^{-1}(\{y\}) = \{x'\}$ and $g(x') = u(\Psi_0^{-1}(x', h(x'))) = u(y)$. So the integral reduces to

$$\int_{\mathbb{R}^d} \left[\sum_{x' \in f^{-1}(\{y\})} g(x') \right] d\mathcal{H}^{d-1}(y) = \int_{\partial\Omega \cap \tilde{U}} u(y)\, d\mathcal{H}^{d-1}(y)$$

$$= \int_{\partial\Omega \cap \tilde{U}} u(y)\, d\sigma(y)$$

$\square$

Now we have everything together to construct traces of Sobolev functions on Lipschitz smooth boundary parts. We follow [Dob10].

Theorem 7.5.17. *Let* $\Omega \subset \mathbb{R}^d$ *open having the segment property,* $d \in \mathbb{N}$. *Let* $\Gamma \subset \partial\Omega$ *be a compact Lipschitz smooth boundary portion. Then for* $1 \leq p < \infty$ *and*

$$q = \begin{cases} \frac{p(d-1)}{d-p} & \text{for } p < d \\ < \infty & \text{for } p = d \end{cases},$$

there exists a linear continuous operator $S : H^{1,p}(\Omega) \to L^q(\Gamma, d\sigma)$ *such that for* $u \in H^{1,p}(\Omega) \cap C_c(\overline{\Omega})$ *it holds* $Su(x) = u(x)$ *for* σ-*a.e.* $x \in \Gamma$. *The operator is unique in the sense that for every other operator* $\tilde{S}$ *with the mentioned properties it holds* $Su(x) = \tilde{S}u(x)$ *for* σ-*a.e.* $x \in \Gamma$, $u \in H^{1,p}(\Omega)$.

Proof. Let $\mathcal{A} := \{(U_i', U_i, \widetilde{U}_i, h_i, r_i, \Psi_i^{(0)}, \Psi_i) \mid 1 \leq i \leq N\}$ be a finite parametrization of Γ as in Definition 7.5.2 and $(\phi_i)_{1 \leq i \leq N}$ a partition of unity subordinate to $\mathcal{A}$. Let $\hat{U}_i := \widetilde{U}_i \cap \Omega$ and $Q_i := U_i' \times (0, r_i)$, $1 \leq i \leq N$. As in the first part of the proof of [Dob10, Satz 6.15] we can define a continuous linear operator $S_i : (C_c^\infty(\mathbb{R}^d)|_{\overline{Q_i}}, \|\cdot\|_{H^{1,p}(Q_i)}) \to L^q(U_i', dx)$ with $S_i u\,(x') := u(x', 0)$, $x' \in U_i'$ for $u \in C_c^\infty(\mathbb{R}^d)$. Since Q_i has the segment property, we have that $C_c^\infty(\mathbb{R}^d)|_{\overline{Q_i}}$ is dense in $H^{1,p}(Q_i)$. So S_i extends to a continuous linear operator $S_i : H^{1,p}(Q_i) \to L^q(U_i', dx)$.

For $u \in C_c^0(\overline{Q_i}) \cap H^{1,p}(Q_i)$ we get by Theorem 7.5.12 a sequence $(u_n)_{n \in \mathbb{N}}$ in $C_c^\infty(\mathbb{R}^d)$ that converges to u both w.r.t. the sup-norm on $\overline{Q_i}$ and in $H^{1,p}(Q_i)$. Dropping to a subsequence we have $S_i u\,(x') = \lim_{n \to \infty} S_i u_n\,(x')$ for σ-a.e. $x' \in U_i'$. Since $S_i u_n\,(x') = u_n(x', 0)$ and $\lim_{n \to \infty} u_n(x', 0) = u(x', 0)$ for $x' \in U_i'$, we get $S_i u\,(x') = u(x', 0)$ for σ-a.e. $x \in U_i'$.

Let T_i as in Lemma 7.5.11 corresponding to Ψ_i. From Lemma 7.5.16 we get for $v \in L^q(U_i, dx)$ that $v \circ \Psi_i \in L^q(\partial\Omega \cap \widetilde{U}_i, d\sigma)$ and the L^q-norm of $v \circ \Psi_i$ can be estimated by the $L^q(U_i, dx)$-norm of v. For $u \in H^{1,p}(\Omega)$, $T_i^{-1} u|_{\hat{U}_i} \in H^{1,p}(U_i \times (0, r_i))$. If $u \in C_c^0(\overline{\Omega})$, then also $T_i^{-1} u \in C_c^0(\overline{U_i \times (0, r_i)})$. Let $u \in C_c^0(\overline{\Omega}) \cap H^{1,p}(\Omega)$ and $x \in \partial\Omega \cap \widetilde{U}_i$. Then $(T_i S T_i^{-1})\,u(x) = (S T_i^{-1} u)(\Psi_i x) = (T_i^{-1} u)(\Psi_i x) = u(x)$. So the operator $S_i' := T_i S_i T_i^{-1}$ defines a continuous linear operator from $H^{1,p}(\Omega)$ to $L^q(\partial\Omega \cap \widetilde{U}_i, d\sigma)$ and $S_i' u\,(x) = u(x)$ for $x \in \partial\Omega \cap \widetilde{U}_i$, $u \in H^{1,p}(\Omega) \cap C_c^0(\overline{\Omega})$. Define $S : H^{1,p}(\Omega) \to L^q(\Gamma, d\sigma)$ by

$$u \mapsto \sum_{i=1}^{N} (\phi_i S_i' u)|_\Gamma.$$

Then S is a continuous linear operator and $Su\,(x) = u(x)$ for σ-a.e. $x \in \Gamma$ for $u \in H^{1,p}(\Omega) \cap C_c^0(\overline{\Omega})$. It is left to prove uniqueness of the trace operator. So assume $\widetilde{S}$ is another trace operator fulfilling the required properties. For $u \in H^{1,p}(\Omega) \cap C_c^\infty(\mathbb{R}^d)$ it holds by assumption $Su = \widetilde{S}u = u$ on Γ. Since Ω has the segment property, we get for $u \in H^{1,p}(\Omega)$ a sequence $(u_k)_{k \in \mathbb{N}}$ in $H^{1,p}(\Omega) \cap C_c^\infty(\mathbb{R}^d)|_\Omega$ such that $u_k \overset{k \to \infty}{\longrightarrow} u$ in $H^{1,p}(\Omega)$. By continuity of both operators we get $Su = \lim_{k \to \infty} Su_k = \lim_{k \to \infty} \widetilde{S}u_k = \widetilde{S}u$ in $L^q(\Gamma, d\sigma)$. Hence $Su(x) = \widetilde{S}u(x)$ for σ-a.e. $x \in \Gamma$. $\qquad\square$

Next we state the divergence theorem.

Theorem 7.5.18. *Let $\Omega \subset \mathbb{R}^d$ open, $d \in \mathbb{N}$, having the segment property. Let $\Gamma \subset \partial\Omega$ be a compact Lipschitz smooth boundary part. Then for a vector* *field $g = (g_1, ..., g_d)$ with $g_i \in H^{1,1}(\Omega)$ and $supp[g] \subset\subset \Omega \cup \Gamma$, $1 \le i \le d$, it holds*

$$\int_\Gamma (\eta, Tr\, g)\, d\sigma = \int_\Omega \nabla \cdot g\, dx$$

where $\nabla \cdot g = \sum_{i=1}^d \partial_i g_i$ is the divergence of g.

Remark 7.5.19. The outward unit normal η can be written in terms of the parametrization of the boundary part Γ from Definition 7.5.2. Let $x \in \Gamma$ and assume that $\Psi_0 = \mathbf{1}$. Then for h as in Definition 7.5.2 we have for $y \in \tilde{U} \cap \Gamma$

$$\eta = \frac{(\nabla h, -1)}{\sqrt{1 + |\nabla h|^2}}.$$

The unit outward normal is unique, see [Alt06, A 6.5].

Proof. Let $K := \operatorname{supp}[g]$. Choose the covering $\mathcal{A}_0 \cup \mathcal{B}$ from Remark 7.5.6 and let $(\phi_i)_{1 \le i \le N'}$ be the corresponding partition of unity. Since Ω has the segment property, we find a sequence $(g_j^{(k)})_{k \in \mathbb{N}}$ of $C_c^\infty(\mathbb{R}^d)$ functions with $g_j^{(k)} \overset{k \to \infty}{\longrightarrow} g_j$ in $H^{1,1}(\Omega)$ for $1 \le j \le d$. Let $1 \le i \le N'$ and assume $\operatorname{supp}[\phi_i] \subset B$ for some cuboid $B \in \mathcal{B}$. Then $\phi_i g$ has compact support in B and partial integration yields

$$\int_B \nabla \cdot \phi_i g^{(k)}\, dx = 0,$$

for all $k \in \mathbb{N}$. The expression is continuous in the $H^{1,1}(\Omega)$-norm so the equation holds also for g. If $\operatorname{supp}[\phi_i] \subset \tilde{U} \cap \overline{\Omega}$ for some $\tilde{U} \in \mathcal{A}_0$, then we get as in [Alt06, A6.8]

$$\int_{\tilde{U} \cap \Omega} \nabla \cdot \phi_i g^{(k)}\, dx = \int_{\Gamma \cap \tilde{U}} (\eta, \phi_i g^{(k)})\, d\sigma = \int_\Gamma (\eta, \phi_i g^{(k)})\, d\sigma.$$

Letting $k \to \infty$ we get

$$\int_{\tilde{U} \cap \Omega} \nabla \cdot \phi_i g\, dx = \int_\Gamma (\eta, \phi_i Tr\, g)\, d\sigma.$$

Summing over i from 1 to N' we obtain since $\sum_{i=1}^{N'} \phi_i = 1$ on $\operatorname{supp}[g]$:

$$\int_\Omega \nabla \cdot g\, dx = \int_\Gamma (\eta, Tr\, g)\, d\sigma.$$

$\square$

Finally we introduce local L^p-spaces and Sobolev spaces.

Definition 7.5.20. Let $p \geq 1$, $M \subset \mathbb{R}^d$, $d \in \mathbb{N}$, and μ a measure on M. Assume that M is endowed with some topology. We define $L^p_{\mathrm{loc}}(M, \mu)$ to consist of all measurable functions f such that for every $K \subset M$, K compact in M, it holds

$$f \in L^p(K, \mu).$$

Definition 7.5.21. Let $p \geq 1$, $G \subset \mathbb{R}^d$, $d \in \mathbb{N}$, with $\overset{\circ}{G} \neq \emptyset$ and $dx(\partial G) = 0$. We define $H^{1,p}_{\mathrm{loc}}(G)$ to consist of all weakly differentiable functions on $\overset{\circ}{G}$, i.e., for $1 \leq i \leq d$ there exists $\partial_i u$ measurable and locally integrable in $\overset{\circ}{G}$ such that

$$\int_{\overset{\circ}{G}} u\, \partial_i \varphi\, dx = - \int_{\overset{\circ}{G}} \partial_i u\, \varphi\, dx \quad \text{for all } \varphi \in C^\infty_c(\overset{\circ}{G})$$

and u, $\partial_i u \in L^p_{\mathrm{loc}}(G)$, $1 \leq i \leq d$.

Note that the weak derivatives are unique on $\overset{\circ}{G}$ due to the local integrability and the fundamental lemma of calculus of variations. Since $dx(\partial G) = 0$, these functions are also unique in the sense of dx-equivalence classes on G.

In common literature only local Sobolev spaces on open sets $\Omega \subset \mathbb{R}^d$ are defined. For example, in [MZ97, p. 21] the space $H^{1,p}_{\mathrm{loc}}(\Omega)$ is defined to consist of all functions u such that $u|_{\Omega_0} \in H^{1,p}(\Omega_0)$ for each $\Omega_0 \subset\subset \Omega$. We will see in Remark 7.5.24 below that in the case of an open set this definition is equivalent to ours.

For our application, however, it is essential to have certain integrability also at boundary parts of Ω. That is why we introduce more general local Sobolev spaces.

Lemma 7.5.22. *Let $p \geq 1$, $G \subset \mathbb{R}^d$, $d \in \mathbb{N}$. Assume that Ω_0 is open in $\mathbb{R}^d$ and $\Omega_0 \subset \overset{\circ}{G}$ and $\overline{\Omega_0} \subset G$ is compact. If $u \in H^{1,p}_{\mathrm{loc}}(G)$, then $u \in H^{1,p}(\Omega_0)$.*

Proof. Let $u \in H^{1,p}_{\mathrm{loc}}(G)$. If $\varphi \in C^\infty_c(\Omega_0)$, then φ can be extended by zero outside Ω_0 to a function in $C^\infty_c(G)$. So the partial integration formula from Definition 7.5.21 is valid for $C^\infty_c(\Omega_0)$ functions. Let $\partial_i u$, $1 \leq i \leq d$, be the weak derivatives of u on $\overset{\circ}{G}$. Since $\overline{\Omega_0} \subset G$ is compact, we get u, $\partial_i u \in L^p(\Omega_0, dx)$ for $1 \leq i \leq d$. $\square$

Lemma 7.5.23. *Let $p \geq 1$, $G \subset \mathbb{R}^d$, $d \in \mathbb{N}$, as in Definition 7.5.21. Assume additionally that $\partial G \cap G$ is open in ∂G and $\partial G \cap G \subset \overline{\overset{\circ}{G}}$. Let u be a measurable function on G. Assume that for every $\Omega_0 \subset \mathbb{R}^d$ open with $\Omega_0 \subset \overset{\circ}{G}$ and compact closure $\overline{\Omega_0} \subset G$ it holds $u \in H^{1,p}(\Omega_0)$. Then $u \in H^{1,p}_{loc}(G)$.*

Proof. For $n \in \mathbb{N}$ set

$$\Omega_n := \left\{ x \in \overset{\circ}{G} \,\Big|\, \mathrm{dist}(x, \partial G \setminus G) > \frac{1}{n} \right\} \cap B_n(0),$$

then

$$\overline{\Omega_n} \subset \left\{ x \in G \,\Big|\, \mathrm{dist}(x, \partial G \setminus G) \geq \frac{1}{n} \right\} \cap \overline{B_n(0)} \subset G.$$

So $\overline{\Omega_n} \subset G$ is compact. Let u fulfill the assumptions of the statement. Then $u \in H^{1,p}(\Omega_n)$ for every $n \in \mathbb{N}$. Since $\overline{\Omega_n} \subset \Omega_{n+1} \cup \overline{\Omega_n} \cap \partial G$, we have $u, \partial_i u \in L^p(\overline{\Omega_n})$ for $1 \leq i \leq d$. Set $u_n := u|_{\Omega_n}$, $n \in \mathbb{N}$. We claim that $\partial_i u_m |_{\Omega_n} = \partial_i u_n$ for every $m \geq n$ and $m, n \in \mathbb{N}$. Indeed, let $m, n \in \mathbb{N}$ with $m \geq n$.

Let $1 \leq i \leq d$ and $\varphi \in C_c^\infty(\Omega_n)$. Then we have $\varphi \in C_c^\infty(\Omega_m)$ for $m \geq n$ by continuing φ by zero outside Ω_n. Using that $u_m|_{\Omega_n} = u_n$ and the support property of φ we get

$$\int_{\Omega_n} u_n \, \partial_i \varphi \, dx = \int_{\Omega_m} u_m \, \partial_i \varphi \, dx = - \int_{\Omega_m} \partial_i u_m \, \varphi \, dx = - \int_{\Omega_n} \partial_i u_m \, \varphi \, dx.$$

Since the weak derivative on Ω_n is unique, we have $\partial_i u_m |_{\Omega_n} = \partial_i u_n$. Define $v_i := \sum_{k=1}^\infty 1_{\Omega_{k+1} \setminus \Omega_k} \partial_i u_k + 1_{\Omega_1} \partial_i u_1$. Then on Ω_n, $n \in \mathbb{N}$, we have $v_i = \partial_i u_n$. Let $\varphi \in C_c^\infty(\overset{\circ}{G})$. Since $\mathrm{supp}[\varphi] \subset \overset{\circ}{G}$ is compact, there exists $n \in \mathbb{N}$ such that $K \subset \Omega_n$. Then we get for $1 \leq i \leq d$

$$\int_{\overset{\circ}{G}} u \, \partial_i \varphi \, dx = \int_{\Omega_n} u_n \, \partial_i \varphi \, dx = - \int_{\Omega_n} \partial_i u_n \, \varphi \, dx = - \int_{\overset{\circ}{G}} v_i \, \varphi \, dx.$$

So u is weakly differentiable with weak derivatives v_i, $1 \leq i \leq d$.

We claim that every compact set K with $K \subset G$ is contained in $\overline{\Omega_n}$ for some $n \in \mathbb{N}$. Indeed. Since $\partial G \cap G$ is open and $K \subset G$, we have $\mathrm{dist}(K, \partial G \setminus G) > 0$. So there exists $n \in \mathbb{N}$ with $\mathrm{dist}(K, \partial G \setminus G) > \frac{1}{n}$ and $K \subset B_n(0)$. So $K \cap \overset{\circ}{G} \subset \Omega_n$. Let $x \in K \cap \partial G \cap G$. Then there exists a sequence $(x_k)_{k \in \mathbb{N}}$ in $\overset{\circ}{G}$ with $\lim\limits_{k \to \infty} x_k = x$. Choosing $K_0 \in \mathbb{N}$ large enough we have $|x_k| < n + 1$ and $\operatorname{dist}(x_k, \partial G \setminus G) > \frac{1}{n+1}$ for $k \geq K_0$. Thus $x \in \overline{\Omega_{n+1}}$.

Since u and its weak derivatives are in $L^p(\Omega_n)$ for every $n \in \mathbb{N}$, we therefore have $u, \partial_i u \in L^p_{\text{loc}}(G)$ for $1 \leq i \leq d$.

$\square$

Remark 7.5.24. Consider the case $G = \Omega$ for $\Omega \subset \mathbb{R}^d$ open. Then $\partial G \cap G = \emptyset$. So the assumptions of Lemma 7.5.25 fulfilled. Together with Lemma 7.5.22 we get that a function u is in $H^{1,p}_{\text{loc}}(\Omega)$ iff $u|_{\Omega_0} \in H^{1,p}(\Omega_0)$ for every $\Omega_0 \subset\subset \Omega$. So in this case our definition and the one of [MZ97, p. 21] coincide.

Lemma 7.5.25. *Let $p \geq 1$, $G \subset \mathbb{R}^d$, $d \in \mathbb{N}$, as in Lemma 7.5.23. Let u be a measurable function on G.*

Assume that: For every $x \in G$ there exists an open neighborhood U of x in $\mathbb{R}^d$ such that $u \in H^{1,p}(U \cap \overset{\circ}{G})$. Then $u \in H^{1,p}_{loc}(G)$.

Proof. Let $\Omega_0 \subset \overset{\circ}{G}$ be open with compact closure in G. Let u fulfill the assumptions. For every $x \in \overline{\Omega_0} \subset G$ there exists U_x open such that $u \in H^{1,p}(U_x \cap \overset{\circ}{G})$. Since $\overline{\Omega_0}$ is compact, we can cover it by finitely many of such U_k, $1 \leq k \leq N$, $N \in \mathbb{N}$. Choose a partition of unity $(\phi_k)_{1 \leq k \leq N}$ for $\overline{\Omega_0}$ subordinate to $(U_k)_{1 \leq k \leq N}$. Since $u \in H^{1,p}(U_k \cap \overset{\circ}{G})$, we have $\phi_k u \in H^{1,p}(\overset{\circ}{G})$. Indeed. Let $1 \leq k \leq N$, $1 \leq i \leq d$ and $\varphi \in C^\infty_c(\overset{\circ}{G})$. Then

$$\int_{\overset{\circ}{G}} u\, \phi_k\, \partial_i \phi\, dx = \int_{\overset{\circ}{G} \cap U_k} u\, \partial_i(\phi_k \phi)\, dx - \int_{\overset{\circ}{G}} u\, \phi\, \partial_i \phi_k\, dx$$

$$= -\int_{\overset{\circ}{G} \cap U_k} \partial_i u\, \phi_k\, \phi\, dx - \int_{\overset{\circ}{G}} u\, \phi\, \partial_i \phi_k\, dx = -\int_{\overset{\circ}{G}} (\partial_i u\, \phi_k + u\, \partial_i \phi_k)\, \phi\, dx.$$

Set $v_i := \sum_{k=1}^N \partial_i(\phi_k u)$, $1 \leq i \leq N$. Let $\varphi \in C^\infty_c(\Omega_0)$ and continue it by zero outside Ω_0. Then we get:

$$\int_{\Omega_0} u\, \partial_i \varphi\, dx = \int_{\Omega_0} \sum_{k=1}^N \phi_k u\, \partial_i \varphi\, dx = \int_{\overset{\circ}{G}} \sum_{k=1}^N \phi_k u\, \partial_i \varphi\, dx$$

$$= -\int_{\overset{\circ}{G}} \sum_{k=1}^N \partial_i(\phi_k u)\, \varphi\, dx = -\int_{\Omega_0} \sum_{k=1}^N \partial_i(\phi_k u)\, \varphi\, dx.$$

So $u \in H^{1,p}(\Omega_0)$ with weak derivatives given by v_i, $1 \leq i \leq d$. So by Lemma 7.5.25 we get $u \in H^{1,p}_{\mathrm{loc}}(G)$. □

7.6 Capacity Estimates

In this section we introduce the notion of the capacity associated with a Dirichlet form and state several results concerning capacities. In the second part of the section we provide details to some capacity estimates used in Section 5.1.

Let $(\mathcal{E}, D(\mathcal{E}))$ be a regular symmetric Dirichlet form on $L^2(E, \mu)$ on the locally compact separable metric space $(E, \mathbf{d})$ with locally finite Borel measure μ.

Denote by $\mathcal{O}$ the family of all open sets in E. For $U \subset E$ open define

$$\mathcal{L}_U := \{u \in D(\mathcal{E}) \,|\, u \geq 1\,\mu - \text{a.e. on } U\,\}$$

and

$$\mathrm{cap}_{\mathcal{E}}(U) := \begin{cases} \inf\limits_{u \in \mathcal{L}_U} \mathcal{E}(u, u) & \text{if } \mathcal{L}_U \neq \emptyset \\ \infty & \text{if } \mathcal{L}_U = \emptyset. \end{cases}$$

For $A \subset E$ arbitrary set

$$\mathrm{cap}_{\mathcal{E}}(A) := \inf_{\substack{A \subset U \\ U \in \mathcal{O}}} \mathrm{cap}_{\mathcal{E}}(U).$$

We call $\mathrm{cap}_{\mathcal{E}}(A)$ the *capacity* of A. This definition is taken from [FOT11, Ch. 2, Sec. 1]. See [MR92, Ch. III, Sec. 2] for a more general definition of capacity for quasi-regular Dirichlet forms. If $U \subset E$ is open and $\mathrm{cap}_{\mathcal{E}}(U) < \infty$ then there exists a unique $e_U \in \mathcal{L}_U$ such that

$$\mathrm{cap}_{\mathcal{E}}(U) = \mathcal{E}_1(e_U, e_U),$$

see [FOT11, Lem. 2.1.1]. For important lemmata concerning capacities see the mentioned works. We state just the results we directly need in this work.

Lemma 7.6.1. *(i) Let $U_n \subset E$, $n \in \mathbb{N}$, such that $\mathrm{cap}_{\mathcal{E}}(U_n) = 0$ for all $n \in \mathbb{N}$. Assume $A \subset \bigcup_{n \in \mathbb{N}} U_n$. Then $\mathrm{cap}_{\mathcal{E}}(A) = 0$.*
(ii) Let $E_1 \subset E$, $(U_n)_{n \in \mathbb{N}}$ be an increasing sequence of open sets in E with $E_1 := \bigcup_{n \in \mathbb{N}} U_n$. Assume $\mathrm{cap}_{\mathcal{E}}(E \setminus E_1) = 0$. Then

$$\lim_{n \to \infty} \mathrm{cap}_{\mathcal{E}}(K \setminus U_n) = 0 \quad \text{for every } K \subset E \text{ compact.}$$

Proof. (i): Follows easily from the definition of the capacity.

(ii): Let $K \subset E$ be an arbitrary compact set. Since U_n is open, we have that $K \setminus U_n$ is compact for $n \in \mathbb{N}$. Furthermore, $K \setminus U_n$, $n \in \mathbb{N}$, is decreasing to $K \setminus E_1$, i.e., $\bigcap_{n \in \mathbb{N}} K \setminus U_n = E \setminus E_1$. By [FOT11, Theo. 2.1.1] we get

$$\lim_{n \to \infty} \operatorname{cap}_{\mathcal{E}}(K \setminus U_n) = \inf_{n \in \mathbb{N}} \operatorname{cap}_{\mathcal{E}}(K \setminus U_n)$$

$$= \operatorname{cap}_{\mathcal{E}}\Big(\bigcap_{n \in \mathbb{N}} K \setminus U_n\Big) = \operatorname{cap}_{\mathcal{E}}(K \setminus E_1) \leq \operatorname{cap}_{\mathcal{E}}(E \setminus E_1) = 0.$$

$\square$

Let us introduce the notion of nests.

Definition 7.6.2. An increasing sequence $(E_k)_{k \in \mathbb{N}}$ of closed subsets of E is called a *nest* if

$$\lim_{k \to \infty} \operatorname{cap}_{\mathcal{E}}(E \setminus E_k) = 0.$$

See [FOT11, Ch. 2, p. 69] or [MR92, Ch. III, Def. 2.1] and [MR92, Ch. III, Theo. 2.11] for a more general definition. An increasing sequence $(E_k)_{k \in \mathbb{N}}$ of closed subsets of E is called a *generalized nest* if

$$\lim_{k \to \infty} \operatorname{cap}_{\mathcal{E}}(K \setminus E_k) = 0 \quad \text{for every } K \subset E \text{ compact.}$$

See [FOT11, Ch. 2, p. 83].

A subset $A \subset E$ is called $\mathcal{E}$-exceptional if $\operatorname{cap}_{\mathcal{E}}(A) = 0$. The definition of capacity yields then also $\mu(A) = 0$. We say that a property holds *$\mathcal{E}$-quasi-everywhere ($\mathcal{E}$-q.e.)* if there exists an $\mathcal{E}$-exceptional set $A \subset E$ such that the property holds for all $x \in E \setminus A$.

Let $(E_k)_{k \in \mathbb{N}}$ be a nest. Define

$$C^0(\{E_k \,|\, k \in \mathbb{N}\}) := \Big\{ f : A \to \mathbb{R} \,\Big|\, \bigcup_{k \in \mathbb{N}} E_k \subset A \subset E \text{ for some } A,$$

$$f|_{E_k} \text{ is continuous for every } k \in \mathbb{N} \Big\}.$$

Definition 7.6.3. Let f be an $\mathcal{E}$-q.e. defined function. We call f *$\mathcal{E}$-quasi-continuous* if there exists an $\mathcal{E}$-nest $(E_k)_{k \in \mathbb{N}}$ such that $f \in C^0(\{E_k \,|\, k \in \mathbb{N}\})$.

See [MR92, Ch. III, Def. 3.2] or [FOT11, Ch. 2, p. 69]. Note that in [FOT11] this continuity is just called quasi-continuous.

The following theorem states that functions in $D(\mathcal{E})$ already have some regularity properties, see [FOT11, Theo. 2.1.3].

Theorem 7.6.4. *For $u \in D(\mathcal{E})$, there exists an $\mathcal{E}$-quasi-continuous modification, i.e., there exists $\widetilde{u}$ with $u = \widetilde{u}$ μ-a.e. and $\widetilde{u}$ is $\mathcal{E}$-quasi-continuous.*

We state results that link probabilistic properties of a process and properties related to capacities.
So we fix a Hunt process $\mathbf{M} = (\mathbf{\Omega}, \mathcal{F}, (\mathcal{F}_t)_{t \geq 0}, (\mathbf{X}_t)_{t \geq 0}, (\mathbb{P}_x)_{x \in E^\Delta})$ with state space a locally compact, separable metric space $(E, \mathbf{d})$. Here $\mathcal{F}$ and $(\mathcal{F}_t)_{t \geq 0}$ are defined as in Section 7.3, i.e., $(\mathcal{F}_t)_{t \geq 0}$ is the natural filtration. Furthermore, assume that $\mathbf{M}$ is μ-symmetric, i.e., for the transition semigroup $(P_t)_{t \geq 0}$ of $\mathbf{M}$ it holds for all $f, g \in \mathcal{B}^+(E)$ that

$$\int_E P_t f(x) g(x) \, d\mu = \int_E f(x) P_t g(x) \, d\mu.$$

See also [FOT11, Sec. 4.1].

We call a subset $B \subset E^\Delta$ nearly Borel if for every probability measure $\nu \in \mathcal{P}(E^\Delta)$ there exists $B_1, B_2 \in \mathcal{B}(E^\Delta)$ with $B_1 \subset B \subset B_2$ and

$$\mathbb{P}_\nu \left(\{\text{There exists } t \geq 0 : \mathbf{X}_t \in B_2 \setminus B_1\} \right) = 0. \tag{7.18}$$

We denote by $\mathcal{B}^n(E^\Delta)$ the set of all nearly Borel sets. Observe that we have $\mathcal{B}(E^\Delta) \subset \mathcal{B}^n(E^\Delta) \subset \mathcal{B}^*(E^\Delta)$. See [FOT11, p. 392] or [BG68, Ch. I, Sec. 10] for this definition. For $B \subset E^\Delta$ define the (first) hitting time by

$$\sigma_B := \inf\{t > 0 \,|\, \mathbf{X}_t \in B\}.$$

Due to [BG68, Ch. I, Theo. 10.7] and the discussion on [BG68, Ch. I, p. 60] we have that σ_B is $\mathcal{F}_t$-adapted for every nearly Borel set $B \subset E^\Delta$. For this it is crucial that $(\mathcal{F}_t)_{t \geq 0}$ is right-continuous and obtained by the augmentation procedure as in (7.8).

We introduce the notion of exceptional sets, see [FOT11, p. 152f].

Definition 7.6.5. A set $N \subset E$ is called *exceptional* if there exists a nearly Borel set $\widetilde{N}$ with $N \subset \widetilde{N}$ such that

$$\mathbb{P}_\mu \left(\sigma_{\widetilde{N}} < \infty \right) = 0.$$

A set $N \subset E$ is called *properly exceptional* if N is nearly Borel, $\mu(N) = 0$ and $E^\Delta \setminus N$ is invariant under $\mathbf{M}$. With invariant we mean here that for $x \in E^\Delta \setminus N$, it holds $\mathbf{X}_t \in E^\Delta \setminus N$ for $t \geq 0$ $\mathbb{P}_x$-a.s.

Note that for an exceptional set we can find a Borel set $\tilde{N}_1$ with $\tilde{N} \subset \tilde{N}_1$ such that $\mathbb{P}_\mu \left(\sigma_{\tilde{N}_1} < \infty \right) = 0$. We say that a property holds *quasi-everywhere* (q.e.) if there exists an exceptional set $N \subset E$ such that the property holds for all $x \in E \setminus N$. From the discussion on [FOT11, p. 152] we get that every exceptional set has μ-measure zero.

We obtain the following theorem, see [FOT11, Theo. 4.1.1].

Theorem 7.6.6. *If $N \subset E$ is exceptional, then there exists a (Borel) set B with $N \subset B$ and B is properly exceptional.*

For $A \subset E$ nearly Borel define

$$p_A^1(x) := \mathbb{E}_x[e^{-\sigma_A}], \quad x \in E.$$

Observe that $p_A^1(x) = 0$ iff $\mathbb{P}_x(\sigma_A < \infty) = 0$.
See [FOT11, Lem. 4.2.1] for the next lemma.

Lemma 7.6.7. *If $U \subset E$ is open with finite $\mathcal{E}$-capacity, then*

$$p_U^1(x) = e_U(x) \quad \mu - a.e.$$

The following theorem links hitting times and capacities. See [FOT11, Theo. 4.2.1].

Theorem 7.6.8.

(i) Let $(U_n)_{n \in \mathbb{N}}$ be a sequence of decreasing open sets with finite capacity. Then

$$\lim_{n \to \infty} cap_{\mathcal{E}}(U_n) = 0 \Leftrightarrow \lim_{n \to \infty} p_{U_n}^1(x) = 0 \quad q.e.$$

(ii) If for $N \subset E$ it holds $cap_{\mathcal{E}}(N) = 0$, then N is exceptional. Assume that every compact set in E has finite capacity. Then the converse is also true, i.e., every exceptional set N has $\mathcal{E}$-capacity zero.

From (i) in the previous theorem we can conclude that for a nest $(A_n)_{n \in \mathbb{N}}$ of closed sets it holds $\lim_{n \to \infty} \sigma_{A_n^c} \geq \mathcal{X}$ $\mathbb{P}_x$-a.s. for quasi-every $x \in E^\Delta$. Here the complements are taken in E^Δ, i.e., $A^c = E^\Delta \setminus A$. The following lemma tells us that this important property is true also for generalized nests.

Lemma 7.6.9. *Let $(E_k)_{k \in \mathbb{N}}$ be a generalized nest. Then*

$$\mathbb{P}_x \left(\lim_{k \to \infty} \sigma_{E_k^c} \geq \mathcal{X} \right) = 1 \quad for \ q.e. \ x \in E.$$

See [FOT11, Lem. 5.1.6].

Next we prove some capacity estimates we used in Section 5.1.

Let $\Upsilon \subset \mathbb{R}^l$, $l \in \mathbb{N}$, be an open set with locally Lipschitz smooth boundary. By cap_1 we denote the capacity of the classical gradient Dirichlet form $(\mathcal{E}, H^{1,2}(\Upsilon))$. For a matrix-valued measurable mapping $A = (a_{ij})_{1 \leq i,j \leq l}$ of symmetric strictly elliptic matrices and measurable density $\varrho \geq 0$ we define the pre-Dirichlet form

$$\mathcal{E}^{A,\varrho}(u,v) = \int_\Upsilon (A \nabla u, \nabla v) \, d\mu,$$

$$u, v \in \mathcal{D}_{A,\varrho} := \{u \in C_c(\overline{\Upsilon}) \,|\, u \in H^{1,1}_{\mathrm{loc}}(\Upsilon), \mathcal{E}^{A,\varrho}(u,u) < \infty\}, \quad (7.19)$$

in the Hilbert space $L^2(\overline{\Upsilon}, \mu)$ where $\mu := \varrho \, dx$. Assume that $(\mathcal{E}^{A,\varrho}, D(\mathcal{E}^{A,\varrho}))$ is closable. Then we denote the closure by $(\mathcal{E}^{A,\varrho}, D(\mathcal{E}^{A,\varrho}))$. The associated capacity we denote by $\mathrm{cap}_{\mathcal{E}^{A,\varrho}}$.

By cap_0 we denote the capacity

$$\mathrm{cap}_0(U) = \inf \left\{ \int_{\mathbb{R}^l} |u|^2 + |\nabla u|^2 dx \,\bigg|\, u \in H^{1,2}(\mathbb{R}^l), U \subset \{u \geq 1\}^\circ \right\}.$$

for a set $U \subset \mathbb{R}^l$. Compare [MZ97, Def. 2.1]. Note that in the case U open, it holds $U \subset \{u \geq 1\}^\circ$ if $u(x) \geq 1$ for *every* $x \in U$.

Theorem 7.6.10. *Let A as in Condition 4.1.1 and 4.1.2. Consider the corresponding Dirichlet form $(\mathcal{E}^{A,1}, D(\mathcal{E}^{A,1}))$ (as in (7.19)) with constant density equal to 1. Denote the corresponding capacity by $\mathrm{cap}_{\mathcal{E}^{A,1}}$. Let ϱ as in (5.2). Denote the closure of the pre-Dirichlet form (5.3), with matrix $\hat{A}$ and density ϱ as in Section 5.1, by $(\mathcal{E}^N, D(\mathcal{E}^N))$ and its capacity by cap_N. Here we take as Ω_0 the set Υ. Assume that $\mathrm{cap}_{\mathcal{E}^{A,1}}(B) = 0$, $B \subset \Upsilon$. Then $\mathrm{cap}_N(\overline{\Upsilon}^k \times B \times \overline{\Upsilon}^{N-k-1}) = 0$ for $1 \leq k \leq N - 1$.*

Proof. We prove the claim for $k = 0$, the case for general k is similar. For $m \in \mathbb{N}$, set $V_m := \overline{\Upsilon}^N \cap B_m(0)$, $K_m := \overline{\Upsilon}^N \cap \overline{B_m(0)}$. Choose $\phi_m \in C_c^\infty(\overline{\Upsilon})$ with $\phi_m = 1$ on K_m and supp $\phi_m \subset\subset V_{m+1}$. Choose a fixed $m \in \mathbb{N}$. Let $u \in \mathcal{D}_{A,1}$, denote the function $\left(\overline{\Upsilon}^N \ni \hat{x} = (x_1, ..., x_N) \mapsto \phi_m(\hat{x}) u(x_1) \in \mathbb{R}\right)$ simply by $\phi_m u$. Then it holds $\phi_m u \in D(\mathcal{E}^N)$ and the norm on $D(\mathcal{E}^N)$ can be estimated by the norm of u on $D(\mathcal{E}^{A,1})$. So the mapping $\mathcal{D}_{A,1} \ni u \mapsto \phi_m u \in D(\mathcal{E}^N)$ extends to a linear continuous mapping from $D(\mathcal{E}^{A,1})$ to $D(\mathcal{E}^N)$.

Since $\mathrm{cap}_{\mathcal{E}^{A,1}}(B) = 0$, there exists a sequence of open sets $U_n \subset \overline{\Upsilon}$ and $u_n \in D(\mathcal{E}^{A,1})$, $n \in \mathbb{N}$, such that $B \subset \bigcap_{n \in \mathbb{N}} U_n$, $u_n \geq 1$, $dx - a.e.$ on U_n

and $\mathcal{E}_1^{A,1}(u_n, u_n) \overset{n\to\infty}{\longrightarrow} 0$. We have $\phi_m u_n \in D(\mathcal{E}^N)$ and $\phi_m u_n \geq 1$ μ-a.e. on $(B \times \overline{\Upsilon}^{N-1}) \cap V_m$.

Since $\mathcal{E}_1^N(\phi_m u_n, \phi_m u_n) \leq C_m \mathcal{E}_1^{A,1}(u_n, u_n)$ for some $C_m < \infty$, we have $\mathrm{cap}_N((B \times \overline{\Upsilon}^{N-1}) \cap V_m) = 0$. From $B \times \overline{\Upsilon}^{N-1} = \bigcup_{m\in\mathbb{N}}(B \times \overline{\Upsilon}^{N-1}) \cap V_m$ and $V_m \subset V_{m+1}$ we get $\mathrm{cap}_N(B \times \overline{\Upsilon}^{N-1}) = 0$. $\qquad\square$

Lemma 7.6.11. *Assume that Υ has the segment property, see Definition 7.5.7. Let A, ϱ as in Condition 4.1.1 and Condition 4.1.2. Let $\phi \in C_c^\infty(\overline{\Upsilon})$. Then for $u \in H^{1,2}(\Upsilon)$ it holds $\phi u \in D(\mathcal{E}^{A,\varrho})$ and the mapping $\iota : H^{1,2}(\Upsilon) \to D(\mathcal{E}^{A,\varrho})$, $u \mapsto \phi u$, is continuous w.r.t. the respective norms.*

Proof. Let $u \in C_c^\infty(\mathbb{R}^d)$. Then $\phi u \in C_c^\infty(\overline{\Upsilon})$ and $\phi u \in \mathcal{D}_{A,\varrho} \subset D(\mathcal{E}^{A,\varrho})$. Since A and ϱ are bounded on $\mathrm{supp}[\phi]$, we find a constant $K < \infty$ such that

$$\mathcal{E}_1^{A,\varrho}(\phi u, \phi u) \leq K \|u\|_{H^{1,2}(\Upsilon)}^2.$$

So we have that the mapping $\iota' : C_c^\infty(\mathbb{R}^d) \to D(\mathcal{E}^{A,\varrho})$, $u \mapsto \phi u$ is a continuous mapping. Since $C_c^\infty(\mathbb{R}^d)|_{H^{1,2}(\Upsilon)}$ is dense in $H^{1,2}(\Upsilon)$, this mapping extends to a continuous mapping $\iota : H^{1,2}(\Upsilon) \to D(\mathcal{E}^{A,\varrho})$. The construction yields that $\iota u = \phi u \, dx$-a.e. for $u \in H^{1,2}(\Upsilon)$. $\qquad\square$

Theorem 7.6.12. *Assume that Υ has the segment property. Let A and ϱ as in Lemma 7.6.11. For $B \subset \overline{\Upsilon}$ with $\mathrm{cap}_1(B) = 0$ it follows $\mathrm{cap}_{\mathcal{E}^{A,\varrho}}(B) = 0$.*

Proof. Let $B \subset \overline{\Upsilon}$ with $\mathrm{cap}_1(B) = 0$. Let $k \in \mathbb{N}$ arbitrary but fixed. Then $B \cap B_k(0)$ has also cap_1 zero. Thus there exist U_n, $n \in \mathbb{N}$, open in $\overline{\Upsilon}$ with $B \cap B_k(0) \subset U_n$ and $\mathrm{cap}_1(U_n) \overset{n\to\infty}{\longrightarrow} 0$. Set $U_n^{(k)} := U_n \cap B_k(0)$, $n \in \mathbb{N}$. This family of sets has the same properties as $(U_n)_{n\in\mathbb{N}}$, i.e., $B \cap B_k(0) \subset U_n^{(k)}$ and $\mathrm{cap}_1(U_n^{(k)}) \overset{n\to\infty}{\longrightarrow} 0$. Of course these sets are also open.

By definition of cap_1, there exists a sequence $(u_n^{(k)})_{n\in\mathbb{N}}$ in $H^{1,2}(\Upsilon)$ with $u_n^{(k)} \geq 1$ on $U_n^{(k)}$, $n \in \mathbb{N}$, and $\|u_n^{(k)}\|_{H^{1,2}(\Upsilon)} \overset{n\to\infty}{\longrightarrow} 0$. Choose a smooth cutoff ϕ for $B_k(0)$ in $B_{k+1}(0)$. Then $\phi u_n^{(k)} \in D(\mathcal{E}^{A,\varrho})$, $\phi u_n^{(k)} \geq 1$ on $U_n^{(k)}$ and $\mathcal{E}^{A,\varrho}(\phi u_n^{(k)}, \phi u_n^{(k)}) \overset{n\to\infty}{\longrightarrow} 0$ by Lemma 7.6.11.

Thus $\mathrm{cap}_{\mathcal{E}^{A,\varrho}}(B \cap B_k(0)) = 0$. This holds for all $k \in \mathbb{N}$ and hence $\mathrm{cap}_{\mathcal{E}^{A,\varrho}}(B) = 0$. $\qquad\square$

Theorem 7.6.13. *For $B \subset \mathbb{R}^l$ with $\mathrm{cap}_0(B) = 0$ it holds $\mathrm{cap}_1(B \cap \overline{\Upsilon}) = 0$.*

Proof. Assume $\mathrm{cap}_0(B) = 0$. Then the definition of cap_0 implies that there exists a sequence $(u_n)_{n\in\mathbb{N}}$ in $H^{1,2}(\mathbb{R}^l)$ with $B \subset \{x \in \mathbb{R}^l \,|\, u_n \geq 1\}^\circ$ and

$\|u_n\|_{H^{1,2}(\mathbb{R}^l)} \xrightarrow{n\to\infty} 0$. Set $U_n := \{x \in \mathbb{R}^l \,|\, u_n \geq 1\}^\circ \cap \overline{\Upsilon}$. These sets are open in the trace topology of $\overline{\Upsilon}$, because $\{x \in \mathbb{R}^l \,|\, u_n \geq 1\}^\circ$ are open in $\mathbb{R}^l$, $n \in \mathbb{N}$. Moreover, $\mathrm{cap}_1(U_n) \leq \|u_n\|_{\Upsilon}\|_{H^{1,2}(\Upsilon)}^2 \leq \|u_n\|_{H^{1,2}(\mathbb{R}^l)}^2$. Since $B \cap \overline{\Upsilon} \subset U_n$ for all $n \in \mathbb{N}$, it follows $\mathrm{cap}_1(B \cap \overline{\Upsilon}) = 0$.   □

By $\mathcal{H}^s$, $0 \leq s < \infty$, we denote the s-dimensional Hausdorff measure, see Definition 7.5.14. Denote by $d^k x$ the Lebesgue measure on $\mathbb{R}^k$, $k \in \mathbb{N}$.

Lemma 7.6.14.   *(i)* $\mathcal{H}^k = d^k x$ *on* $\mathbb{R}^k$, $k \in \mathbb{N}$.

(ii) Let $f : \mathbb{R}^k \to \mathbb{R}^l$, $k, l \in \mathbb{N}$, *be Lipschitz and* $0 \leq s < \infty$. *Then*

$$\mathcal{H}^s(f(B)) \leq (Lip(f))^s \mathcal{H}^s(B) \quad for\, B \subset \mathbb{R}^k.$$

Of course, $\mathrm{Lip}(f)$ denotes the Lipschitz constant of f. For the proof see [EG09, Sec. 2.2, Theo. 2] and [EG09, Sec. 2.4, Theo. 1]. From this lemma we can conclude that the Lipschitz smooth image of a compact subset of a k-dimensional subspace has finite k-dimensional Hausdorff measure.

For the following theorem see [MZ97, Theo. 2.52].

Theorem 7.6.15. *For* $B \subset \mathbb{R}^l$ *with* $\mathcal{H}^{l-2}(B) < \infty$ *it holds* $\mathrm{cap}_0(B) = 0$.

Combining this with Theorem 7.6.13 and Theorem 7.6.12 we obtain with Υ, A and ϱ as in Theorem 7.6.12:

Corollary 7.6.16. *For* $B \subset \mathbb{R}^l$ *with* $\mathcal{H}^{l-2}(B) < \infty$ *it holds* $\mathrm{cap}_{\mathcal{E}^{A,\varrho}}(B \cap \overline{\Upsilon}) = 0$.

7.7 Integration with respect to Functionals

In this section we give details concerning the definition and properties of integrals with respect to positive, increasing and continuous $\mathcal{F}_t$-adapted processes. In the case of additive functionals, additional properties will be given.

Without loss of generality we consider just strict AF (in the sense of Definition 6.1.2). In the case of a non-strict AF with exceptional set $N \subset E$ the results in this chapter are applied by considering the restricted process $\mathbf{M}^{E^\Delta \setminus N}$, see Definition 6.1.2 for the notation.

Let $F : [0, \infty) \to \mathbb{R}_0^+$ be an increasing and continuous function with $F(0) = 0$. For $0 \leq a < b < \infty$ define $\mu_F((a, b]) = F(b) - F(a)$. This map extends to a measure on $\mathcal{B}(\mathbb{R}_0^+)$, the so-called *Lebesgue-Stieltjes measure* denoted also by μ_F. Observe that $\mu_F(\{c\}) = 0$ for all $c \geq 0$. The integral defined with respect to this measure is called the *Lebesgue-Stieltjes integral*, see e.g. [Kle06, Def. 1.57]. Note that the measure is uniquely defined by its behavior on the set of all semi-open intervals.

Fix now a Markov process $\mathbf{M} = (\Omega, \mathcal{F}, (\mathcal{F}_t)_{t\geq 0}, (\mathbf{X}_t)_{t\geq 0}, (\mathbb{P}_x)_{x \in \hat{E}^\Delta})$ with càdlàg paths and state space a metric space $(\hat{E}, \mathbf{d})$. Let $(P_t)_{t\geq 0}$ be the transition semigroup which we may consider as linear operator acting both on $\mathcal{B}_b(\hat{E})$ or $\mathcal{B}_b^*(\hat{E})$. Let $A := (A_t)_{t\geq 0}$, $A_t : (\Omega, \mathcal{F}) \to \mathbb{R}_0^+$ be an increasing, continuous and finite functional. Then for every $\omega \in \Omega$ the mapping $[0, \infty) \ni t \mapsto A_t(\omega)$ induces a Lebesgue-Stieltjes measure denoted by $\mu_{A(\omega)}$.

Proposition 7.7.1. *Let $A : \Omega \to \mathbb{R}_0^+$ be an $\mathcal{F}_t$-adapted, increasing, continuous and finite process. Let $\mu_{A(\omega)}$, $\omega \in \Omega$, be the measure induced by the mapping $t \mapsto A_t(\omega)$. For $f \in \mathcal{B}_b^+([0, \infty) \times \hat{E})$ define $(f \cdot A)$ by*

$$[0, \infty) \times \Omega \ni t \times \omega \mapsto \int_0^t f(s, \mathbf{X}_s(\omega)) \, d\mu_{A(\omega)} =: \int_0^t f(s, \mathbf{X}_s) \, dA_s =: (f \cdot A)_t.$$

Then $(f \cdot A)_{t\geq 0}$ is an $\mathcal{F}_t$-adapted, positive, continuous and increasing process. For $f \in \mathcal{B}_b([0, \infty) \times \hat{E})$ the integral exists and $|(f \cdot A)_t| \leq (|f| \cdot A)_t$. The mapping $f \mapsto (f \cdot A)_t$ is linear.

Proof. By right-continuity of the paths of $(\mathbf{X}_s)_{t\geq 0}$ we have that the mapping $s \mapsto f(s, \mathbf{X}_s)$ is $\mathcal{B}([0, \infty))$-measurable. Note that for $\omega \in \Omega$ the mapping $t \mapsto A_t(\omega)$ fulfills all the properties required to define the measure $\mu_{A(\omega)}$ as Lebesgue-Stieltjes measure. For every $0 \leq t < \infty$ the integral of course exists since $\mu_{A(\omega)}([0, t]) = A(\omega)(t) < \infty$ and $f(s, \mathbf{X}_s)$ is bounded on $[0, t]$. Positivity follows from the properties of the integral. For $t_2 \geq t_1$ we have

$$(f \cdot A)_{t_2} = \int_0^{t_2} f(s, \mathbf{X}_s) dA_s \geq \int_0^{t_1} f(s, \mathbf{X}_s) dA_s = (f \cdot A)_{t_1}.$$

Now let $f \in \mathcal{B}_b([0, \infty) \times \hat{E})$. Then $\int_0^t f^+ dA_s$ and $\int_0^t f^- dA_s$ exists and hence $\int_0^t f dA_s := \int_0^t f^+ dA_s - \int_0^t f^- dA_s$ is well-defined and we have $|(f \cdot A)| \leq (|f| \cdot A)$. The linearity follows then from the linearity of the integrals with respect to $\mu_{A(\omega)}$ for $\omega \in \Omega$.

If $f \in C_b([0, \infty) \times \hat{E})$, the integral $(f \cdot A)_t$ can be approximated pathwisely by Riemann sums for each $0 \leq t < \infty$. This implies that $(f \cdot A)$ is $\mathcal{F}_t$-adapted. So adaptedness for general $f \in \mathcal{B}_b([0, \infty) \times \hat{E})$ follows using the functional monotone class theorem, see Corollary 2.2.4.

It is left to prove continuity in t. Let $T > 0$. For $0 \leq t < T$ we have

$$\int_0^t f(s, \mathbf{X}_s)\, dA_s = \int_0^T 1_{[0,t]}(s)\, f(s, \mathbf{X}_s)\, dA_s.$$

For a sequence $(t_n)_{n \in \mathbb{N}}$ in $[0, T]$ converging to t we have $1_{[0,t_n]}(s) f(s, \mathbf{X}_s) \overset{n \to \infty}{\longrightarrow} 1_{[0,t]}(s) f(s, \mathbf{X}_s)$ for all $s \in [0, \infty)$ with $s \neq t$. Since $\mu_{A(\omega)}(\{t\}) = 0$, we have $\mu_{A(\omega)}$-a.e. convergence. Thus the claim follows by Lebesgue dominated convergence on $[0, T]$.

$\square$

Corollary 7.7.2. *Let $(A_t)_{t \geq 0}$ be as in Proposition 7.7.1 but assume that $A_t < \infty$ only for $t < \mathcal{X}$ and $A_t = A_{\mathcal{X}}$ for $t \geq \mathcal{X}$. For $f \in \mathcal{B}_b^+([0, \infty) \times \hat{E})$ define $(f \cdot A)$ by*

$$[0, \infty) \times \mathbf{\Omega} \ni t \times \omega \mapsto \lim_{n \to \infty} \int_0^{t \wedge (\mathcal{X} - \frac{1}{n}) \vee 0} f(s, \mathbf{X}_s(\omega))\, d\mu_{A(\omega)}$$

$$:= \int_0^t f(s, \mathbf{X}_s)\, dA_s =: (f \cdot A)_t.$$

Then $(f \cdot A)_{t \geq 0}$ is an $\mathcal{F}_t$-adapted, positive, continuous and increasing process.

Proof. For $n \in \mathbb{N}$ define $A_\cdot^n := A_{\cdot \wedge (\mathcal{X} - \frac{1}{n}) \vee 0}$. Then A^n fulfills the assumption of Proposition 7.7.1 and

$$\int_0^{t \wedge (\mathcal{X} - \frac{1}{n}) \vee 0} f(s, \mathbf{X}_s(\omega))\, d\mu_{A(\omega)} = (f \cdot A^n)_t, \quad t \geq 0.$$

Since $(f \cdot A^n)$, $n \in \mathbb{N}$, fulfills all claimed properties and the sequence is increasing, we get the claim for $(f \cdot A)$. $\square$

Proposition 7.7.1 shows that for $f \in \mathcal{B}_b^+([0, \infty) \times \hat{E})$ and A as in the proposition the process $(f \cdot A)$ fulfills again the assumptions of the proposition. Thus we can consider integrals with respect to $(f \cdot A)$ in the sense of the proposition.

Lemma 7.7.3. *Let A as in Proposition 7.7.1 and $f, g \in \mathcal{B}_b^+([0, \infty) \times \hat{E})$.*

(i) It holds $(g \cdot (f \cdot A)) = (gf \cdot A)$, i.e.,

$$\int_0^t g(s, \mathbf{X}_s)\, d(f \cdot A)_s = \int_0^t g(s, \mathbf{X}_s)\, f(s, \mathbf{X}_s)\, dA_s \quad for\ t \geq 0.$$

(ii) For $t \geq 0$ it holds

$$f(t, \mathbf{X}_t)A_t = \int_0^t f(s, \mathbf{X}_s)dA_s + \int_0^t A_s f(s, \mathbf{X}_s)ds.$$

Proof. (i): One easily calculates that $\mu_{(h \cdot A)(\omega)} = h(\,\cdot\,, \mathbf{X}.)\,\mu_{A(\omega)}$ for $\omega \in \Omega$. So for $\omega \in \Omega$

$$\mu_{g \cdot (f \cdot A)(\omega)} = g(\,\cdot\,, \mathbf{X}.)\,\mu_{(f \cdot A)(\omega)} = g(\,\cdot\,, \mathbf{X}.)\,f(\,\cdot\,, \mathbf{X}.)\,\mu_{A(\omega)}.$$

(ii): First, let $f \in C^0([0,t])$. Then both $\int_0^t f(s)dA_s$ and $\int_0^t A_s f(s)ds$ can be approximated by Riemannian sums. So we get from [Tay65, Theo. 9-5 I]:

$$f(t)A_t = \int_0^t f(s)dA_s + \int_0^t A_s f(s)ds. \tag{7.20}$$

For general $f \in \mathcal{B}_b([0,t])$ we get (7.20) by a functional monotone class argument. Let $f \in \mathcal{B}_b^+([0,\infty) \times E)$. For $\omega \in \Omega$ set $F_\omega(s) := f(s, \mathbf{X}_s(\omega))$, $s \in [0,\infty)$. Applying (7.20) to $F_\omega(s)$ we get the claim. □

Under the assumption that $(A_t)_{t \geq 0}$ is additive and f does not depend explicitly on t, the functional $(f \cdot A)$ is also additive.

Proposition 7.7.4. *Let $(A_t)_{t \geq 0}$ as in Proposition 7.7.1. Assume additionally that $(A_t)_{t \geq 0}$ is additive with additivity set $\Lambda \in \mathcal{F}$. Let $f \in \mathcal{B}_b(\hat{E})$. Then $(f \cdot A)$ is additive on Λ. So together with Proposition 7.7.1, $(f \cdot A)$ is a CAF.*

Proof. For $t, s \geq 0$, $\omega \in \Lambda$, we have to show

$$\int_0^{t+s} f(\mathbf{X}_r)dA_r\,(\omega) = \int_0^t f(\mathbf{X}_r)dA_r\,(\omega) + \int_0^s f(\mathbf{X}_r)dA_r\,(\theta_t\omega).$$

Note that the pathwise definition of the integral yields

$$\int_0^s f(\mathbf{X}_r)\,dA_r\,(\theta_t\omega) = \int_0^s f(\mathbf{X}_r(\theta_t\omega))\,d\mu_{A(\theta_t\omega)}(r)$$

$$= \int_0^s f(\mathbf{X}_{t+r}(\omega))\,d\mu_{A(\theta_t\omega)}(r). \tag{7.21}$$

Let $\omega \in \Lambda$. We show first that for $g \in \mathcal{B}_b([0,T])$, $t + s \leq T < \infty$,

$$\int_0^{t+s} g(r)\,d\mu_{A(\omega)}(r) = \int_0^t g(r)\,d\mu_{A(\omega)}(r) + \int_0^s g(t+r)\,d\mu_{A(\theta_t\omega)}(r). \tag{7.22}$$

Note that the family of all functions fulfilling (7.22) is a linear vector space which is closed under monotone convergence. Furthermore, for the constant one function the equality also holds. So using the functional monotone class theorem, see Theorem 2.2.3, it is enough to prove the claim for indicator functions of semi-open intervals. Let $0 \le a < b \le T$. We have to show

$$\int_0^{t+s} 1_{(a,b]}(r)\, d\mu_{A(\omega)}(r) = \int_0^t 1_{(a,b]}(r)\, d\mu_{A(\omega)}(r)$$
$$+ \int_0^s 1_{(a,b]}(t+r)\, d\mu_{A(\theta_t\omega)}(r). \quad (7.23)$$

If $b \le t$, then the second integral on the right-hand side is zero, and the first integral of the right-hand side coincides with the integral of the left-hand side. So it is left to consider the case $b \ge t$. Then

$$\int_0^{t+s} 1_{(a,b]}(r)\, d\mu_{A(\omega)}(r) = \int_0^t 1_{(a,t]}(r)\, d\mu_{A(\omega)}(r) + \int_t^{t+s} 1_{(a,b]}(r)\, d\mu_{A(\omega)}(r)$$
$$= \int_0^t 1_{(a,t]}(r)\, d\mu_{A(\omega)}(r) + A_{(t+s)\wedge b} - A_{a \vee t}.$$

Since $b \ge t$, we have

$$A_{(t+s)\wedge b} - A_{a \vee t} = (A_{s \wedge (b-t)} - A_{(a-t) \vee 0})(\theta_t \cdot) = \int_0^s 1_{(a-t,b-t]}(r)\, d\mu_{A(\theta_t\omega)}(r).$$

Thus

$$\int_0^{t+s} 1_{(a,b]}(r)\, d\mu_{A(\omega)}(r) = \int_0^t 1_{(a,b]}(r)\, d\mu_{A(\omega)}(r)$$
$$+ \int_0^s 1_{(a,b]}(t+r)\, d\mu_{A(\theta_t\omega)}(r).$$

Define $g(r) := f(\mathbf{X}_r(\omega))$ for $r \in [0,\infty)$. Then (7.22) together with (7.21) yields

$$\int_0^{t+s} f(\mathbf{X}_r)\, dA_r\,(\omega) = \int_0^{t+s} g(r)\, d\mu_{A(\omega)}(r) =$$
$$\int_0^t g(r)\, d\mu_{A(\omega)}(r) + \int_0^s g(t+r)\, d\mu_{A(\theta_t\omega)}(r)$$
$$= \int_0^t f(\mathbf{X}_r)\, dA_r\,(\omega) + \int_0^s f(\mathbf{X}_r)\, dA_r\,(\theta_t\omega).$$

$\square$

Definition 7.7.5. Let $(A_t)_{t\geq 0}$ be a strict finite PCAF of $\mathbf{M}$ with additivity set $\Lambda \in \mathcal{F}$. Let $\mu_{A(\cdot)}$ be the corresponding pathwise Lebesgue-Stieltjes measure. We call a $\mathcal{B}(\hat{E})$-measurable function f locally A-*integrable* if for every $0 < T < \infty$ and $x \in \hat{E}$

$$\int_0^T |f(\mathbf{X}_s)| \, d\mu_{A(\cdot)}(s) < \infty \quad \mathbb{P}_x - a.s.$$

Define

$$\Lambda_f := \bigcap_{N\in\mathbb{N}} \left\{ \omega \in \Lambda \,\middle|\, \int_0^N |f|(\mathbf{X}_s) \, d\mu_{A(\cdot)}(s) \, (\omega) < \infty \right\}.$$

Then $\mathbb{P}_x(\Lambda_f) = 1$ for $x \in \hat{E}$. Define for $\omega \in \Omega$ and $0 \leq t < \infty$

$$\int_0^t f(\mathbf{X}_s) \, dA_s := \begin{cases} \int_0^t f^+(\mathbf{X}_s) \, dA_s - \int_0^t f^-(\mathbf{X}_s) \, dA_s & \text{if } \omega \in \Lambda_f \\ 0 & \text{else.} \end{cases}$$

Proposition 7.7.6. *Let $(A_t)_{t\geq 0}$ be a strict finite PCAF with additivity set Λ. Let f $\mathcal{B}(\hat{E})$-measurable and locally A-integrable. Define $(f \cdot A)$ by*

$$[0,\infty) \times \Omega \ni t \times \omega \mapsto \int_0^t f(\mathbf{X}_s) \, dA_s, \tag{7.24}$$

in the sense of Definition 7.7.5. Then $(f \cdot A)$ is a finite CAF with additivity set Λ_f.

Proof. Note that (7.24) is well-defined and $\mathbb{P}_x(\Lambda_f) = 1$ for $x \in \hat{E}$ since f is locally A-integrable. We have for $\omega \in \Lambda_f$

$$\int_0^N |f|(\mathbf{X}_s) \, dA_s \, (\theta_t\omega) = \int_0^N |f|(\mathbf{X}_s(\theta_t\omega)) \, d\mu_{A(\theta_t\omega)}(s)$$

$$= \int_t^{N+t} |f|(\mathbf{X}_s(\omega)) \, d\mu_{A(\omega)}(s) < \infty.$$

So Λ_f is again shift-invariant. Moreover, $f^+(\mathbf{X}_s) \wedge n \uparrow f^+(\mathbf{X}_s)$ and $f^-(\mathbf{X}_s) \wedge n \uparrow f^-(\mathbf{X}_s)$.

Thus $(f \cdot A)_t = 1_{\Lambda_f}(\lim_{n\to\infty}(f^+ \wedge n \cdot A)_t - (f^- \wedge n \cdot A)_t)$. Note that $\Lambda_f \in \mathcal{F}_t$ for $t \geq 0$ since $\mathbb{P}_x(\Lambda_f) = 1$ for every $x \in \hat{E}$. So from Proposition 7.7.1 we get that $(f \cdot A)$ is $\mathcal{F}_t$-adapted. That $(f \cdot A)$ is additive on Λ_f follows from $(f^+ \wedge n \cdot A)$ and $(f^- \wedge n \cdot A)$ being additive on Λ and shift-invariance of Λ_f.

It is left to show continuity on $[0, T]$. We have $(f \cdot A)_t = (1_{[0,t]} f \cdot A)_T$ and $|1_{[0,t]} f| \leq |1_{[0,T]} f|$. So the continuity follows as in the proof of Proposition 7.7.1 using Lebesgue's dominated convergence. □

We state the following well-known lemma for further reference.

Lemma 7.7.7. *Let $(A_t)_{t \geq 0}$ be a strict finite PCAF with additivity set Λ. Assume $\mathbb{E}_x[A_t] < \infty$ and $\mathbb{E}_x\left[\mathbb{E}_{\mathbf{X}_t}[A_s]\right] < \infty$ for $0 \leq s, t < \infty$ and every $x \in \hat{E}$. Then for $x \in \hat{E}$ it holds*

$$\mathbb{E}_x\left[A_{t+s} | \mathcal{F}_t\right] = \mathbb{E}_{\mathbf{X}_t}\left[A_s\right] + A_t \quad \mathbb{P}_x - a.s. \text{ for } t, s \geq 0.$$

For $0 \leq t \leq a \leq b < \infty$, $f \in \mathcal{B}_b(E)$ it holds

$$\mathbb{E}_x\left[f(\mathbf{X}_a)(A_b - A_a) \,|\, \mathcal{F}_t\right] = \mathbb{E}_{\mathbf{X}_t}\left[f(\mathbf{X}_{a-t})(A_{b-t} - A_{a-t})\right] \quad \mathbb{P}_x - a.s.$$

Proof. Let $t, s \geq 0$ and $x \in E$. Since A_s is $\mathcal{F}$-measurable, we have that the mapping $E \ni y \mapsto \mathbb{E}_y[A_s]$ is $\mathcal{B}^*(E)$-measurable. This follows from the Markov property of $\mathbf{M}$, more precisely from Definition 7.3.2(iii). By Lemma 7.3.4, $\mathbf{X}_t$ is $\mathcal{F}_t/B^*(E)$-measurable. So altogether, the mapping $\Omega \ni \omega \mapsto \mathbb{E}_{\mathbf{X}_t(\omega)}[A_s]$ is $\mathcal{F}_t$-adapted.

So using the properties of the conditional expectation and the Markov property of $\mathbf{M}$ (in particular, Lemma 7.3.7(iv)), we get

$$\mathbb{E}_x\left[A_{t+s} \,|\, \mathcal{F}_t\right] = \mathbb{E}_x\left[1_\Lambda(A_t + A_s(\theta_t \cdot)) \,|\, \mathcal{F}_t\right] = A_t + \mathbb{E}_{\mathbf{X}_t}\left[A_s\right]$$

and

$$\begin{aligned}\mathbb{E}_x\left[f(\mathbf{X}_a)(A_b - A_a) \,|\, \mathcal{F}_t\right] &= \mathbb{E}_x\left[f(\mathbf{X}_a)(A_{b-t}(\theta_t \cdot) - A_{a-s}(\theta_t \cdot)) \,|\, \mathcal{F}_t\right] \\ &= \mathbb{E}_{\mathbf{X}_t}\left[f(\mathbf{X}_{a-t})(A_{b-t}(\cdot) - A_{a-t}(\cdot))\right].\end{aligned}$$

□

For the next lemma we use the notion of restricted processes, see Definition 7.3.18 and Definition 6.1.2.

Lemma 7.7.8. *Let $E_1 \subset E^\Delta$, $N \subset E_1$ Borel, such that $E^\Delta \setminus E_1$ and $E_1 \setminus N$ are properly exceptional. Set $\widetilde{E}_2 := (E_1 \cup \{\Delta\}) \setminus N$ and $\widetilde{E}_1 := E_1 \cup \{\Delta\}$. Let $\mathbf{M}^{\widetilde{E}_1}$ and $\mathbf{M}^{\widetilde{E}_2}$ be the corresponding restricted processes. Let $(A_t)_{t \geq 0}$ be a strict finite PCAF of $\mathbf{M}^{\widetilde{E}_2}$ with additivity set $\Lambda \subset \mathbf{\Omega}^{\widetilde{E}_2}$. Assume*

$\mathbb{E}_x^{\widetilde{E}_1}[\mathbb{E}_{\mathbf{X}_t}^{\widetilde{E}_2}[A_s]] < \infty$ for $0 \le s < \infty$ and $P_t^{\widetilde{E}_1}(x, \widetilde{E}_1 \setminus \widetilde{E}_2) = 0$ for $0 < t < \infty$ and every $x \in \widetilde{E}_1$. Let $\Lambda_0 := \Omega^{\widetilde{E}_1} \cap \bigcap_{n \in \mathbb{N}} \theta_{1/n}^{-1}(\Lambda)$. For $\varepsilon > 0$ define $A_t^\varepsilon := 1_{\Lambda_0} A_{t-\varepsilon}(\theta_\varepsilon \cdot)$ Let $\varepsilon < T < \infty$. Then for all $f \in \mathcal{B}_b([0,T])$ it holds

$$\mathbb{E}_x^{\widetilde{E}_1}\left[\int_\varepsilon^T f(s)\, dA_s^\varepsilon\right] = \mathbb{E}_x^{\widetilde{E}_1}\left[\mathbb{E}_{\mathbf{X}_\varepsilon}^{\widetilde{E}_2}\left[\int_0^{T-\varepsilon} f(\varepsilon + s)\, dA_s\right]\right] \quad \text{for } x \in \widetilde{E}_1.$$
(7.25)

Proof. Note that the class of functions for which (7.25) holds is a linear vector space and closed under monotone convergence. So by the functional monotone class theorem, see Theorem 2.2.3, it is enough to prove the equality for indicator functions of semi-open intervals. Let $x \in \widetilde{E}_1$ and $0 \le a < b \le T$. Define $f(s) := 1_{(a,b]}(s)$. If either $a, b \le \varepsilon$ or $a, b \ge T$ then both expressions in (7.25) are zero. If $a \le \varepsilon$ and $b \ge \varepsilon$ we can replace a by ε. If $b \ge T$ and $a \le T$ we can replace b by T. In both cases the values in (7.25) are not affected. So we may assume $\varepsilon \le a < b \le T$. Note that $\mathbb{P}_x^{\widetilde{E}_1}(\Lambda_0) = 1$ for $x \in \widetilde{E}_1$ by Lemma 7.3.22. For $\omega \in \Lambda_0$ it holds

$$\int_\varepsilon^T f(s)\, dA_s^\varepsilon = \int_\varepsilon^T 1_{(a,b]}(s)\, dA_s^\varepsilon = A_b^\varepsilon(\omega) - A_a^\varepsilon(\omega)$$

$$= A_{b-\varepsilon}(\theta_\varepsilon \omega) - A_{a-\varepsilon}(\theta_\varepsilon \omega). \quad (7.26)$$

From Lemma 7.3.23 we get for $x \in \widetilde{E}_1$

$$\mathbb{E}_x^{\widetilde{E}_1}\left[1_{\Lambda_0}(A_{b-\varepsilon}(\theta_\varepsilon \cdot) - A_{a-\varepsilon}(\theta_\varepsilon \cdot))\right] =$$

$$\mathbb{E}_x^{\widetilde{E}_1}\left[\mathbb{E}_{\mathbf{X}_\varepsilon}^{\widetilde{E}_2}[A_{b-\varepsilon} - A_{a-\varepsilon}]\right] = \mathbb{E}_x^{\widetilde{E}_1}\left[\mathbb{E}_{\mathbf{X}_\varepsilon}^{\widetilde{E}_2}\left[\int_0^{T-\varepsilon} 1_{(a-\varepsilon, b-\varepsilon]}(s)\, dA_s\right]\right]$$

$$= \mathbb{E}_x^{\widetilde{E}_1}\left[\mathbb{E}_{\mathbf{X}_\varepsilon}^{\widetilde{E}_2}\left[\int_0^{T-\varepsilon} 1_{(a,b]}(\varepsilon + s)\, dA_s\right]\right] = \mathbb{E}_x^{\widetilde{E}_1}\left[\mathbb{E}_{\mathbf{X}_\varepsilon}^{\widetilde{E}_2}\left[\int_0^{T-\varepsilon} f(\varepsilon + s)\, dA_s\right]\right].$$
(7.27)

Combining (7.26) and (7.27) we get for $x \in \widetilde{E}_1$

$$\mathbb{E}_x^{\widetilde{E}_1}\left[\int_\varepsilon^T f(s)\, dA_s^\varepsilon\right] = \mathbb{E}_x^{\widetilde{E}_1}\left[\mathbb{E}_{\mathbf{X}_\varepsilon}^{\widetilde{E}_2}\left[\int_0^{T-\varepsilon} f(\varepsilon + s)dA_s\right]\right].$$

$\square$

The next lemma shows that taking the conditional expectation with respect to $\mathcal{F}_s$ „shifts" integration with respect to an additive functional to the left.

Lemma 7.7.9. *Let $(A_t)_{t\geq 0}$ be a strict finite PCAF with additivity set Λ. Assume $\mathbb{E}_x[A_t] < \infty$ and $\mathbb{E}_x[\mathbb{E}_{\mathbf{X}_t}[A_s]] < \infty$ for $0 \leq s,t < \infty$ and every $x \in E$. Let $f \in \mathcal{B}_b([0,\infty) \times \hat{E})$. For $0 \leq s \leq a,b < \infty$ and $x \in \hat{E}$ it holds*

$$\mathbb{E}_x\left[\int_a^b f(r,\mathbf{X}_r)\, dA_r \,\Big|\, \mathcal{F}_s\right] = \mathbb{E}_{\mathbf{X}_s}\left[\int_{a-s}^{b-s} f(s+r,\mathbf{X}_r)dA_r\right] \quad \mathbb{P}_x - a.s.$$
(7.28)

and

$$\mathbb{E}_x\left[\int_a^b f(r,\mathbf{X}_r)\, dA_r\right] = P_s.\mathbb{E}\left[\int_{a-s}^{b-s} f(s+r,\mathbf{X}_r)dA_r\right](x).$$
(7.29)

In particular,

$$\mathbb{E}_x\left[\int_s^\infty e^{-r}dA_r\right] = e^{-s} P_s.\mathbb{E}\left[\int_0^\infty e^{-r}dA_r\right](x).$$
(7.30)

Proof. We prove (7.28) first for continuous bounded functions on $[0,\infty) \times \hat{E}$. Let $f \in C_b([0,\infty) \times \hat{E})$. Let $x \in \hat{E}$. Let $M \in \mathbb{N}$. Define $r_k^M := a + \frac{b-a}{M}k$, $k = 0,...,M$. For $\omega \in \Omega$ define

$$F^M(r,\omega) := \sum_{i=0}^{M-1} f\big(r_{i+1}^M, \mathbf{X}_{r_{i+1}^M}(\omega)\big) 1_{(r_i^M, r_{i+1}^M]}(r).$$

and $F(r,\omega) := f\big(r,\mathbf{X}_r(\omega)\big)$. Then it holds for $r \in (a,b]$, $F^M(r,\omega) \overset{M\to\infty}{\longrightarrow} F(r,\omega)$. Thus with $F_s^M(r,\cdot) := \sum_{i=0}^{M-1} f\big(r_{i+1}^M, \mathbf{X}_{r_{i+1}^M - s}\big) 1_{(r_i^M - s, r_{i+1}^M - s]}(r)$, we have

$$\mathbb{E}_x\left[\int_a^b f(r,\mathbf{X}_r)\, dA_r\right] = \lim_{M\to\infty} \mathbb{E}_x\left[\int_a^b F^M(r,\cdot)\, dA_r\right],$$
(7.31)

$$\mathbb{E}_x\left[\int_a^b f(r,\mathbf{X}_r)\, dA_r \,\Big|\, \mathcal{F}_s\right] = \lim_{M\to\infty} \mathbb{E}_x\left[\int_a^b F^M(r,\cdot)\, dA_r \,\Big|\, \mathcal{F}_s\right],$$

and

$$\mathbb{E}_{\mathbf{X}_s}\left[\int_{a-s}^{b-s} f(s+r,\mathbf{X}_r)\, dA_r\right] = \lim_{M\to\infty} \mathbb{E}_{\mathbf{X}_s}\left[\int_{a-s}^{b-s} F_s^M(r,\cdot)\, dA_r\right] \quad \mathbb{P}_x - a.s.$$

So to prove (7.28) it is enough to show

$$\mathbb{E}_x\left[\int_a^b F^M(r,\cdot)\,dA_r\,\Big|\,\mathcal{F}_s\right] = \mathbb{E}_{\mathbf{X}_s}\left[\int_{a-s}^{b-s} F_s^M(r,\cdot)\,dA_r\right].$$

Let M be fixed. By Lemma 7.7.7 we have

$$\mathbb{E}_x\left[f\big(r_{i+1}^M,\mathbf{X}_{r_{i+1}^M}\big)\big(A_{r_{i+1}^M} - A_{r_i}\big)\,\Big|\,\mathcal{F}_s\right]$$
$$= \mathbb{E}_{\mathbf{X}_s}\left[f\big(r_{i+1}^M,\mathbf{X}_{r_{i+1}^M-s}\big)\big(A_{r_{i+1}^M-s} - A_{r_i^M-s}\big)\right].$$

So

$$\mathbb{E}_x\left[\int_a^b F^M(r,\cdot)\,dA_r\,\Big|\,\mathcal{F}_s\right] = \sum_{i=0}^{M-1} \mathbb{E}_x\left[f\big(r_{i+1}^M,\mathbf{X}_{r_{i+1}^M}\big)\big(A_{r_{i+1}^M} - A_{r_i^M}\big)\,\Big|\,\mathcal{F}_s\right]$$
$$= \sum_{i=0}^{M-1} \mathbb{E}_{\mathbf{X}_s}\left[f\big(r_{i+1}^M,\mathbf{X}_{r_{i+1}^M-s}\big)\big(A_{r_{i+1}^M-s} - A_{r_i^M-s}\big)\right]$$
$$= \mathbb{E}_{\mathbf{X}_s}\left[\int_{a-s}^{b-s} F_s^M(r,\cdot)\,dA_r\right].$$

Altogether, we get that (7.28) holds for $f \in C_b([0,\infty) \times \hat{E})$. Using the functional monotone class theorem, more precisely Corollary 2.2.4, we get the equation for all $f \in \mathcal{B}_b([0,\infty) \times \hat{E})$. Equation (7.29) follows now by taking the expectation in (7.28). Equation (7.30) follows by taking $a = s$, $f(r) := \exp(-r)$ and monotone convergence to replace b by ∞.

$\square$

Bibliography

[AD75] R. A. Adams. *Sobolev spaces.* Pure and Applied Mathematics, 65. New York-San Francisco-London: Academic Press, Inc. XVIII, 1975.

[AKR03] S. Albeverio, Y. Kondratiev and M. Röckner *Strong Feller properties for distorted Brownian motion and applications to finite particle systems with singular interactions.* Finite and infinite dimensional analysis in honor of Leonard Gross. Providence, RI: American Mathematical Society (AMS). Contemp. Math. 317, 15-35 (2003).

[ASZ09] L. Ambrosio, G. Savaré, and L. Zambotti. Existence and stability for Fokker-Planck equations with log-concave reference measure. *Probab. Theory Relat. Fields*, 145(3-4):517–564, 2009.

[CB06] D. A. Charalambos and K. C. Border. *Infinite dimensional analysis. A hitchhiker's guide. 3rd ed.* Berlin: Springer. xxii, 2006.

[Alt06] H. W. Alt. *Lineare Funktionalanalysis. Eine anwendungsorientierte Einführung. 5th revised ed.* Berlin: Springer. xiv, 2006.

[AR12] S. Andres and M.-K. von Renesse. Uniqueness and regularity for a system of interacting Bessel processes via the Muckenhoupt condition. *Trans. Am. Math. Soc.*, 364(3):1413–1426, 2012.

[BH00] R. F. Bass and E. P. Hsu. Pathwise uniqueness for reflecting Brownian motion in Euclidean domains. *Probab. Theory Relat. Fields*, 117(2):183–200, 2000.

[BH90] R. F. Bass and P. Hsu. The semimartingale structure of reflecting Brownian motion. *Proc. Am. Math. Soc.*, 108(4):1007–1010, 1990.

[BH91] R. F. Bass and P. Hsu. Some potential theory for reflecting Brownian motion in Hölder and Lipschitz domains. *Ann. Probab.*, 19(2):486–508, 1991.

[Bau78] H. Bauer. *Wahrscheinlichkeitstheorie und Grundzüge der Maßtheorie. 3., neubearb. Aufl.* de Gruyter Lehrbuch. Berlin - New York: Walter de Gruyter., 1978.

[BG13] B. Baur and M. Grothaus. Construction and strong Feller property of distorted elliptic diffusion with reflecting boundary. *Potential Analysis*, Online First, doi:10.1007/s11118-013-9355-8, 2013.[1]

[BG13b] B. Baur and M. Grothaus. Skorokhod decomposition for a reflected $\mathcal{L}^p$-strong Feller diffusion with singular drift. *submitted for publication*, 2013.[1]

[BGS13] B. Baur, M. Grothaus and P. Stilgenbauer. Construction of $\mathcal{L}^p$-strong Feller Processes via Dirichlet Forms and Applications to Elliptic Diffusions. *Potential Analysis*, 38(4):1233–1258, 2013.

[BG68] R. M. Blumenthal and R. K. Getoor. *Markov processes and potential theory*. Pure and Applied Mathematics, 29. A Series of Monographs and Textbooks. New York-London: Academic Press. X, 1968.

[BKR97] V. I. Bogachev, N. V. Krylov and M. Röckner. Elliptic regularity and essential self-adjointness of Dirichlet operators on $\mathbb{R}^n$. *Ann. Sc. Norm. Super. Pisa, Cl. Sci., IV. Ser.*, 24(3):451–461, 1997.

[BKR01] V. I. Bogachev, N. V. Krylov and M. Röckner. On regularity of transition probabilities and invariant measures of singular diffusions under minimal conditions. *Commun. Partial Differ. Equations*, 26(11-12):2037–2080, 2001.

[Che93] Z. -Q. Chen. On reflecting diffusion processes and Skorokhod decompositions. *Probab. Theory Relat. Fields*, 94(3):281–315, 1993.

[PR02] G. Da Prato and M. Röckner. Singular dissipative stochastic equations in Hilbert spaces. *Probab. Theory Relat. Fields*, 124(2):261–303, 2002.

[DM82] C. Dellacherie and P. -A. Meyer. *Probabilities and potential. B: Theory of martingales. Transl. from the French and prep. by J. P. Wilson.* , 1982.

[DN07] J. -D. Deuschel and T. Nishikawa. The dynamic of entropic repulsion. *Stochastic Processes Appl.*, 117(5):575–595, 2007.

[Dob10] M. Dobrowolski. *Angewandte Funktionalanalysis. Funktionalanalysis, Sobolev-Räume und elliptische Differentialgleichungen.* Springer-Lehrbuch Masterclass. Berlin: Springer. xi, 2010.

---

[1] As of January 2014

[Doh05] J. M. Dohmann. Feller-type properties and path regularities of Markov processes. *Forum Math.*, 17(3):343–359, 2005.

[Doo53] J. L. Doob. *Stochastic processes.* New York: Wiley, 1953.

[EN00] K. -J. Engel and R. Nagel. *One-parameter semigroups for linear evolution equations.* Berlin: Springer, 2000.

[EG09] L. C. Evans and R. F. Gariepy. *Measure theory and fine properties of functions.* Studies in Advanced Mathematics. Boca Raton: CRC Press. viii, 1992.

[FG07] T. Fattler and M. Grothaus. Strong Feller properties for distorted Brownian motion with reflecting boundary condition and an application to continuous N-particle systems with singular interactions. *J. Funct. Anal.*, 246(2):217–241, 2007.

[FG08] T. Fattler and M. Grothaus. Construction of elliptic diffusions with reflecting boundary condition and an application to continuous N-particle systems with singular interactions. *Proc. Edinb. Math. Soc., II. Ser.*, 51(2):337–362, 2008.

[Fat08] T. Fattler. *Construction and analysis of elliptic diffusions and applications to continuous particle systems with singular interactions.* Dr. Hut, München, 2008.

[FOT94] M. Fukushima, Y. Oshima and M. Takeda. *Dirichlet forms and symmetric Markov processes. 2nd revised and extended ed.* de Gruyter Studies in Mathematics 19. Berlin: Walter de Gruyter. x, 1994.

[Fuk85] M. Fukushima. Energy forms and diffusion processes. In *Mathematics + physics. Vol. 1*, pages 65–97. World Sci. Publishing, Singapore, 1985.

[FOT11] M. Fukushima, Y. Oshima and M. Takeda. *Dirichlet forms and symmetric Markov processes. 2nd revised and extended ed.* de Gruyter Studies in Mathematics 19. Berlin: Walter de Gruyter. x, 2011.

[FT95] M. Fukushima and M. Tomisaki. Reflecting diffusions on Lipschitz domains with cusps – analytic construction and Skorohod representation. *Potential Anal.*, 4(4):377–408, 1995.

[FT96] M. Fukushima and M. Tomisaki. Construction and decomposition of reflecting diffusions on Lipschitz domains with Hölder cusps. *Probab. Theory Relat. Fields*, 106(4):521–557, 1996.

[Fun03] T. Funaki. Hydrodynamic limit for $\nabla\varphi$ interface model on a wall. *Probab. Theory Relat. Fields*, 126(2):155–183, 2003.

[Fun05] T. Funaki. Stochastic interface models. Dembo, Amir et al. Berlin: Springer. Lecture Notes in Mathematics 1869, 105-274 (2005)., 2005.

[FO01] T. Funaki and S. Olla. Fluctuations for $\nabla\varphi$ interface model on a wall. *Stochastic Processes Appl.*, 94(1):1–27, 2001.

[GT77] D. Gilbarg and N. S. Trudinger. *Elliptic partial differential equations of second order. Reprint of the 1998 ed.* Classics in Mathematics. Berlin: Springer. xiii, 2001.

[IW81] N. Ikeda and S. Watanabe. *Stochastic differential equations and diffusion processes.* , 1981.

[KS91] I. Karatzas and S. E. Shreve. *Brownian motion and stochastic calculus. 2nd ed.* Graduate Texts in Mathematics, 113. New York etc.: Springer-Verlag. xxiii., 1991.

[Kle06] A. Klenke. *Probability theory. (Wahrscheinlichkeitstheorie.).* Berlin: Springer. xii, 2006.

[KR05] N. V. Krylov and M. Röckner Strong solutions of stochastic equations with singular time dependent drift. *Probab. Theory Relat. Fields*, 131(2):154–196, 2005.

[LS84] P. -L. Lions and A. S. Sznitman. Stochastic differential equations with reflecting boundary conditions. *Commun. Pure Appl. Math.*, 37:511–537, 1984.

[LS96] V. A. Liskevich and Yu. A. Semenov. Some problems on Markov semigroups. Demuth, Michael (ed.) et al., Schrödinger operators, Markov semigroups, wavelet analysis, operator algebras. Berlin: Akademie Verlag. Math. Top. 11, 163-217, 1996.

[Lun95] A. Lunardi. *Analytic semigroups and optimal regularity in parabolic problems.* Progress in Nonlinear Differential Equations and their Applications. 16. Basel: Birkhäuser. xvii, 1995.

[MR92] Z. Ma and M. Röckner *Introduction to the theory of (non-symmetric) Dirichlet forms.* Universitext. Berlin: Springer-Verlag. viii, 1992.

[MZ97] J. Malý and W. P. Ziemer. *Fine regularity of solutions of elliptic partial differential equations.* Mathematical Surveys and Monographs. 51. Providence, RI: American Mathematical Society (AMS). xiv, 1997.

[McK63] H. P. jun. McKean. A. Skorohod's stochastic integral equation for a reflecting barrier diffusion. *J. Math. Kyoto Univ.*, 3:85–88, 1963.

[Miy03] Y. Miyazaki. The L^p resolvents of elliptic operators with uniformly continuous coefficients. *J. Differ. Equations*, 188(2):555–568, 2003.

[Miy06] Y. Miyazaki. Higher order elliptic operators of divergence form in C^1 or Lipschitz domains. *J. Differ. Equations*, 230(1):174–195, 2006.

[Mor66] C. B. Morrey. *Multiple integrals in the calculus of variations. Reprint of the 1966 original.* Classics in Mathematics. Berlin: Springer. ix, 2008.

[PW94] E. Pardoux and R. J. Williams. Symmetric reflected diffusions. *Ann. Inst. Henri Poincaré, Probab. Stat.*, 30(1):13–62, 1994.

[Paz83] A. Pazy. *Semigroups of linear operators and applications to partial differential equations.* Applied Mathematical Sciences, 44. New York etc.: Springer-Verlag. VIII, 1983.

[RS06] M. Röckner and Z. Sobol. Kolmogorov equations in infinite dimensions: well-posedness and regularity of solutions, with applications to stochastic generalized Burgers equations. *Ann. Probab.*, 34(2):663–727, 2006.

[Rud70] W. Rudin. *Real and complex analysis.* McGraw-Hill Series in Higher Mathematics. New York etc.: McGraw-Hill Book Company. xi, 1966.

[Sai87] Y. Saisho. Stochastic differential equations for multi-dimensional domain with reflecting boundary. *Probab. Theory Relat. Fields*, 74:455–477, 1987.

[Sch06] F. Schwabl. *Statistical mechanics. (Statistische Mechanik.) 3rd revised ed.* Berlin: Springer, 2006.

[Sha06] S. V. Shaposhnikov. On Morrey's estimate of the Sobolev norms of solutions of elliptic equations. *Math. Notes*, 79(3):413–430, 2006.

[Sim63] G. F. Simmons. *Introduction to topology and modern analysis.* International Series in Pure and Applied Mathematics. New York etc.: McGraw-Hill Book Company. XV, 1963.

[Sko61] A. V. Skorokhod. Stochastic equations for diffusion processes in a bounded region. *Teor. Veroyatn. Primen.*, 6:287–298, 1961.

[Sti10] P. Stilgenbauer. *Elliptic diffusions on general open sets with singular matrix coefficients.* University of Kaiserslautern, 2010.

[Sto83] L. Stoica. On the construction of Hunt processes from resolvents. *Z. Wahrscheinlichkeitstheor. Verw. Geb.*, 64:167–179, 1983.

[SV71] D. W. Stroock and S. R. S. Varadhan. Diffusion processes with boundary conditions. *Commun. Pure Appl. Math.*, 24:147–225, 1971.

[Stu98a] K. -T. Sturm. How to construct diffusion processes on metric spaces. *Potential Anal.*, 8(2):149–161, 1998.

[Stu98b] K. T. Sturm. Diffusion processes and heat kernels on metric spaces. *Ann. Probab.*, 26(1):1–55, 1998.

[Tan79] H. Tanaka. Stochastic differential equations with reflecting boundary condition in convex regions. *Hiroshima Math. J.*, 9:163–177, 1979.

[Tay65] A. E. Taylor. *General theory of functions and integration.* New York-Toronto-London: Blaisdell Publishing Company, a division of Ginn and Company. XVI, 1965.

[Tri78] H. Triebel. *Interpolation theory, function spaces, differential operators.* North-Holland Mathematical Library. Vol. 18. Amsterdam - New York - Oxford: North-Holland Publishing Company., 1978.

[Tru03] G. Trutnau. Skorokhod decomposition of reflected diffusions on bounded Lipschitz domains with singular non-reflection part. *Probab. Theory Relat. Fields*, 127(4):455–495, 2003.

[Wer11] D. Werner. *Functional analysis. (Funktionalanalysis.) 7th revised ed.* Springer-Lehrbuch. Berlin: Springer. xiii, 2011.

[WZ90] R. J. Williams and W. A. Zheng. On reflecting Brownian motion - a weak convergence approach. *Ann. Inst. Henri Poincaré, Probab. Stat.*, 26(3):461–488, 1990.

[Zam08] L. Zambotti. Fluctuations for a conservative interface model on a wall. *ALEA, Lat. Am. J. Probab. Math. Stat.*, 4:167–184, 2008.

Index